Page, Thornton
Beyond the Milky Way

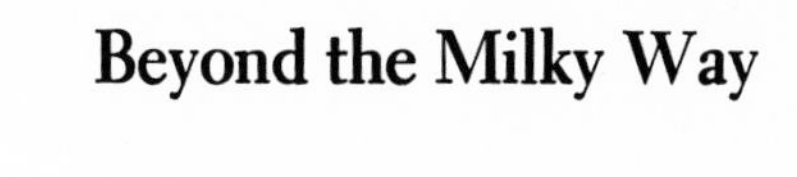
Beyond the Milky Way

THE MACMILLAN SKY AND TELESCOPE LIBRARY OF ASTRONOMY

With the advent of space exploration the science of astronomy enters a new phase—that of practical application significant not only to scientists but also to the public at large. The rapid and critical developments in astronomy that preceded this new phase are presented in this unique library of astronomy made up of articles that first appeared in the prominent journals *Sky and Telescope, The Sky*, and *The Telescope*.

Volume 1. Wanderers in the Sky
Volume 2. Neighbors of the Earth
Volume 3. The Origin of the Solar System
Volume 4. Telescopes
Volume 5. Starlight
Volume 6. The Evolution of Stars
Volume 7. Stars and Clouds of the Milky Way
Volume 8. Beyond the Milky Way

Beyond the Milky Way

GALAXIES, QUASARS, AND THE NEW COSMOLOGY

EDITED BY THORNTON PAGE & LOU WILLIAMS PAGE

VOLUME 8 *Sky and Telescope* Library of Astronomy

Illustrated with over 100 photographs, drawings, and diagrams

The Macmillan Company, New York, New York

Collier-Macmillan Limited, London

The articles in this book were originally published in slightly different form in *The Sky*, copyright American Museum of Natural History 1937, 1940; in *Sky and Telescope* copyright © Sky Publishing Corporation 1944, 1945, 1948, 1950, 1951, 1953, 1954, 1956, 1957, 1960, 1961, 1963, 1964, 1965, 1966, 1967, 1968 and in *The Telescope*.

Library of Congress Catalog Card Number: 69-10504

SECOND PRINTING 1971

The Macmillan Company
866 Third Avenue
New York, N. Y. 10022
Collier-Macmillan Canada Ltd., Toronto, Ontario

Printed in the United States of America

Contents

3. Evolution

4. Peculiar Galaxies

5. Quasars

6. Masses, Luminosities, and Models

7. Cosmology

Illustrations

Figure

Tables

Preface

Strictly speaking, there can be only one universe. But each man has his own concept of the outside world. For centuries astronomers have studied the sun, moon, planets, stars, and other things in the sky, piecing together a consistent picture of the astronomical universe, which extends on every side to distances far larger than any experienced on earth. Previous volumes in this series have shown how distances are measured in the solar system out to billions of miles from the earth, and distances to stars in the Milky Way out to many thousands of light years.

In this book we will go well beyond the Milky Way, to other galaxies, each of them comparable to our Milky Way in size and content, some as far as 8 billion light years from us. Although these vast distances are dramatic, and characteristic of the astronomers' universe, we are here more concerned with how they have been determined, and with the observations and reasoning that have been used to learn about conditions in such remote places. A large part of this research has taken place during the past forty years, and many of the results have been described in the magazine Sky and Telescope or its predecessors, The Sky and The Telescope.

The history of Sky and Telescope is told in Appendix I. The magazine is available at most public libraries in bound volumes (13,000 pages total from 1941 to 1968). Today Sky and Telescope is read by amateur astronomers as well as by professionals, by students of astronomy, and by their professors. The articles appearing in any one year reflect the topics of greatest interest and the state of astronomical knowledge then. After reviewing all the many articles that have appeared since 1931, the editors selected the ones that mark major developments in eight different areas; each group of articles provided the theme for a volume in this series, Macmillan's Sky and Telescope Library of Astronomy. Most of the articles reprinted in the present volume are recent ones, reflecting the growing interest in studies of galaxies and quasars, but to show what was known, say, ten or twenty years ago, we have selected some earlier articles, several of them written by now-famous astronomers.

Each article appears here essentially as it was originally printed, although some minor changes have been made for consistency in style, and a few deletions to maintain continuity. Omissions of one or more paragraphs are indicated by ellipses, and additions by square brackets. The editors' comments between articles, identified by the initials TLP, note contrasts between past views and those accepted today. As in previous volumes, these comments are not intended to be critical; they simply emphasize the progress of astrophysics and cosmology.

THORNTON PAGE
LOU WILLIAMS PAGE

January 1969

About the Editors

Thornton Page is now Fisk Professor of Astronomy at Wesleyan University, Middletown, Connecticut, and director of the Van Vleck Observatory. After studying at Yale University he went to England as a Rhodes Scholar; he received a Ph.D. from Oxford University in 1938. He then spent two years at the Yerkes Observatory and the University of Chicago. Starting in 1940 he worked for the U.S. Navy on magnetic sea mines, first in the Naval Ordnance Laboratory and then as a junior officer in the Pacific, ending up in Japan studying the effects of atomic-bomb explosions. After the war he returned to the University of Chicago and observed galaxies with the 82-inch Struve telescope of the McDonald Observatory in western Texas. In 1950 he was called back to military service with the Army's Operations Research Office; he spent two years as scientific advisor in the U.S. Army Headquarters, Heidelberg, Germany, and several months in Korea during the war there. Since 1958 he has spent four summers at the Lick and Mount Wilson Observatories in California, taught astronomy courses at the University of Colorado, UCLA, and Yale. During 1965, 1966, and 1967 he was NAS Research Associate at the Smithsonian and Harvard Observatories, and during 1968 and 1969 at the NASA Manned Spacecraft Center. In February 1969 Dr. Page was given an honorary doctorate degree by the University of Córdoba, Argentina.

Mrs. Page received her Ph.D. in geology from the University of Chicago. She is the author of *A Dipper Full of Stars* and has collaborated with Mr. Page and others in writing a three-volume text, *Introduction to the Physical Sciences*. More recently, she has served as editor of the Connecticut Geological and Natural History Survey, taught a science course at Wesleyan University, and has written a book on geology for young readers called *The Earth and Its Story*. The Pages also collaborated with others in writing the high school text for an earth-science course sponsored by the National Science Foundation and have, individually, produced several research papers in their widely separated fields.

Introduction

Although astronomy is probably the oldest science, and men have been wondering about the stars since 3000 B.C. or earlier, only in the last few hundred years has our understanding of the universe developed rapidly. Part of this recent development has been due to technical inventions—the telescope about 1600, the spectrograph and photographic plates in the nineteenth century, cosmic-ray detectors fifty years ago, radio telescopes twenty years ago, and space probes ten years ago (see Volume 4 of this series, Telescopes). Equally important, however, are theoretical developments—mechanics in the seventeenth century, atomic physics in the nineteenth, relativity and nuclear physics in the twentieth.

Astronomers have used these sciences to expand their picture of the universe at a rate that greatly exceeds the velocity of light—even though no material object (or light itself) can exceed that speed. One hundred years ago we knew that the stars are hundreds of light years distant and that our Milky Way is an enormous galaxy of stars. Then, in rapid succession, astronomers measured the diameter of this Galaxy as about 100,000 light years, the distances to other, outside galaxies as millions of light years, and most recently, the distances of very luminous small galaxies ("quasars") as probably billions of light years. It is fair to say that the astronomers' understanding of the universe expanded five billion light years beyond the Milky Way during the past fifty years, reaching out at 100 million times the speed of light (one light year per year).

For the reader unfamiliar with these rapid developments, this Introduction presents a rough overview of the astronomers' methods of measuring distances, a general picture of the Milky Way Galaxy and its nearest neighbors (the Magellanic Clouds), and a basic question about the extent of the entire universe. Chapter 1 examines distance measurements in greater detail. Chapters 2 and 3 are devoted to "normal" galaxies, their contents (stars and nebulae), and how they are expected to change over billions of years. Chapter 4 deals with peculiar galaxies and leads to Chapters 5 and 6, which discuss the extreme case of the quasars, *probably the most distant objects yet observed. Chapter 7, on cosmology, shows the efforts of astronomers and physicists to fit all this together into a self-consistent picture of the universe.*

The estimated distances of stars, clusters of stars, galaxies, clusters of galaxies, and quasars all depend on accurate measurements of photographs showing tiny changes in the directions of some stars—that is, tiny changes in their positions in the sky.—TLP

Man and His Expanding Universe (I)

PERCY W. WITHERELL

(*Sky and Telescope*, September, October, and December 1944; January 1945)

Astronomers for many centuries observed, listed, and made charts of the positions of the stars, and recorded their apparent magnitudes or relative brightnesses. To measure the stellar distances was a more difficult problem. Patterning their observations after the surveyor's trigonometric method, astronomers a century ago first used the distance across the earth's orbit, or a portion thereof, as a baseline. In some cases it is possible to compare positions of a star at six-month intervals, when the earth has been displaced the diameter of its orbit, 186 million miles; in other instances, depending on the location of the star and of the observatory, smaller intervals than this must be used. In any case, however, the change in position of a bright star against the background of much fainter stars (presumably much farther away on the average) gave the necessary information to determine the star's distance in terms of the radius of the earth's orbit.

The angle at the star that is subtended by the earth's orbital radius is the stellar *parallax*. By modern methods, developed by the late Frank Schlesinger of Yale University at the turn of the century, observatories possessing long-focus refractors, such as Yale's southern station, Allegheny, and Leander McCormick, measure ten to twenty or more plates of a region on very precise measuring machines. To procure the plates requires a number of years, and the measurements need skill and patience, so that today directly measured parallaxes are counted only in the thousands.

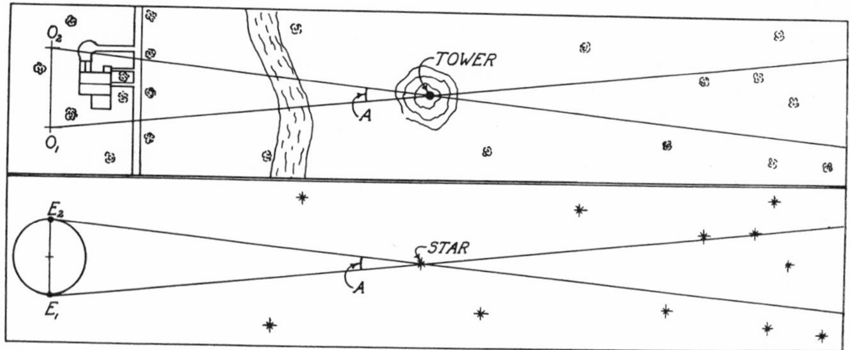

FIG. 1. The principle of triangulation is shown above as applied to a terrestrial problem and below as applied to the distance of a star from its heliocentric parallax. Observations from O_1 and O_2 or E_1 and E_2 determine the parallax angle A and hence the distance of the tower from O and the star from earth (E).

Sirius, the brightest star, is some 54 million million miles from us, while Alpha Centauri, the nearest known star to the sun, is about 25 million million miles away, so that its light takes about 4¼ years to reach us. At its distance, the astronomical unit (distance from earth to sun) makes an angle of only about 3/4 of a second of arc, or about 1/1,728,000 of a whole circle!

One of the chief advantages of trigonometric parallaxes, which are fairly accurate for stars within 200 light years (1200 million million miles), is that they provide a way to check the accuracy of other ways of ascertaining the distances of the stars. If other methods[1] show comparable results for the nearer stars, the astronomer can use them for the more distant objects with greater confidence.

From time immemorial the Milky Way attracted the attention of mankind, but it was not until the telescope revealed some of the individual stars that any conception of its true nature was realized. Sir William Herschel was one of the first to show by actual observations

[1] The "other methods" include comparison of measured star brightnesses: fainter stars are expected to be more distant. Distances determined from trigonometric parallax, however, showed that this is not true in all cases. Some farther stars are brighter than nearer ones. There are wide differences in intrinsic luminosities—the brightnesses stars would have if they were all at the same distance from us. Luckily, it was found that a star's luminosity can be estimated from its spectrum, and in the case of pulsating *Cepheid variables*, the luminosity can be determined quite accurately from the pulsation period. Thus rough estimates (made, first, by Sir William Herschel) of the distances of faint stars in the Milky Way (based simply on how faint they appeared to be) have been refined to yield the more accurate distances given later in this article.—TLP

and counts of stars that the existence of a watch-shaped arrangement of stars located in the plane of the Milky Way would explain the immense concentration of stars in every direction in that plane. The comparative thinness of the system would explain the much smaller number of stars we see when looking at right angles to the plane of the Milky Way (galactic equator)—that is, when we look toward the north pole of the Galaxy, in Coma Berenices, or toward the south pole, in Sculptor.

Careful study of the movements of stars across the sky (*proper motions*) antedated the measurement of their distances by more than a century. It was Edmund Halley, of comet fame, who first showed that Arcturus and Sirius were changing their positions among the stars, so that stars could no longer be regarded as "fixed." Although rapid strides were made in the measurement of stellar motions, it was not until the present century that sufficient information became available to prove that stars in the Milky Way Galaxy are in rotation around a common center. It had been thought, of course, that such rotation was necessary for the dynamic equilibrium of such a flattened system. The study of the motions of the stars, including their motions in the line of sight (radial velocities determined from shifts in the lines in stars' spectra), finally showed that the center of rotation is in the direction of the stars in Sagittarius. This also explained the density of the stars in that direction, where they are more abundant than in any other part of the sky, for we are looking toward the region of greatest concentration about the galactic center. It is now thought that the *cosmic year*, or time for the sun and its planets to turn once around the galactic center is about 200 million years; the distance to be covered is so great that the sun must travel about 150 miles per second along its galactic orbit.

How big is our Galaxy? This problem has been studied for many years. The evidence seems to show a probable diameter in excess of 100,000 light years[2] and a thickness of about 10,000 light years. There seems to be a central nucleus extending over about a quarter of this area. Harlow Shapley has recently shown "the existence of a spherical haze of stars surrounding the ellipsoidal system" mentioned above. Our sun is about 30,000 light years from the center and a little above the central galactic plane. This latter position is revealed by the fact that the Milky Way in the sky seems to be slightly displaced to the south of the galactic equator, which is itself, of course, a great circle on the sky.

[2] One light year, the distance that light travels in one year, is about 10^{13} km or 6×10^{12} mi. Astronomers prefer the larger unit, parsec, which is the distance of a star with parallax 1″. One parsec (psc) is 3.26 light years, over 200,000 times the distance to the sun.—TLP

Incidentally, it may help our inferiority complex to learn that our sun is not quite so insignificant in the star family as formerly supposed. Of all the known stars located within 16 light years of us, our sun is surpassed in absolute brightness by only four and is superior in candle-power to the other known forty-two stars within this sphere of space. There is also the possibility that the discovery of more invisible dwarf companions and of other faint stars may increase the number of those in the latter group.

Charles Messier (1730–1817), a Frenchman searching for comets, found that some of the hazy objects in the sky did not change their positions in relation to their neighboring stars and so were not comets. To eliminate the work of repeating observations on objects which he had previously determined were not comets, he recorded these starless nebulosities (*nébuleuse sans étoiles*) in his now famous catalogue [see p. 28]. The objects in this catalogue are commonly designated by the letter M followed by a number, there being 109 in all. This list, made over 160 years ago, has since been classified as containing 30 open clusters, 29 globular clusters, four planetary nebulae, seven diffuse nebulae, 35 galaxies, and two unconfirmed objects, either comets or stars mistaken for nebulae by Messier. His small telescope was unable to show any individual stars even in the mighty Hercules globular cluster, M 13.

Photographic plates [such as that reproduced in Fig. 2] taken by modern telescopes, have shown the number of stars in M 13 that are bright enough to be recorded on the plates exceeds 60,000, and it is

FIG. 2. The globular cluster, M 13, in Hercules photographed with the 200-inch Hale Telescope. (Mount Wilson and Palomar Observatories photograph)

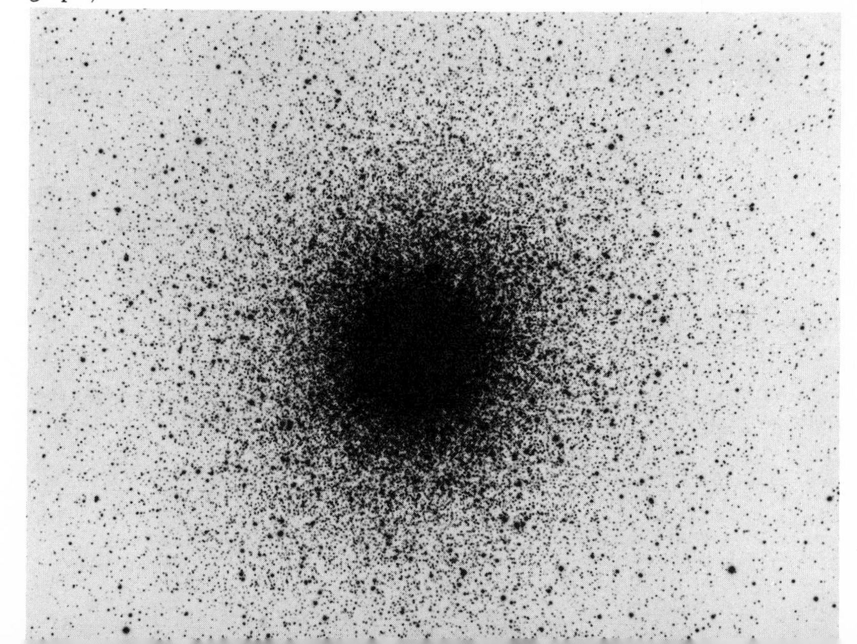

estimated that there are probably over 100,000 stars in the cluster. Photographs of the surrounding background of the sky record many faint galaxies, which shows that this region is not obscured by dust clouds, and we can rely on the apparent luminosity of the stars in this cluster in determining its distance and its actual dimensions.

Fortunately, there are Cepheid variables in M 13. The period of the variation of a Cepheid tells us its absolute magnitude. This has been proved by thousands of observations of Cepheids in such remote objects as the Magellanic Clouds. If we know the absolute magnitude of a star, it is easy to compute its distance, using also its apparent magnitude. The distance of the Hercules cluster is found to be about 36,000 light years, and its diameter is about 320 light years, including the fainter, less dense outlying portions. A count of the number of stars of different magnitudes per unit area that can be resolved in the outer regions of the cluster, combined with study of the apparent density of the unresolved stars in the central portion, makes possible the above estimates of the total population. M 13 is similar in size and star density to most other globular clusters. The "Giant of Palomar" [200-inch telescope, under construction in 1944] may help resolve the central mass of this beautiful naked-eye cluster and give us a better knowledge of its true structure.

In Canes Venatici is M 3, whose two hundred cluster-type Cepheids, with periods of less than a day, have been used to establish standards for measuring distances to other clusters. M 4, in Scorpius, may be the nearest globular cluster, but it is obscured by intervening dust clouds and is not conspicuous. The southern regions of the sky are favored with 47 Tucanae, a giant globular cluster appearing near the Small Magellanic Cloud but not physically associated with it. There is also the famous Omega Centauri, 21,000 light years distant, and in Sagittarius is M 22, a bright object near the Milky Way center.

Globular clusters are located north and south of the plane of the Milky Way, to about 50° north and south galactic latitudes, and while they are more strongly concentrated near the central plane, they avoid completely the galactic equator. This peculiar distribution may be accounted for by the absorption by dark matter in the galactic plane, for the center of the system of globular clusters is near galactic longitude 327°, latitude 0°. This is close to the meeting place of Scorpius, Ophiuchus, and Sagittarius, where the star clouds of the Milky Way are brightest [see Fig. 3].

A quarter of a century ago Harlow Shapley proposed that the spatial arrangement of the ninety-three globular clusters then known indicated the central point around which our Galaxy of stars, dust, nebulae, and

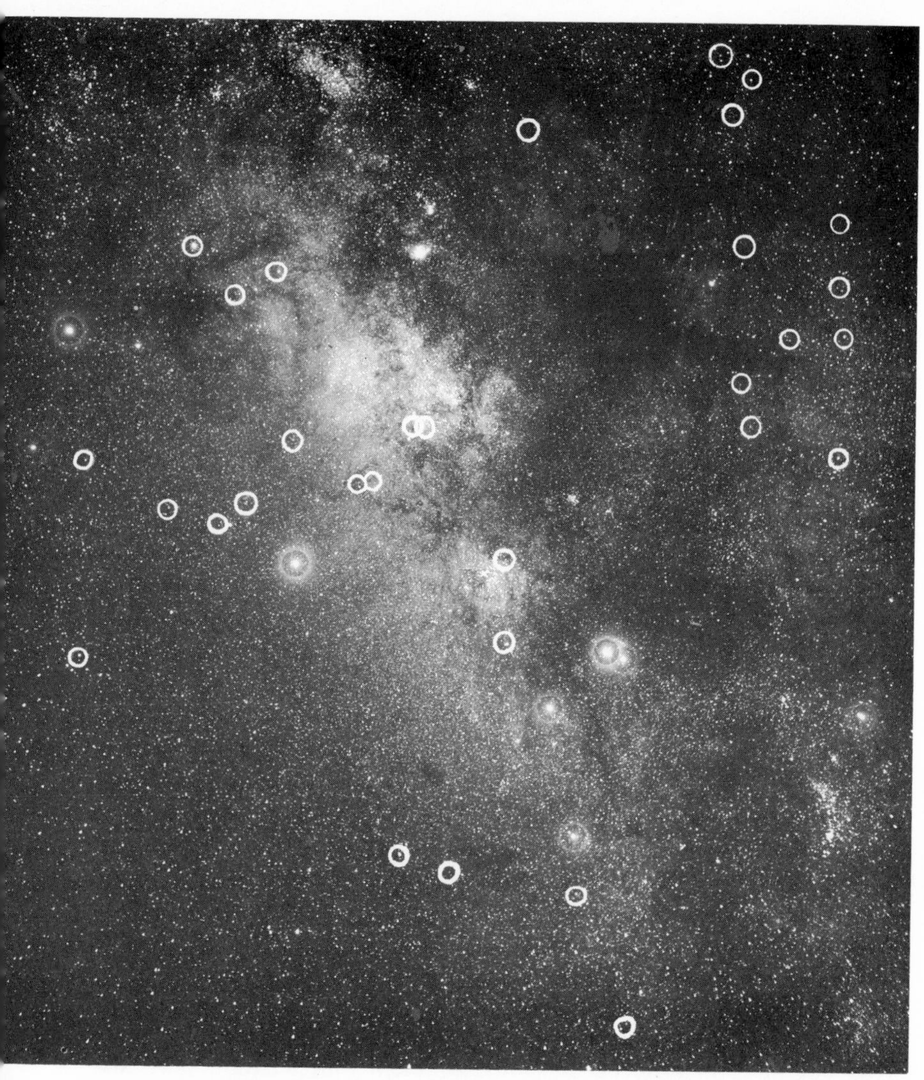

FIG. 3. The distribution of the globular clusters (circled). One third of all globulars are shown here, centering on the Milky Way in Sagittarius. (Harvard College Observatory photograph)

clusters is also arranged. As the globulars are found mostly in the half of the sky centered on Sagittarius, it was apparent that our sun was rather remote from the galactic center. Other methods of investigation have amply confirmed the original assumption, and the globular clusters are found to form a kind of spherical framework for the more flattened system of stars.

Thus has man realized the immense size of his own Galaxy. But one should not be too puffed up with its importance, for "pride goeth before a fall," and a humble spirit will be in order when we consider our neighboring galaxies.

Ferdinand Magellan, in planning his trip to the Spice Islands [1519–1521] by a southwesterly course, was assisted by Ruy Faleiro, an astronomer. When the latter, after casting his own horoscope and finding it unfavorable, decided not to go, the world lost a chance for an early technical description of the Large and Small Magellanic Clouds. These objects were named in honor of the Great Circumnavigator, whose sudden death in the Philippines prevented his writing his own memoirs. The Clouds were recorded, however, by Antonio Pigafetta, a volunteer associate. Their location, at two corners of an equilateral triangle which they form with the south celestial pole, was useful to navigators, but little was known of their structure for many centuries. Certainly, no one dreamed how useful they would become in our time as laboratories in which to establish standards for the measurement of the universe.

The Englishman Thomas Wright wrote in 1750 that the Milky Way is only one of many systems of stars, though the others may not resemble the Galaxy or each other in structure. "That this in all probability may be the case is in some degree made evident by the many cloudy spots just perceivable by us . . . in which, although visibly luminous spaces, no one star or particular constituent body can possibly be distinguished. These, in all likelihood, may be external creations, bordering upon the known one, too remote for even our telescopes to reach."

In 1873 a *Manual of Astronomy* by Henry Kiddle described the "Greater and Lesser Cloudlets." "The former is in Dorado," it said; "the latter in Toucan. These objects are distinguished for their great extent, the larger covering a space of about 42 square degrees, and the smaller being of about one fourth that extent, but of greater brightness. The telescope of John Herschel decomposed them into separate stars, star clusters, and numerous distinct nebulae. In the larger cloud [see Fig. 4] Herschel counted 582 single stars, 46 star clusters, and 291 nebulae; in the smaller cloud, 200 single stars, 7 star clusters, and 37 nebulae. In the immensity of their extent and diversity of objects which they present, they are only comparable to that apparently the greatest of all clusters, the Milky Way."

In 1900 the Bruce 24-inch photographic refractor at the Harvard station, then at Arequipa, Peru, began to record the southern skies on large-size plates. The Magellanic Clouds showed tens of thousands of stars and hundreds of star clusters and gaseous nebulae. Then Henrietta Leavitt, at the Harvard Observatory, started the series of events which has so amazingly increased our knowledge of the size of the known

FIG. 4. The Large Magellanic Cloud. (Harvard College Observatory photograph)

universe. She began to record the apparent magnitudes of the stars on the Bruce plates, publishing in 1906 a list of 808 faint variables in the Large Cloud and 969 in the Small. Later she showed that the period of each variable (time between maxima) had a distinct relation to its luminosity.

Ejnar Hertzsprung of the Netherlands showed that the Cepheid variables of the Milky Way are giant (highly luminous) stars, and if the Cepheids in the Magellanic Clouds were considered to be giants also, as seemed likely, they must be very distant to appear so faint. Then Harlow Shapley gave additional proof by a study of the short-period Cepheid variables in the globular clusters, and finally produced a diagram relating the period of a Cepheid to its absolute luminosity.

Here, then, was a standard to measure the distances to the faint objects. Assuming that all Cepheid variables of the same period are of equal absolute luminosity, the distance can be computed by a simple relationship between the absolute and apparent magnitudes. The results were startling. The distance to the Large Magellanic Cloud is 75,000 light years (a light year is 6 million million miles), and that to the Small Cloud is 84,000 light years. Their centers are about 30,000 light years apart.

Recent plates taken in red light have shown a haze of stars around the main clouds, indicating the possibility that the two systems are nearly in contact or even contained in a common network. The Magellanic Clouds and our own Galaxy have a mutual gravitational relationship, and the haze of stars surrounding each system may be actually in contact with the others.

If a man from a primitive tribe of central Asia were taken in a wooden case on board the *Queen Mary*, and then allowed to look in several directions through holes bored in the sides of the box, he would have a very vague idea of the shape and functions of an ocean liner. To one who had seen large ships such glimpses would reveal much more, and to a captain of such a vessel many details would be easily recognizable in spite of the limitations of the view through the holes in the box.

Let us, then, take a look outside of our own Galaxy and study the appearance of other systems, from which we may get a better idea of the probable structure of our own. Our nearest "sister" among the galaxies is the Great Nebula in Andromeda, M 31, which may be seen easily with the naked eye as an elongated patch of light not far from the star Beta in Andromeda.

An amateur's telescope confirms the elongation of this galaxy as we see it, but long-exposure photographs through large instruments are required to show the wealth of detail, and it is only recently that the 100-inch telescope [at the Mount Wilson Observatory] has been used to resolve into stars the very bright nucleus of this galaxy. The Andromeda system is probably a flattened, disklike spiral, which would appear to us (as many other galaxies do) as a thin wafer were its principal plane directly in our line of sight. That plane is, however, tilted 15° from the line of sight, so the galaxy appears as a rather elongated ellipse. It has a greatest diameter, in ordinary photographs, of about 35,000 light years, but studies by means of the microdensitometer, an instrument for measuring the intensity of a photographic image, show it to be surrounded by a three-dimensional haze nearly twice this size. The stars which compose this haze are not bright enough to be seen as individuals, but their presence constitutes an interesting analogy to the spherical haze of stars surrounding our own Galaxy. In size and general appearance, then, M 31 is similar to the Milky Way system.

The Andromeda galaxy is rich in novae, supernovae, and Cepheid variable stars. By means of the periods and apparent magnitudes observed for the Cepheids, the galaxy's distance is estimated at about 750,000 light years,[3] with an uncertainty of about 10 per cent. The intervening space between us and M 31 is reasonably clear of absorbing gas and dust, as shown by the presence of an average number of stars in our own Galaxy visible in this part of the sky as well as by the observation of an average number of much more distant external galaxies in the area around the Andromeda system.

FIG. 5. The triple system in Andromeda: M 31, M 32 (just left of center), and NGC 205 (below center, ⅝ inch from bottom). The circle indicates a region with surface brightness equal to that in our Galaxy at the position of the sun relative to the galactic center. (Photograph by Hans Vehrenberg)

Closely associated with M 31 [see Fig. 5] are M 32 and NGC 205. M 32 is a spheroidal galaxy, 1800 light years in diameter and without supergiant stars. NGC 205 is even smaller than M 32 [and was not catalogued by Messier (see p. 30)]. These two systems and the nucleus of M 31 have recently been resolved on photographs showing individual stars.

[3] As shown in later chapters, the more accurate modern estimate of the distance to M 31 is 2 million light years. The estimate of 1945, as well as the distances to other galaxies estimated then, was smaller than present-day measurements because of an error in measuring the intrinsic luminosities of Cepheid variables, which were used as distance indicators (see p. 49). The way in which this error was discovered is described in Volume 7 of this series, *Stars and Clouds of the Milky Way*.—TLP

M 33, a neighbor of M 31, situated in the adjoining constellation of Triangulum, is a disk of about 12,500 light years' diameter, tilted at an angle of some 30°. It has a sphere of faint stars around it and is on the large side as spiral nebulae go, but it contains only a fraction as many stars as M 31. It is a much fainter object as we see it, although a few observers have reported observing it without optical aid.

M 31 and M 33 each have about fifty known Cepheids and many supergiant stars; in addition, about twenty-five novae are seen in M 31 each year. This is many times more than are visible to us in our own Galaxy, because of our point of disadvantage amid the dust and dark clouds along the plane of the Milky Way.

In another direction in space, in Sagittarius, at a distance of about half a million light years, lies NGC 6822, an irregular patch known as Barnard's galaxy. Another neighbor is IC 1613, located between Pisces and Cetus, but at a distance of about 900,000 light years.

Recently two other galaxies have been discovered to be neighbors of ours. One near the south galactic pole, in Sculptor, is in the form of a globular star cluster but with the diameter of a galaxy, 5500 light years. It is about 250,000 light years away, and has no apparent supergiant stars, no open clusters or gaseous nebulae. Another such system is in Fornax, 500,000 light years distant; however, it has a few globular clusters associated with it and is several times brighter intrinsically than the Sculptor system.

Thus our family, or Local Group, of galaxies includes eleven known systems which we have enumerated: the Milky Way Galaxy, the Andromeda spiral, both in the giant class; M 33, a medium-sized spiral; the small NGC 205 and M 32; the irregular systems of the Large and Small Magellanic Clouds; the small irregulars NGC 6822 and IC 1613; and the small symmetrical systems in Sculptor and Fornax.

In addition to these there are other candidates for admission to the Local Group: spirals NGC 6946, IC 10, and IC 342, and some six others not yet catalogued. Their status is not determined because of the uncertainties in distance estimates; they all, however, seem to be within a million light years of our Galaxy and so are near neighbors, astronomically speaking.

The Local Group of Galaxies[1]

GEORGE O. ABELL

(*Sky and Telescope*, January 1964)

Our own Galaxy is now recognized to belong to a small cluster—a congregation of seventeen members called the Local Group. The Local Group, extending over a diameter of some 3 million light years, contains three spiral galaxies—our own, the Andromeda galaxy (M 31), and the Spiral in Triangulum (M 33). It also contains four irregular galaxies, four rather smallish elliptical galaxies, and six dwarf elliptical galaxies. These latter are stellar systems only a few thousand light years across and which contain only a few million stars. Our Galaxy, in contrast, is about 100,000 light years across and contains roughly 100 billion member stars. There are a few other galaxies whose membership in the Local Group is possible but not certain.

In the Local Group, at least, it is seen that the most common kind of galaxy is an elliptical of dwarfish dimensions. The same is observed to be true in the few other groups and clusters of galaxies that are near enough for such dwarfs to be identified.

The accompanying table lists the known members of the Local Group and some of their properties. The visual magnitudes have not been measured for all of the galaxies, and not all of those that have been measured are known with much precision. The distances and diameters listed represent the author's judgment of the best data available today, but some of these figures are quite provisional and can be expected to undergo future revision. The absolute magnitudes are more uncertain yet; some (given in parentheses) are merely educated guesses. Masses have been calculated from the rotation rates of four of the galaxies. Masses for three dwarf ellipticals (in parentheses) are based on the assumption that the radii of those galaxies are limited by the tidal forces produced on them by our own Galaxy.

[1] The material here is reprinted from the August 1963 *Griffith Observer* by permission of its publisher, Griffith Observatory and Planetarium, Los Angeles, California.

TABLE 1. MEMBERS OF THE LOCAL GROUP OF GALAXIES

Galaxy	Type[1]	Right ascension (1950) h m	Decli-nation (1950) ° ′	Visual mag. (m_v)	Distance kilo-pscs.	Distance thous. lt. yrs.	Diameter kilo-pscs.	Diameter thous. lt. yrs.	Absolute mag. (M_v)	Radial velocity km/sec	Mass (solar masses)
Milky Way	Sb						30	100	(−21)		2×10^{11}
Large Magel. Cloud	Irr I	5 26	−69	0.9	48	160	10	30	−17.7	+276	2.5×10^{10}
Small Magel. Cloud	Irr I	0 50	−73	2.5	56	180	8	25	−16.5	+168	
Ursa Minor System	E4 (d)	15 08.2	+67 18		70	220	1	3	(−9)		
Sculptor System	E3 (d)	0 57.5	−33 58	8.0	83	270	2.2	7	−11.8		(2 to 4×10^6)
Draco System	E2 (d)	17 19.4	+57 58		100	330	1.4	4.5	(−10)		
Fornax System	E3 (d)	2 37.5	−34 44	8.3	190	600	6.6	22	−13.3	+39	(1.2 to 2×10^7)
Leo II System	E0 (d)	11 10.8	+22 26	12.04	230	750	1.6	5.2	−10.0		(1.1×10^6)
Leo I System	E4 (d)	10 05.8	+12 33	12.0	280	900	1.5	5	−10.4		
NGC 6822	Irr I	19 42.1	−14 54	8.9	460	1500	2.7	9	−14.8	−32	
NGC 147	E6	0 30.4	+48 13	9.73	570	1900	3	10	−14.5		
NGC 185	E2	0 36.1	+48 04	9.43	570	1900	2.3	8	−14.8	−305	
NGC 205	E5	0 37.6	+41 25	8.17	680	2200	5	16	−16.5	−239	
NGC 221 (M 32)	E3	0 40.0	+40 36	8.16	680	2200	2.4	8	−16.5	−214	
IC 1613	Irr I	1 00.6	+ 1 41	9.61	680	2200	5	16	−14.7	−238	
NGC 224 (M 31)[2]	Sb	0 40.0	+41 00	3.47	680	2200	40	130	−21.2	−266	4×10^{11}
NGC 598 (M 33)[2]	Sc	1 31.0	+30 24	5.79	720	2300	17	60	−18.9	−189	8×10^9

[1] (d) indicates a dwarf system. [See p. 172.]

[2] M 31 and M 33 are the bright objects in Andromeda and Triangulum, respectively.

The Local Group is now considered to have the seventeen members listed in Table 1. However, there may be others hidden behind nearby clouds of obscuring dust in the Milky Way. In fact, these dust clouds are so numerous in the disk of our Galaxy that the band around the sky centered on the Milky Way and extending 10° to 20° on either side of the galactic equator is called the "zone of avoidance." With a few exceptions, no galaxies are seen in this zone. The exceptions occur in "windows," or "holes," between the dust clouds.

The "galactic equator," a circle extending completely around the sky following the center of the Milky Way, is used by astronomers as the basis of coordinates in the sky: "galactic latitude, b" and "galactic longitude, l." These two angles, analogous to latitude and longitude on the spherical earth, serve to describe the location of a star, or cluster, or galaxy on the sphere of the sky. Galactic latitude is measured in degrees of arc north (positive) or south (negative) of the galactic equator. Galactic longitude is measured (also in degrees) eastward along the galactic equator from a specified point. Before 1961, this point ($l = 0$) was where the galactic equator crosses the celestial equator in the constellation Aquila. In 1961 galactic longitude was redefined (and designated l^{II}); it is now measured from the center of the Milky Way Galaxy in Sagittarius. The galactic equator was also changed slightly, and the new galactic latitude is written b^{II}. The reason for these changes, fully described in Volume 7 of this series, was that measures of hydrogen 21-cm radio radiation show an accurate ring around the sky at $b^{II} = 0$. This is caused by hydrogen in the disk of the Galaxy.—TLP

A Northern Window in the Milky Way

(*The Telescope*, September–October 1938)

It has been well known for a hundred years or so that spiral nebulae avoid the Milky Way. The word "avoid" is, of course, used in a special sense, indicating that we find spirals in high galactic latitudes and intermediate latitudes, but that in low latitudes, within 10° or 20° of the galactic circle, we rarely if ever see or photograph these external systems. Years ago this apparent avoidance was frequently considered to be some fundamental detail in the structure of the universe. It was thought that perhaps fields of stars and nebulae were for some reason, again unknown, mutually exclusive. But with the demonstration that the spiral nebulae and their kind are actually external galaxies, the avoidance was immediately seen to be apparent only—the result, in fact, of obscuring matter along the plane of the Milky Way. The external galaxies no doubt exist in more or less equal abundance in all latitudes, but in and near the Milky Way star stratum the light from the external systems is so greatly dimmed by the absorbing and scattering gases and dust of interstellar space that the apparent avoidance results.

Fortunately the great cosmic clouds of dust are pretty well localized along the plane of the Galaxy, and the spaces in high latitudes, toward the poles of the galactic system, are essentially transparent. In these high latitudes we can study the external galaxies with confidence that the observed luminosities and the deduced distances are not seriously disturbed by space absorption.

If external galaxies were scattered with absolute uniformity throughout space, we could use the observed distribution as a rigorous indication of the location of absorbing clouds and their potency in dimming light. Even with the considerable departure from uniformity, the galaxies still provide us a useful tool for outlining the zone of avoidance along the Milky Way. One of the extensive nebular surveys under way at the Harvard Observatory emphasizes the distribution of galaxies in the two belts of galactic latitude, each 10° wide, which lie between the latitudes 20° and 30° on the northern and the southern sides of the Milky Way. The irregularities in the distribution of the faint galaxies in these belts will indicate both true irregularities in the distribution of galaxies and apparent irregularities that arise from interposed absorption. But if any galaxies whatever are found in a given region within these belts, we have evidence that the absorption is not very great. In the direction of the rich star clouds toward the center of our own Galaxy, in Sagittarius, Scorpio, and Ophiuchus, no external galaxies whatever are found over areas of many square degrees. But in a few places along the Milky Way, rather near the galactic equator, a few systems shine through. When many are found we call the particular low-latitude region in which they occur a galactic window—an opening, more or less transparent, in the clouds of absorbing matter. A southeastern galactic window, in the constellations of Pavo, Ara, and Telescopium, has been extensively studied in recent years at Harvard. Not only has a census been made of the external galaxies, but the variable stars have been measured because they can throw light on the extent of the Milky Way in this particular direction. It is only in this south-

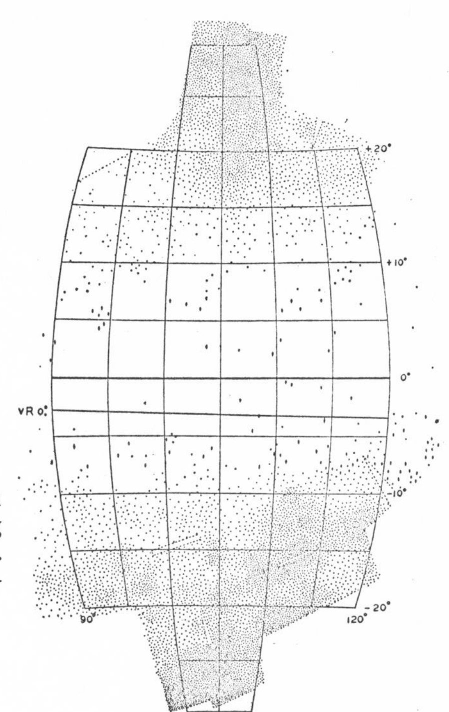

FIG. 6. Distribution of galaxies in Perseus and Cassiopeia, showing the galactic window near l = 120°. (Harvard College Observatory diagram)

eastern window that we have definitely located large numbers of stars in the Milky Way that lie beyond the galactic center. In the northern part of the Milky Way there is something of a galactic window, which is indicated by recognized external galaxies near the galactic equator. Figure 6 illustrates one result of the research by Harlow Shapley and Rebecca Jones on a large region in the constellations of Cassiopeia and Perseus. The area studied covers about 1400 square degrees. The coordinate system of the figure gives galactic latitude [b^I] and galactic longitude [l^I]. The slanted line below center [labeled *vR*o°] shows the galactic equator as derived by the Dutch astronomer P. J. van Rhijn from the distribution of faint stars. (Note that the semitransparent window comes much nearer to this than to the galactic equator in its conventional position.) Each dot in the diagram indicates an external galaxy. In low latitudes, near the galactic equator, there are not many galaxies shining through; however, in comparison with the average population along the Milky Way the number is very large. In this direction the absorbing material of the Milky Way is relatively thin. It is, in fact, about 150° from the direction to the center of the galactic system. . . . The distance to the rim of the Galaxy in the direction of this northern galactic window may not be very large.

Man and His Expanding Universe (II)

PERCY W. WITHERELL

(*Sky and Telescope*, February and March 1945)

Observations by many astronomers, made during more than a century, resulted in the discovery of thousands of hazy objects, galaxies, nebulae, star clusters, and the like, which are recorded in Dreyer's *New General Catalogue* (1890) and its supplements. A plot of the galaxies in this list showed a very uneven distribution in different directions. Whether this was real, or due only to the nonsystematic arrangement of the regions that had been under observation, was not definitely established until the Shapley-Ames catalogue was compiled at Harvard College Observatory in 1932.

In this catalogue, the positions and photographic magnitudes of all such objects down to magnitude 12.8 were determined for the entire celestial sphere. A plot of these 1249 galaxies shows 823 on the northern side of the Milky Way and 426 on the southern side. There are almost none recorded along the Milky Way, because the interstellar dust along our galactic plane completely obscures or seriously decreases the brightness of objects beyond that dust, so that such objects appear fainter than the limit of the catalogue. Giant galaxies away from the Milky Way are recorded to a distance of 20 million light years, dwarfs to about 4 million, with an average of 10 million for the list.

In the northern part of the constellation Virgo, and stretching over into Coma Berenices, about midway between Arcturus and Regulus, is a concentration of one hundred bright galaxies known as the Virgo cluster. None of these is visible to the naked eye, as they are 8 million light years away. The Shapley-Ames list includes only the giant Virgo galaxies. If fainter magnitudes, to the fifteenth, are included, the number in the cluster is doubled. In the vicinity are many more galaxies, scattered through Lynx, Coma, and Ursa Major, in a region 40° in length. [A small part—about one square degree—is shown in Figure 7.] Fortunately, this group is located in a position for favorable observation, in that space absorption is not especially troublesome. About three fourths of the galaxies are spirals, the rest spheroidal, with a few irregular members similar to the Magellanic Clouds. The cluster as a

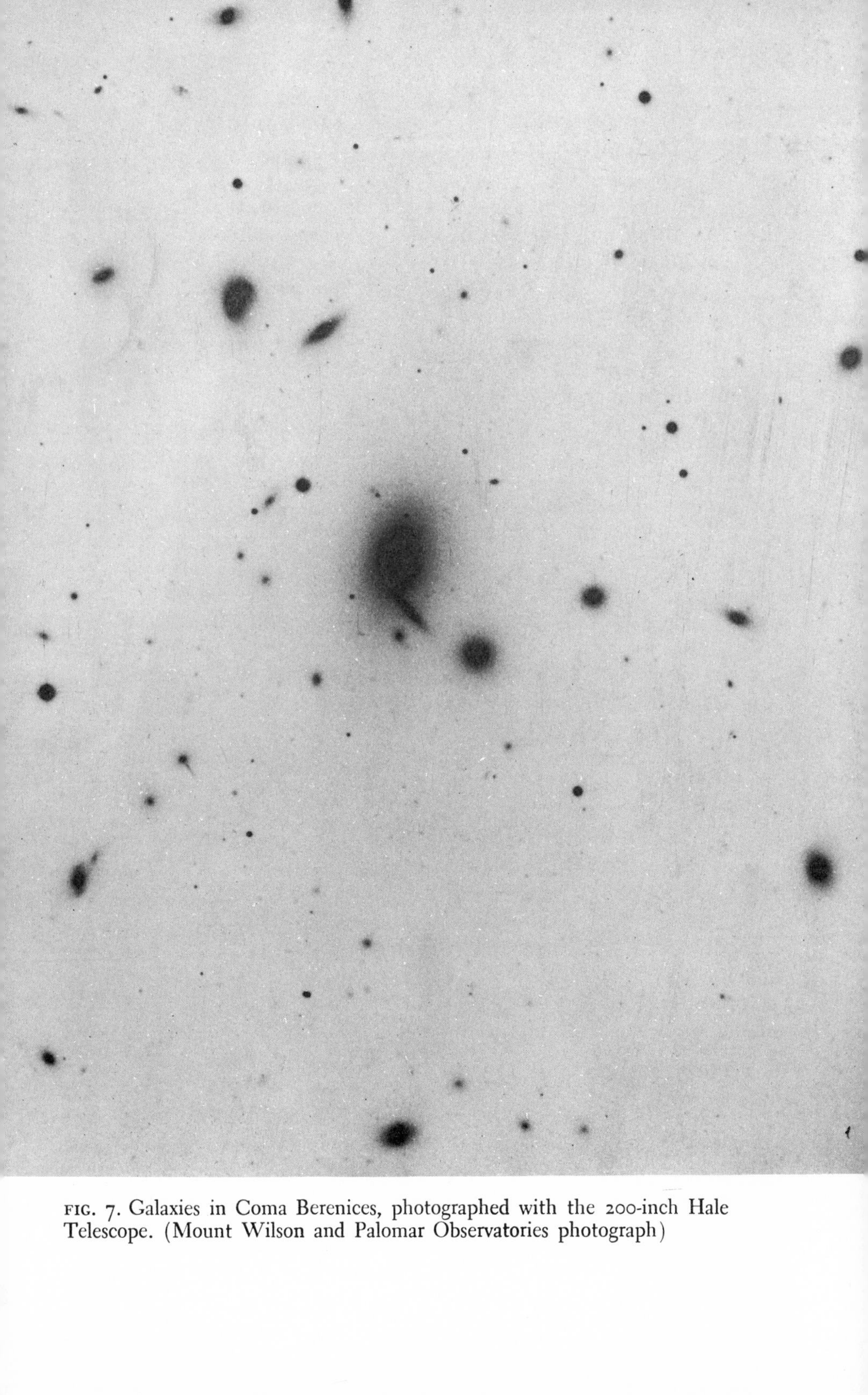

FIG. 7. Galaxies in Coma Berenices, photographed with the 200-inch Hale Telescope. (Mount Wilson and Palomar Observatories photograph)

whole is receding from us at the rate of 700 miles per second. The individual members are very active, some moving at speeds of 1500 miles per second inside the group.

The total mass of this aggregation of galaxies is so enormous that the individual members average 200 billion suns. As this seems too large for the observed luminosity of the group as a whole, it is possible that a considerable amount of the mass is composed of nonluminous matter in the intergalactic spaces. Although the spectra of some of the galaxies show the presence of groups of hot blue stars or very bright nebulae, the average spectrum is like the sun's.

Southward of the main group of the Virgo cluster, there are many galaxies that extend over 30° toward the constellation of Centaurus. Northward are many more galaxies, as mentioned, and these are of about the same luminosity and distance. Are all these part of a great cloud of galaxies, of which our Local Group is a small condensation and the Virgo group a much larger concentration? If so, the cloud extends over 3 million light years.

In the southern skies, in the constellation of Fornax, are seventeen galaxies which undoubtedly are interrelated. The brightest member, NGC 1316, is of the spheroidal type, and nine of the seventeen galaxies are spheroidal. Other galaxy groups are similar in having a spheroidal as their brightest member.

We have noted the existence of the gigantic galaxies of Andromeda and our own Milky Way system, surrounded by the medium and dwarf galaxies in our neighborhood group. Although the smaller galaxies are not observed in the distant clusters, it is a fair assumption that they exist among the large types which we do observe. The evidence seems to be that the brightest and faintest galaxies in a group are about 2.5 magnitudes brighter and fainter, respectively, than the average. We observe the brightest portion, and use this hypothesis to estimate the total luminosity of the group, and so its distance. It is found on the average that the absolute magnitude of the fifth brightest galaxy in a group is −16.4, and with this assumption and that galaxy's apparent magnitude, the average distance of the group can be determined with fair accuracy. It is better to use the fifth brightest than the first brightest galaxy in a cluster, for then the chances are much less that a bright foreground object has been accidentally chosen.

By this method, twenty-five clusters of galaxies with as many as or more members than the Virgo system have been measured. There are a hundred groups with numbers similar to our Local Group. There are thousands of irregular condensations that suggest mutual association.

Double and triple galaxies in close gravitational relationship are common. Some of these are of similar type; others are as dissimilar as a spheroidal alongside one of the Magellanic Clouds would be.

We have previously suggested that a survey of the outside systems would help us understand our own Galaxy. Conversely, the range from double stars through all degrees of association up to the famous globular clusters that we find in our own system has counterparts among the distant objects of the metagalaxy.

Not satisfied with the knowledge gained of the arrangement of the universe to a distance of 10 to 20 million light years, and with curiosity aroused by hazy images not quite clear enough to classify on the plates of the thirteenth-magnitude survey, astronomers at Harvard were anxious to extend their study to greater distances. Meanwhile, test photographs at Mount Wilson, made with the 100-inch reflecting telescope, and its ability to reach very faint magnitudes, had indicated that the inner metagalaxy might not be a typical sample of the whole universe, which made further study by survey methods imperative.

While the 100-inch continues to probe space in selected but limited areas of the sky, at Oak Ridge, in the town of Harvard, Massachusetts, the 16-inch Metcalf refractor is surveying the northern sky with 3-hour exposures on fast plates, each plate covering 30 square degrees. These plates record stars a little fainter than the eighteenth magnitude and galaxies to 17.6 in a satisfactory manner. A similar systematic survey of the southern sky is being made with the 24-inch Bruce refractor at Harvard's Bloemfontein station.

The combined program has been in progress for over fifteen years, and plates for nearly the whole sky are now on hand for study. The microscopic examination and recording of each of the thousands of images on each plate takes weeks of patient labor. The reward is the discovery of thousands of galaxies not previously recorded.

Some 500,000 new galaxies have been found already during examination of plates covering more than half of the sky. The magnitudes of galaxies are not easy to determine accurately, but thousands of these new systems have also been so classified. The limiting magnitude of the survey corresponds to an average distance of 100 million light years, and many more years of study are required before the census of galaxies in this volume of space can be completed.

The extreme limit of distance reached by the Mount Wilson observations is now about 500 million light years, or five times the distance of the eighteenth-magnitude survey; this represents a volume of space about 125 times as great. In probing this vast region, the 100-inch telescope can obtain only samples here and there, but these are of great value in determining the overall distribution of galaxies in what must be a good-sized portion of the universe.

We have noted the effect of interstellar dust and gas in shutting off the observation of galaxies along the plane of the Milky Way. Probably similar causes create some apparent irregularities in the distribution of galaxies in other parts of the sky. However, in certain high latitudes which are free of absorbing material, there are actual irregularities of distribution. Large clouds of galaxies seem to demonstrate that in the nearer regions of the metagalaxy these systems are not uniformly distributed.

When we described the results of the thirteenth-magnitude survey, we noticed that the number of systems photographed in the northern hemisphere was greater than that photographed in the southern. The eighteenth-magnitude survey does not show as large a difference, but there are still more galaxies in the north than in the south. Even omitting the Virgo and Fornax clusters does not eliminate this effect. It is not now known whether there is a gradual increase of density of population from south to north or whether there may be a great cloud of galaxies in the northern part of the metagalaxy beyond the 100-million-light-year limit of this survey.

Additional evidence of irregular distribution is found in the thousands of faint galaxies in Andromeda, Triangulum, Pegasus, and Pisces. In the southern constellations of Pictor and Dorado there are masses of galaxies more than 100 million light years away.

Thus the evidence to date shows that there are irregularities of distribution of the galaxies in many sections of the sky. Nevertheless, when the sphere under consideration is enlarged to a radius of 500 million light years by use of the Mount Wilson observations, the average space density of the galaxies appears to be nearly uniform. There is definitely no evidence that we are able to observe any part of the universe near its edge; nor, for that matter, have we any indication of the existence of a metagalactic center.

The problem of the size of the universe is inextricably tied up with that of the expanding universe. Spectroscopic study of distant galaxies, successfully carried to a distance of 250 million light years at Mount Wilson, shows that all but the few galaxies in our local group are receding from us. At least, this is the conventional interpretation of the observed shift of the spectra of these galaxies toward the red. The amount of Doppler shift (p. 316) is proportional to their distances; as shown in Figure 8, galaxies 1 million light years away recede at 100 miles

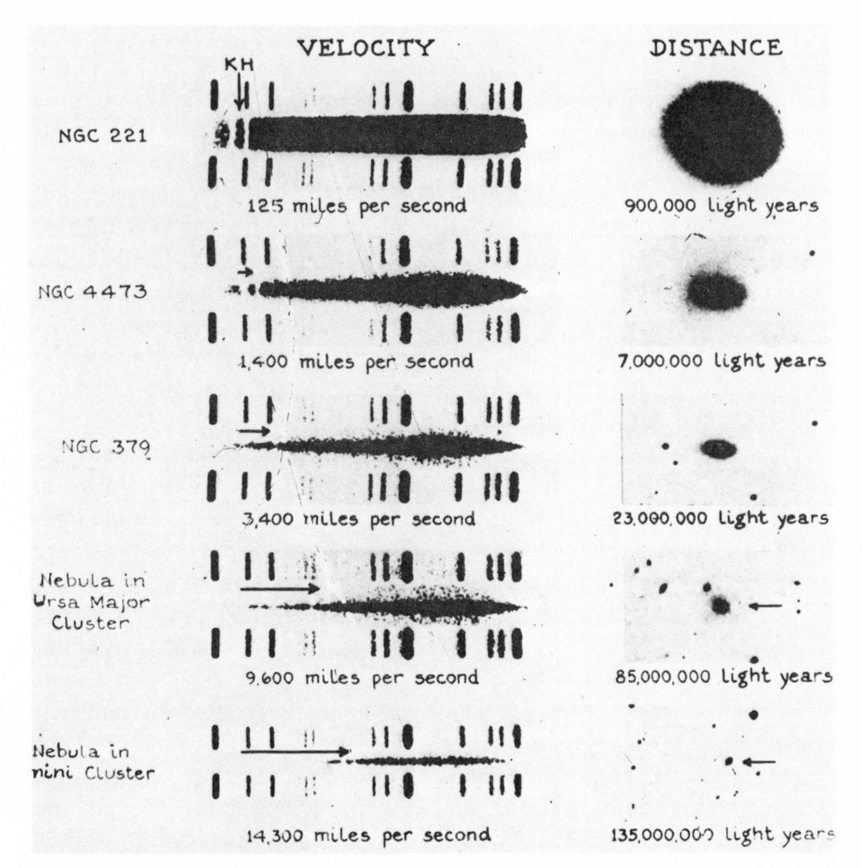

FIG. 8. The right-hand pictures show how, with increasing distance, galaxies appear fainter and fainter. (The numerical distances have since been corrected.) The corresponding red shifts in their spectra provide evidence for an expanding universe. The arrows indicate the conspicuous calcium lines and the amount of the shift in each case. The comparison spectrum is of helium. (Chart by M. L. Humason, from *The Realm of the Nebulae* by Edwin Hubble, Yale University Press, New Haven, Connecticut, 1936)

a second;[1] at 5 million light years the recession is at 500 miles a second; at 25 million it is 2500 miles a second. At 100 million light years, the rate of expansion has a value of 10,000 miles a second; at 250 million light years it is more than one eighth the speed of light!

[1] Because of errors in the distance estimates, this number and the following ones are incorrect. Edwin P. Hubble, who did this work at the Mount Wilson Observatory, discovered the correct "law of red shifts," $v = HD$, where H is the "Hubble constant," at first thought to be 100 miles per second for every million light years. It is now considered to be 18 mi/sec per million light years (or 100 km/sec per million parsecs, since 1 mile = 1.6 kilometers, and 1 parsec = 3.26 light years). The red shift, from which v is measured, is always larger in more distant galaxies, but v never exceeds the speed of light, as shown in Chapter 6.—TLP

The expansion thus seems to give a linear increase in the speed, so that if galaxies are ever observed at distances of some 2 billion light-years, they should appear to be going away from us faster than the speed of light! This is incredible, for modern physics considers that nothing can move faster than light. Perhaps the 200-inch telescope at Mount Palomar [then being assembled] will help solve this problem, for it is expected to reach galaxies as far as 1 billion light years distant. If the linear relation fails to hold, perhaps the universe can still be considered as expanding. If, however, it is indicated that the speed of light will be apparently exceeded, then some other interpretation of the excessive Doppler shifts of the nebulae may take precedence.

Already, many hypotheses have been advanced as alternate explanations of the red shift. The radiations of the distant galaxies have been traveling for many millions of years, through space filled with dust, gas, electrons, atomic particles and building blocks, and subject to the radiation flooding through space in all directions from other stars and nebulae. It is possible that the light has become "tired" and lost some of its original energy. The energy of a quantum of radiation is proportional to its frequency, so that red light represents less energy than blue light. Many astronomers are even now ready to accept this explanation for the red shift of the galaxies.

Incidentally, an expanding universe is a logical result of relativity theory, and this in turn leads to a determination of the present age of the universe. By probing farther into space and giving us spectra of more distant objects, the new Mount Palomar telescope is expected to help solve such problems or at least to limit some of our conjectures. Schmidt cameras, with their ability to photograph fairly large areas to faint magnitudes, are the survey telescopes of the present and future.

Thus, with the improvement of the technique of observations and their interpretation, mankind's knowledge of the universe will continue to expand. As the frontiers are explored, new vistas will appear, and the seeker for the truth will continue to be entranced with the fascinating study of man's expanding universe.

The errors in distance (D), noted in footnotes on p. 11 and p. 23, led, in 1945, to an estimate of the age of the universe (t_0) about five times too small. Hubble's law of red shifts states that the radial velocity of a galaxy is proportional to its distance, or $v = HD$, where H is Hubble's constant. A simple estimate of t_0 is based on the idea that each galaxy has moved away from us at its present velocity v (without any change in v) since it was near us t_0 years ago. That is, the distance it has moved is $D = vt_0$. Since $v = HD$, $vt_0 = v/H$, and

$t_0 = 1/H$. Using the modern value, H = 100 km/sec per million parsecs, t_0 = 1,000,000 parsecs/100 km/sec, or the time needed to go 10^6 parsecs (3.1×10^{19} km) at 100 km/sec, or 3.1×10^{17} seconds = 10^{10} (10 billion) years. The values given in Figure 8 (omitting the nearby NGC 221 and its negative v) are H = 1280, 790, 590, and 560 km/sec per megaparsec, corresponding to t_0 less than 2 billion years. It is now known that the sun and the earth are older than 4 billion years, and the stars in some clusters are as old as 8 to 10 billion years, as shown in Volume 6 of this series. So the new, larger estimates of galaxy distances, giving t_0 = 10 billion years, are more consistent with other age determinations.

In 1945 the observations of radio emission, which we now know to come from stars, nebulae, and galaxies had just begun. An American, Grote Reber, had built a 30-ft. radio telescope (described in Volume 7 of this series) and measured the radio "noise" coming from the Milky Way. Australians, British, and Dutch astronomers led the way in radio astronomy, and discovered that other galaxies beside our own can be "seen" by radio telescopes.—TLP

Radio Energy from M 31

(*Sky and Telescope*, August 1951)

At Jodrell Bank Experimental Station of the University of Manchester, England, R. Hanbury Brown and C. Hazard have completed the first successful detection of radio noise from an external galaxy, M 31, the Andromeda nebula. They employed the world's largest antenna, a paraboloid built of wire mesh with an aperture of 218 feet and a focal length of 126 feet. Although this apparatus is necessarily fixed in position, pointing to the zenith, the primary receiving unit can be swung back and forth at the focus to receive energy at least 15½° on either side of the zenith.

According to the account in *Nature* (November 25, 1950), the experiment consisted of fixing the beam (2° wide) at a number of different elevations, so that in each 24-hour period it swept over a strip of sky 2° wide. Six such strips centered between declinations 38°47′ and 43°00′ north indicated a conspicuous source of radiation at 0^h40^m, + 40°55′, as shown in Figure 9. The coordinates of the Andromeda nebula are 0^h40^m, + 40°59′.

At the radio frequency to which the giant radio telescope was tuned, 158.5 megacycles, the dimensions of the source of radio energy appear to be ¾° by ½°, of the proper size and orientation to be identified with M 31. This size was checked with records of celestial radio "point" sources, in particular the bright radio star in Cygnus, which is known to have a diameter of less than six minutes of arc—small in comparison with the beam width.

From the intensity of the radiation, the effective black-body temperature is about 1000° K, which compares with Grote Reber's similar determination for the nucleus of our Galaxy, 1500°K.

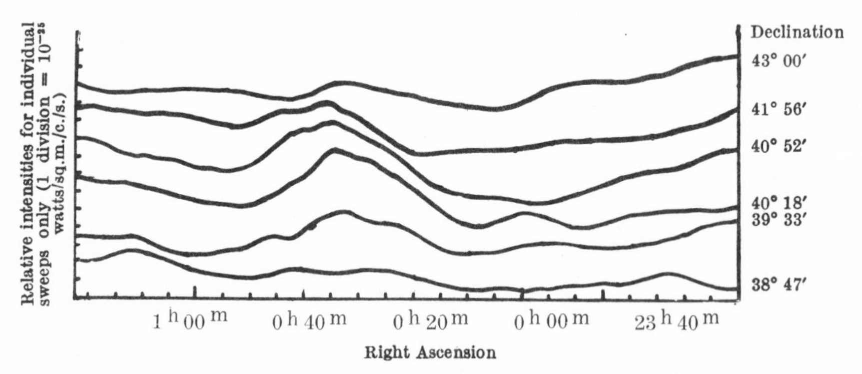

FIG. 9. Coinciding humps in four of the six sweeps, plotted here with arbitrary vertical displacement, reveal the radio emission from M 31. (Courtesy *Nature*)

At about this time Dutch astronomers discovered that large, tenuous clouds of interstellar hydrogen gas "broadcast" at 1431 megacycles frequency, or 21 centimeters wavelength—an "emission line" in the radio spectrum predicted by the theory of the hydrogen atom. The Doppler red shifts illustrated in Figure 8 are changes in optical wavelength ($\Delta\lambda$) from which the radial velocity (v) can be calculated; the change in wavelength (λ) is $\Delta\lambda = \lambda v/c$, where c is the velocity of light, 3×10^5 km/sec. In the radio spectrum, as expected, the Doppler red shift is also observed.—TLP

Radio 21-Centimeter Line in NGC 5668

GEORGE S. MUMFORD

(*Sky and Telescope*, February 1966)

At the National Radio Astronomy Observatory in West Virginia, the 300-foot radio telescope is being used in a general survey of the 21-cm radiation of hydrogen from galaxies. So far eighty-eight systems have been studied, according to Morton S. Roberts in *The Astrophysical Journal* [**142**, 148, July 1965]. Among them, NGC 5668, an Sc spiral [see p. 64] about 15 million light years distant, has the largest radial velocity yet detected at this wavelength.

There is much interest in knowing over how wide a range of wavelengths the simple textbook formula for deriving radial velocities from Doppler shifts of spectral lines applies. And over what range of radial velocities do optical and 21-cm determinations agree?

A comparison made at Harvard in 1962 of the optical and radio line-of-sight velocities for twenty-nine galaxies showed that they have virtually the same values in each case. Thus, the simple Doppler formula is valid over a 500,000-to-1 range of wavelengths, from 21 cm to about 4000 angstroms.

The twenty-nine galaxies have radial velocities from -343 kilometers per second to $+660$. (Positive velocities indicate recession; negative, approach.) And now Roberts reports that NGC 5668 recedes at 1577 kilometers per second, in reasonable agreement with the optical result. This doubles the span in which 21-cm and optical velocities are known to be equivalent.

This consistency of red shifts in optical and radio spectra, and the consistency with cosmological theory (described in Chapters 5, 6, and 7) have so increased confidence in Hubble's law of red shifts that it is now commonly used to derive distances from red shifts measured in the spectra of faint galaxies. The red shift, $\Delta\lambda/\lambda$, is often designated by the symbol z, so that $v = cz$ by the simple Doppler formula, or $D = cz/H = 3 \times 10^9 z$ parsecs, or about $10^{10} z$ light years (using $c = 3 \times 10^5$

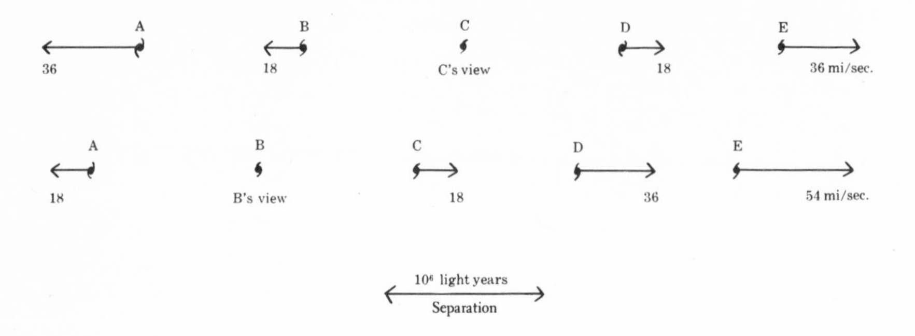

FIG. 10. The Hubble law, $v = HD$, is illustrated here for five galaxies in a line, each 10^6 light years from the next. The upper line shows radial velocities measured from C, the middle galaxy; A and B are moving to the left, D and E to the right, A and E twice as fast as B and D. The lower line shows radial velocities measured by someone on galaxy B who considers that he is at rest. He also sees the other galaxies moving away from him at velocities proportional to distance; hence, Hubble's law implies no center of the universe.

km/sec, $H = 100$ km/sec per million parsecs, and 1 parsec = 3.26 light years). However, if z is larger than about 0.1 (v larger than 30,000 km/sec), the Doppler formula must be modified to fit relativity theory, as shown in Chapter 5. Although values of z larger than 2 have recently been measured, the corrected Doppler formula never gives v larger than c (300,000 km/sec), an upper limit to the speeds of material objects in relativity theory.

It may seem that the measured radial velocities of distant galaxies—away from us on all sides—implies that we are in the center of the universe (and unpopular!). However, the form of Hubble's law, $v = HD$, is such that an observer in each galaxy would get the same impression as we do (see Fig. 10). Another thing that might indicate a center of the universe of galaxies is how they are spread out around us. If most of them were on one side of the sky (like the globular clusters in Figure 3), we would suspect that the center is in that direction. This possibility has been investigated several times by visual and photographic surveys of galaxies all around the sky.

Early surveys led to lists, or "catalogues," such as Messier's, which was published in four parts: M 1 through M 45 in the Histoire de l'Académie Royale des Sciences for 1774, M 46 through M 68 in the Connaissance des Temps for 1781, M 69 through M 103 in the same almanac for 1784, and M 104 through M 109 by his colleague Pierre Méchain in the Berliner Astronomisches Jahrbuch for 1786. (Of these 109 objects, M 40, M 47, M 48, M 91, and M 102 are erroneous—not at the correct positions of nebulae or galaxies or clusters.) The second (much larger) catalogue was the New General Catalogue, published in 1888, and based in part on surveys by Sir William Herschel in England and Sir John Herschel in South Africa. The former contained 109 and the latter 7840 "nebulae" and star clusters which could be seen as fuzzy objects in the telescope eyepiece. Galaxies are known by their numbers in these catalogues—thus, for example, M 31, M 32, and NGC 205 (shown in Fig 5). The overlap between Messier numbers and NGC numbers of galaxies is given in Table 2. The true nebulae (clouds of interstellar gas) are omitted from this list, as are the star clusters.

More recent surveys led to maps, or atlases, of galaxies. One, done by Edwin Hubble with the 100-inch reflector at Mount Wilson Observatory, was based on "sample" photographs separated by about 5°. This established the "zone of avoidance" (see p. 15) and showed that distant galaxies are about equally distributed over the sky, after allowance is made for obscuring dust clouds nearby in the Milky Way.

Three other sets of photographs, exposed longer in order to show fainter galaxies, helped to establish the "number law," $\log N = 0.6m + K$, where N is the average number of galaxies brighter than magnitude m in one square degree, and K is a constant. This important result of Hubble's surveys is consistent with a roughly uniform density of galaxies throughout the spherical volume of space surveyed. That is, when we look twice as far (photographing galaxies one quarter as bright) we see eight times as many. The volume of a sphere is proportional to the

FIG. 11. The Palomar 48-inch Schmidt telescope with Edwin P. Hubble holding the control box and looking through the guide telescope. Photographic plates are loaded in a special holder in the middle of the 20-foot-long tube. (Courtesy Mount Wilson and Palomar Observatories)

TABLE 2. GALAXIES IN MESSIER'S CATALOGUE[1]

		Visual magnitude	*Constellation (Location in the sky)*	*Type*[2]	*Distance*[3]
M 31	NGC 224	4	Andromeda ("great nebula")	Sb	2
M 32	221	9	Andromeda (companion)	E2	2
M 33	598	6	Triangulum	Sc	3
M 49	4472	9	Virgo	E2	40
M 51	5194	9	Canes Ven. ("Whirlpool")	Sc	15
M 58	4579	10	Virgo	Sb	40
M 59	4621	10	Virgo	E5	40
M 60	4649	10	Virgo	E2	40
M 61	4303	10	Virgo	Sc	40
M 63	5055	9	Canes Venatici	Sc	16
M 64	4826	8	Coma	Sb	12
M 65	3623	10	Leo	Sa	20
M 66	3627	9	Leo	Sb	20
M 74	628	10	Pisces ("Pinwheel")	Sc	25
M 77	1068	10	Cetus	Sb	40
M 81	3031	7	Ursa Major	Sb	7
M 82	3034	9	Ursa Major ("Exploding")	Irr	7
M 83	5236	8	Hyades	Sb	15
M 84	4374	10	Virgo	So	40
M 85	4382	10	Coma	So	40
M 86	4406	10	Virgo	E3	40
M 87	4486	10	Virgo (Jet)	Eop	40
M 88	4501	10	Coma	Sc	40
M 89	4552	11	Virgo	Eo	40
M 90	4569	10	Virgo	Sa	40
M 94	4736	9	Canes Venatici	Sb	15
M 95	3351	10	Leo	SBb	25
M 96	3368	10	Leo	Sa	25
M 98	4192	10	Coma	Sb	40
M 99	4254	10	Coma	Sc	40
M 100	4321	10	Coma	Sc	40
M 101	5457	8	Ursa Major	Sc	15
M 104	4594	8	Virgo (Sombrero)	Sb	40
M 105	3379	10	Leo	Eo1	24
M 106	4258	9	Canes Venatici	Sb	15
M 108	3556	10	Ursa Major	Sc	25
M 109	3992	10	Ursa Major	SBb	40

[1] From *Leaflet* 40 of the Astronomical Society of the Pacific, "Messier's Clusters and Nebulae," by Owen Gingerich, October 1967. *NGC* stands for the *New General Catalogue*, a list of 7840 nebulae and clusters published in 1888 by J. L. E. Dreyer, *Memoirs* of the Royal Astronomical Society, **49**.

[2] See p. 64 for description of types.

[3] In millions of light years.

cube of the radius (D^3), and the brightnesses of distant galaxies are inversely proportional to the square of their distances ($1/D^2$). This combination, together with the definition of magnitude, gives the 0.6m in the number law, and shows that there can be no additional visible dimming by obscuring dust between galaxies.

The most recent survey, done in the early 1950's with the 48-inch Schmidt telescope (Fig. 11) at Palomar, led to the Palomar Atlas, or Sky Survey, a collection of 1870 plates covering two thirds of the whole sky (north of declination −40°) to magnitude 20. Copies of these plates have been distributed to most of the active observatories, where they are being analyzed and used to select individual galaxies (among other objects) for detailed study with larger telescopes. The survey is now being extended to the south celestial pole with a telescope in Chile.—TLP

Telescopes for the Future

(*Sky and Telescope*, October 1964)

There are only nine telescopes in the United States with apertures over 50 inches, and already they fall far short of providing the observing time needed by active astronomers.

This shortage was stressed by Ira S. Bowen, recently retired director of Mount Wilson and Palomar Observatories, when he gave the 1964 Henry Norris Russell lecture before the American Astronomical Society [in Flagstaff, Arizona, June 25, 1964].

According to Bowen, the problem will become much more acute before relief is gained, since five to twenty years are needed to design and build a major telescope. The costs of constructing and operating large instruments are considerable. At present-day prices, duplicating the 200-inch Hale telescope would require over $15 million. Allowing 5 per cent interest and 3 per cent depreciation, operating such an instrument would cost at least $3000 a night. A further bottleneck is the small number of individuals who combine the engineering and astronomical knowledge necessary in such a project.

The basic question in planning for the future is whether a single huge telescope should be built, or a number of identical smaller ones.

Bowen is confident that a 400-inch telescope could be constructed, although many basic new problems would have to be solved. On the other hand, a battery of four 200-inch telescopes would accumulate information at the same rate as the 400-inch, when used for direct photography or photoelectric photometry. For spectroscopic work, the four would collect twice as much information per year.

Either solution to the basic question would probably cost $50 million to $100 million. However, the four identical instruments should be somewhat less expensive than the 400-inch.

But there are far cheaper ways of greatly extending the usefulness of existing telescopes. Plans are under way to provide the 200-inch Palomar telescope with an auxiliary lens that will treble its effective focal length. Permitting longer exposures, this accessory should allow twenty-fourth-magnitude stars to be photographed, instead of the present limit of 23.

A very simple procedure for recording elusive details is multiple photography, in which several negatives are superimposed to make a print. This intensifying technique is being increasingly used in planetary photography and in studying the spectra of faint galaxies.

Image converter tubes offer exciting prospects of making small telescopes do the work of much larger ones. Such an intensification device converts incoming light into electrons; the electron flux is amplified and projected on a screen to give an image that can be viewed or photographed. Bowen stressed the importance of making such improvements in auxiliary instrumentation, as contrasted with building larger telescopes.

By using larger telescopes, longer exposures, "faster" photographic plates, and image tubes, astronomers have pushed their observations farther and farther out into space, finding more and more galaxies. There is no sign of an "edge" to the universe, or of any decrease in the number of galaxies per unit volume. Can the galaxies be infinite in number, spread out uniformly in an infinite universe? If so, we would see galaxies covering the sky completely—small, distant ones between the large, nearer ones—and the sky would not be dark at night.—TLP

Some Thoughts on Olbers' Paradox

OTTO STRUVE

(*Sky and Telescope*, March 1963)

While delivering the Halley Lecture at Oxford University on May 16, 1962, Hermann Bondi of London University told his audience:

"Many of you will know that for quite a little while I have had a bee in my bonnet about the darkness of the night sky—that the simple fact that it is dark at night seems to me to give a considerable clue to the structure of the universe. Many of you will be familiar with Olbers' old argument in which you simply suggest that if we do live in a uniform universe, then the number of objects at distance R goes up like R^2, the intensity of light received from each of them goes down like $1/R^2$, and therefore from every thin spherical shell we get the same amount of light, which should add up to an infinite amount, or if we are rather more cautious, to a very large finite amount. The main way out, nowadays, is to ascribe the actual result, the darkness of the sky, to the expansion of the universe; the distant sources move away from us at a speed so high that it materially diminishes the intensity of light that we receive from them, so that instead of diverging, or converging to a very large sum, the amount of light from distant matter in fact converges to a very small sum."

In recent years Bondi and other British astronomers have repeatedly called attention to the importance of this paradox of Olbers. In a stationary universe of infinite extent, uniformly strewn with stars, our line of sight would always end at the surface of a star, and the entire sky should therefore be bright—with a surface brightness like the sun's! So familiar is the darkness of the sky at night that it is easy to forget its fundamental importance.

My personal interest in Olbers' paradox goes back some years. I was familiar with the fact that nineteenth-century astronomers, including Heinrich Olbers himself, believed it could be explained by assuming the presence of absorbing material in interstellar space that dimmed the light of distant stars. Early in the present century it was realized that this explanation is not adequate. As Bondi aptly put it, "What

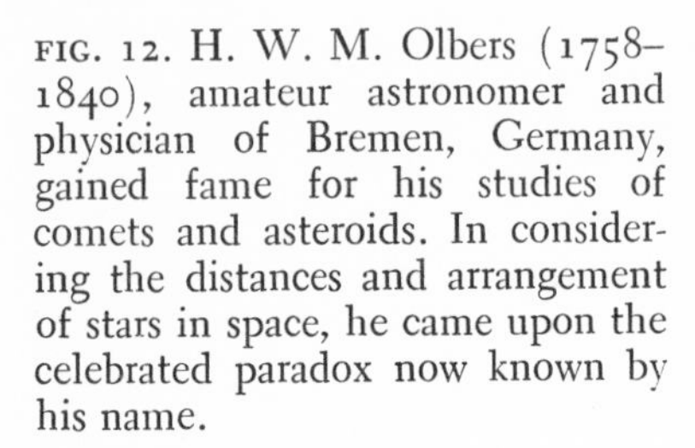

FIG. 12. H. W. M. Olbers (1758–1840), amateur astronomer and physician of Bremen, Germany, gained fame for his studies of comets and asteroids. In considering the distances and arrangement of stars in space, he came upon the celebrated paradox now known by his name.

happens to the energy absorbed by the [interstellar matter]? It clearly must heat the [matter] until it reaches such a temperature that it radiates as much as it receives, and hence it will not reduce the average density of radiation."

Since Olbers' explanation had proved unsatisfactory, and no other one had gained acceptance, I kept reading all I could about this problem. . . . Especially interesting were some comments by F. G. W. Struve [Otto Struve's great-grandfather] in his *Études d'astronomie stellaire,* St. Petersburg, 1847. There it was noted that precisely the same argument proposed by Olbers in 1823 had already been advanced by the Swiss astronomer Philippe Loys de Chéseaux, of Lausanne, in an appendix to his 1744 publication *Traité de la comète qui a paru en décembre 1743.* . . . In an article in the 1888 edition of the *Encyclopaedia Britannica,* the physicist P. G. Tait stated, "Chéseaux and Olbers endeavoured to show that because the sky is not all over as bright as the sun, there is absorption in interstellar space . . . an idea ingeniously developed by Struve." But Tait took the paradox as evidence for a finite universe. . . .

Philippe Loys (or Louis) de Chéseaux was born in 1718, the son of a wealthy landowner near Lausanne. . . . At a relatively early age he received various scientific honors. He was elected to foreign membership in the Paris Academy of Sciences, the Royal Society, and the scientific societies of Stockholm and Göttingen. The St. Petersburg Academy not only offered him membership but invited him to Russia as director of its observatory. Chéseaux, whose health had been delicate since childhood, could not avail himself of this opportunity. Instead, he went to

Paris in August 1751 to present to the academy his results on the motions of the sun and moon, the shape of the earth, and more. Soon after his return to Lausanne he became very ill, and died on November 30, 1751, at the age of only thirty-three.

The appendix to his 1744 memoir has the title "On the Force of Light, Its Propagation through the Ether, and on the Distance of the Fixed Stars." The discussion resembles Olbers'. . . . Chéseaux assumed that first-magnitude stars were of the same intrinsic brightness as the sun and estimated that they were about 240,000 astronomical units distant. The apparent diameter of such a star would then be about 1/125 second of arc, if it were the same size as the sun. Obviously all the first-magnitude stars together would occupy only a tiny fraction of the entire celestial sphere. A second set of stars, appearing a quarter as bright, would be twice as far, and each of them 1/250 second in diameter. But since there would be four times as many of them, they would occupy just as much sky area as the disks of the first-magnitude stars.

By including stars at greater and greater distances, Chéseaux concluded that the universe of stars need not even be infinite for the sky to be as bright as the sun. If the fixed stars were uniformly spread in space out to 760 billion astronomical units [3.7×10^6 parsecs], every point on the celestial sphere would be covered by a star disk, and we would receive from the hemisphere of the sky that is above the horizon nearly 100,000 times as much light and heat as the sun alone gives us.

Olbers used essentially the same argument. . . . To account for the darkness of the night sky, he assumed that "space is transparent only to this extent, that of 800 rays emitted by Sirius 799 reach to a distance equal to our distance from Sirius." Olbers then computes that stars 554 times as remote as Sirius lose half their light by interstellar absorption before we see them. At 5500 times as far, the surface brightness is reduced to 0.001 that of Sirius. The German astronomer concluded that stars more than 30,000 times the distance of the Dog Star do not make any contribution to the brightness of the night sky.

As we have already seen, the hypothesis of interstellar absorption fails to resolve the paradox. Bondi analyzed Olbers' assumptions, and noted that the amount of light we receive from an object that is, say, 500 million light years distant does not depend upon the present luminosity of the object, but on its luminosity 500 million years ago, when the rays we now see were being emitted.

According to Bondi, Olbers' assumptions can be restated: (1) the average density of stars and their average luminosity do not vary throughout space; (2) the same quantities do not vary with time; (3) there are no large systematic movements of the stars; (4) space is Euclidean; (5) the known laws of physics apply.

At least one of these premises must be incorrect if we are to avoid the erroneous conclusion that the night sky is bright. Bondi suggests that 2 or 3, or both, may have to be dropped. If we drop 3 and accept the red shifts of galaxies as observed by Hubble, Humason, and others, then we can account for the darkness of the night sky without invoking the hypothesis of interstellar absorption. For if distant galaxies are receding rapidly, the energy content of the radiation we receive from them is reduced.

Bondi's summary of his rather elaborate discussion is worth reading: "This little argument may well serve as a prototype of scientific arguments. We start with a theory, the set of assumptions that Olbers made. We have deduced from them by a logical argument consequences that are susceptible to observation, namely, the brightness of the sky. We have found that the forecasts of the theory do not agree with observation, and thus the assumptions on which the theory is based must be wrong. We know, as a result of Olbers' work, that whatever may be going on in the depths of the universe, they cannot be constructed in accord with his assumptions. By this method of empirical disproof, we have discovered something about the universe and so have made cosmology a science. . . . Thus the darkness of the night sky, the most obvious of all astronomical observations, leads us almost directly to the expansion of the universe, this remarkable and outstanding phenomenon discovered by modern astronomy."

We might ask whether the expansion of the universe could have been discovered solely as the result of Bondi's interpretation of the darkness of the night sky. Historically, of course, the red shift in the spectra of distant galaxies was already known long before Bondi wrote his books on cosmology. Most observers would probably give a negative answer to our question, while many theoreticians might seriously argue that no spectra were needed to deduce the expansion. In actual fact, I believe, progress comes from the simultaneous efforts of observers and theoreticians. Most important, the work of Chéseaux and Olbers drew the attention of later astronomers to the possibility of interstellar matter, and led to the demonstration of interstellar absorption by Robert Trumpler at Lick Observatory in 1930.

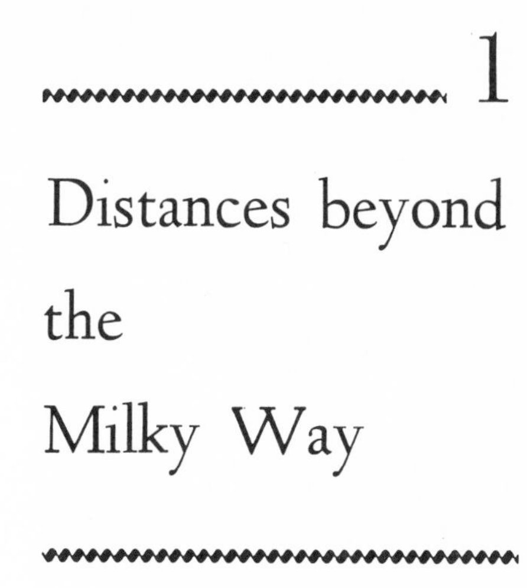

1 Distances beyond the Milky Way

For two reasons, the Milky Way Galaxy must enter this discussion: (1) we must know how far it extends from our position within it before we can identify as "extragalactic" the objects beyond its edge; (2) as "the Galaxy" (a name derived from galaktos, *the Greek word for milk), it is a prototype for the other galaxies found outside it—similar in size, mass, contents, and in some cases, structure.*

A detailed description of the Galaxy is given in Volume 7 of this series, Stars and Clouds of the Milky Way. *Avoiding the complexities, it may be said that the observed rotation of the Milky Way, with a circular speed of about 250 km/sec at our sun's position, depends on the mass of the Galaxy and the distance of the sun from its center. The rotational speed changes with distance from the center, and can be measured by radio or optical measures of Doppler shifts in the spectra of gas clouds or stars at known distances from the sun. These distances can be estimated from the brightnesses of pulsating variable stars like RR Lyrae and Delta Cephei, and of globular clusters (see p. 6). However, corrections must be made for interstellar obscuration that dims these stars and clusters, as well as for the effect of distance.*—TLP

Size of the Galaxy

(*Sky and Telescope*, July 1957)

"The problem of the scale of the Galaxy is one of obvious importance and currently appears to be particularly on the minds of astronomers," said Peter van de Kamp of Sproul Observatory, who was one of six speakers at the May 10 [1957] conference on this subject sponsored by the National Science Foundation. . . .

At present the generally used value of the distance from the sun to the center of the Milky Way Galaxy is 8200 parsecs (about 27,000 light years), from two accordant determinations. One of these was by Walter Baade of Mount Wilson and Palomar Observatories, in 1953, using the apparent brightnesses of thirty-six RR-Lyrae-type variable stars nearly in the direction of the galactic nucleus. The other result came from an analysis by Dutch radio astronomers in 1954 of the rotation of the Galaxy as indicated by interstellar neutral hydrogen clouds.

A critical examination of both results revealed that they may be more uncertain than is often supposed, the conference was told by Harold Weaver of Leuschner Observatory. Both methods are indirect, he pointed out, and are affected in an involved manner by uncertainties in the intrinsic brightnesses of stars and in the amount of dimming of starlight by interstellar dust.

For example, Baade in his 1953 work believed that the absolute magnitude of the variable stars used had a mean error of ±0.2 magnitude, which by itself could affect his distance determination by 10 per cent. When other sources of error are taken into account, Weaver said, the uncertainty of the method employed by Baade must, at present, be at least as great as 1000 parsecs and may be much greater.

The speed of rotation of the Galaxy in the sun's vicinity, which is a factor in interpreting the radio observations, may be larger than the usually cited values of around 200 kilometers per second, Weaver believes. He suggested that the inconsistency which thus appears in our picture of galactic rotation could be removed if the sun were farther from the center of our Milky Way system than generally accepted.

Another conference speaker was Maarten Schmidt, [then] a Carnegie fellow at Mount Wilson and Palomar Observatories, who discussed the distribution of neutral hydrogen gas inside our Galaxy, on the basis of measurements of 21-cm radiation. The gas clouds are believed to move in practically circular orbits around the galactic center. The rotational speeds depend on the total mass of the Galaxy and the manner in which that mass is distributed.

He found that the total mass of the Milky Way system is between 110 billion and 540 billion suns, depending on the assumptions made as to the size of the Galaxy and the nature of its rotation. These two extreme values bracket the currently adopted figure of 340 billion suns for the mass of our neighboring spiral galaxy, Messier 31 in Andromeda.

D. W. N. Stibbs of England's Oxford University Observatory spoke on a fundamental question in the evaluation of galactic rotation and the size of our system—the intrinsic colors of the Cepheid variable stars. Only if their true colors are known can the space reddening of these stars be evaluated, yet such evaluation must be made if we are to find the true distances of these variables by comparing their apparent and absolute magnitudes. Stibbs described four possible approaches to the problem, including one in which the reddening of each Cepheid is estimated by means of the color excesses (reddening) of blue stars in its immediate surroundings.

Cecilia Payne-Gaposchkin of Harvard College Observatory made use of several hundred RR Lyrae variable stars recently discovered by Baade in the Sagittarius star clouds. These faint and distant variables are evidently symmetrically distributed around the nucleus of the Galaxy. Periods and apparent magnitudes were derived for these variable stars by Sergei Gaposchkin, also of Harvard, who used Baade's photographs. If we compare the apparent magnitudes (around 17.5 for the most part) of these stars with their absolute magnitudes, then we obtain the distance to the center, provided proper allowance for interstellar absorption can be made by the astronomer.

The important innovation in Mrs. Gaposchkin's calculation is discarding the traditional view that all RR Lyrae variables have absolute magnitudes close to 0.0. Growing observational evidence for a change comes from the fact that precise color-magnitude arrays for globular clusters cannot be exactly superimposed. Moreover, Allan Sandage of Mount Wilson and Palomar Observatories has shown theoretically that RR Lyrae stars of younger globular clusters should be intrinsically more luminous than those in older globulars (where shorter periods prevail).

The direct determination of the absolute magnitude of RR Lyrae stars as 0.0 depended on the nearby stars of this type in the general galactic field, but these stars are no longer to be considered as identical with the RR Lyrae variables in globular clusters, nor with those at the galactic center. The last group has an exaggerated preference for the very short periods (around 0.3 day instead of half a day) exhibited by variables in old globular clusters. Therefore Mrs. Gaposchkin has adopted +0.76 for the absolute magnitude of the galactic center variables.

On this basis, the distance of the nucleus may be as great as 11,200 parsecs (36,500 light years). This value is an upper limit, for space between us and the center of the Milky Way could be somewhat less transparent than implied by Mrs. Gaposchkin's adoption of 1.5 magnitudes for the effect of interstellar absorption.

It was with just this problem, space reddening and absorption, that the sixth speaker at the conference, A. E. Whitford, [then of] Washburn Observatory, was concerned. He pointed out that Baade's final value for the center's distance depended on measures made in 1946 of the reddening of the globular cluster NGC 6522, which lies in the center of Baade's original field of variable stars, and at a distance slightly greater than that of the variables themselves.

Whitford has re-examined the 1946 measures of NGC 6522 in view of our present knowledge of space reddening and of the characteristics of globular clusters. The total photographic absorption of 2.7 to 2.9 magnitudes adopted by Baade is probably too high. Whitford said, "It would not be surprising if the photographic absorption in front of the bulge variables is as low as 1.5 magnitudes." (This is the same figure Mrs. Gaposchkin used.) The consequence would be a large upward revision in Baade's 8200-parsec distance to the center.

The general consensus of the conference appeared to be that our Milky Way system is larger than had been supposed, by perhaps 25 per cent; it was agreed, however, that the difficult problem of establishing its size precisely requires much further study.

The distance from our sun to the center of the Galaxy is not yet accurately determined, but is probably close to 10,000 parsecs (33,000 light years), giving the Galaxy a mass close to that of M 31.

The pulsating Cepheid variables, so useful in estimating distances, can also be seen in the nearer galaxies outside the Milky Way. In fact, it was for Cepheids in the Magellanic Clouds that the relation between pulsation period and intrinsic luminosity was first noticed. Again, the obscuration by interstellar dust clouds requires a correction.—TLP

Extragalactic Cepheids Studied

ROBERT R. COLES

(*The Sky*, July 1940)

In a paper presented at the Eighth American Scientific Congress, in Washington, May 15 [1940], Harlow Shapley, director of Harvard Observatory, discussed some of the factors influencing the apparent magnitudes of Cepheid variable stars. He has just finished a survey of about three hundred of these stars in the Small Magellanic Cloud, the first study of absorption in a galaxy outside our own Milky Way system. Shapley finds that the light of such a star is absorbed as it passes through the Cloud, so the star may appear as much as half a magnitude fainter than it otherwise would, although the average decrease seems to be about one quarter of a magnitude.

The position of a star on the near or far side of the Cloud therefore affects the star's apparent brightness; also, in some cases "doubling" resulted from two stars in the same line of sight. Because of these factors, and because of uncertainties in the period-luminosity relationship for long-period Cepheids, Shapley estimates that the distances of some galaxies found by the Cepheid variable method may be in error by as much as 30 per cent. No major change in the distances of the Magellanic Clouds is required.

Further study of the period-luminosity relationship for Cepheids of very short and very long periods is required, and also, the establishment of more dependable magnitude standards in the southern sky, where important studies of external galaxies are being made.

The Clouds of Magellan

OTTO STRUVE

(*Sky and Telescope*, December 1954)

Every astronomer who has been south of the equator has noted the extraordinary richness of the southern sky. Not only is the southern Milky Way far more brilliant and full of structure than its northern

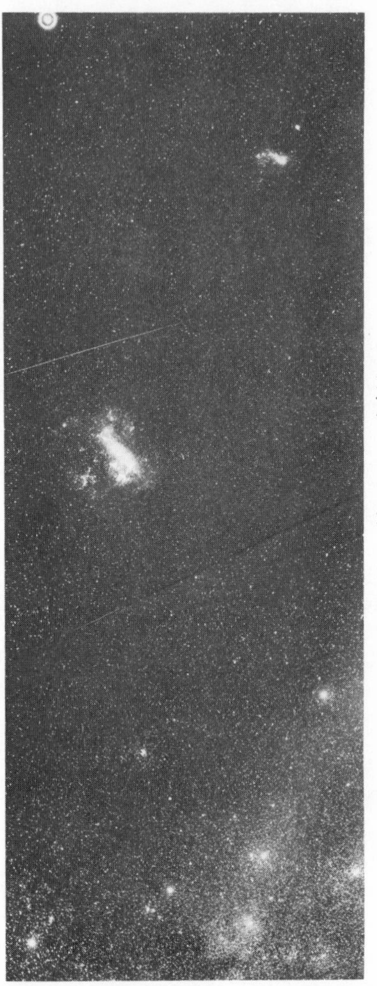

FIG. 13. The Large Magellanic Cloud is left of center in this picture, and the Small Magellanic Cloud is near the upper right. The overexposed star image in the upper left corner is of Achernar, a first-magnitude star in the constellation Eridanus. Just above the Small Cloud, looking like a fuzzy bright star, is the great globular cluster 47 Tucanae. In the lower right is part of the Milky Way. This photograph was taken by T. Houck with the 3-inch Ross-Tessar camera at Harvard's Boyden Station on December 26, 1953; exposure 3 hours on a blue-sensitive plate. The south celestial pole is located about 2 inches to the right of the Large Cloud. (Harvard College Observatory photograph)

counterpart, but the two great naked-eye galaxies in the constellations of Dorado and Tucana lend the southern heavens a mysterious interest that is not shared by the northern sky.

Those of us who have never seen the Magellanic Clouds (Fig. 13) have all too often minimized the importance of these extragalactic neighbors of ours. Astronomy owes much to Harvard Observatory, whose southern stations, from 1890 to 1925 at Arequipa, Peru, and later in South Africa, have photographed the Clouds of Magellan for two generations. Almost all of our early knowledge of them comes from the work of Harvard astronomers: Henrietta S. Leavitt published in 1907 a list of 808 variable stars in the Large Cloud and 969 in the Small Cloud, and by 1912 she had established the period-magnitude relation of Cepheid variables in the clouds; Annie J. Cannon studied the spectra of the brightest stars in the clouds; and Harlow Shapley and his collaborators collected and systematized the pertinent information. (See Shapley's popular book *Galaxies,* and *Introduction to Astronomy* by Cecilia Payne-Gaposchkin.)

The two clouds are located some 30° and 45° from the Milky Way. Long ago the British physicist A. S. Eddington pointed out that these large galactic latitudes make it unlikely that the Magellanic Clouds are parts of our Galaxy, even though they look like isolated scraps of the Milky Way.

The distances of the two clouds from us are nearly the same, about 50,000 parsecs, as inferred from the apparent magnitudes of the Cepheid and cluster-type variables that they contain. The relative distances with respect to the center of our Galaxy are shown schematically in Figure 14. There is no doubt that the clouds are independent galaxies and not merely condensations within the Milky Way.

They are, in fact, classified as irregular galaxies [see p. 66], with total absolute magnitudes of —17.5 and —16, rather more luminous than an average small galaxy of absolute magnitude —14. Yet, the clouds are about 2 and 3.5 magnitudes intrinsically fainter than giant spirals like the Andromeda nebula (M 31) and our own Milky Way. It is therefore reasonable to expect tht the total mass of either cloud will also be considerably smaller than the mass of the Milky Way.

The two clouds are not alike, and neither of them is a miniature replica of the Milky Way Galaxy. The Large Cloud contains blue and red supergiant stars, as well as obscuring dust, which blots more distant

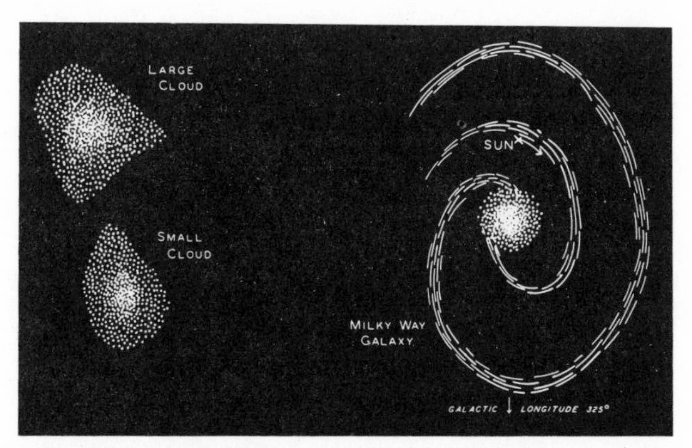

FIG. 14. In this schematic diagram, the sizes and distances of the Magellanic Clouds and of our Galaxy are shown to approximate scale.

galaxies from view. It has much ionized hydrogen gas in the form of luminous nebulosities—the Tarantula nebula is only one of a great many. There are also several thousand typical Cepheid variables. All these kinds of objects are what we find in the spiral arms of our Milky Way, and are representative of Baade's Population I [see p. 321]. There are, however, several globular clusters in the Large Cloud, and at least one of them contains RR Lyrae variables. These are typical Population-II objects. Hence the Large Cloud, though made up mostly of Population-I objects, has a small admixture of Population II. [The stellar Populations I (young) and II (old) are described in Volumes 6 and 7 of this Series.—TLP]

The Small Cloud is almost free of obscuring dust; distant galaxies shine through it undimmed. This suggests that the Small Cloud is relatively richer in Population-II objects than the Large Cloud. Yet there are some hydrogen emission nebulae in the Small Cloud, so we cannot assign to it a pure Type-II population. Cecilia Payne-Gaposchkin of Harvard College Observatory suggests that the two populations are about equally represented. . . .

Not only are the Magellanic Clouds an important source of information about the stellar contents of galaxies, but recently they have added to our knowledge of how two neighbor galaxies may interact. . . . In the Large Cloud, there is a pronounced "axis," which suggested to the Australian astronomer H. C. Russell, in 1890, a resemblance to barred spirals; he noted the cloud's "incipient spiral structure."

Although the Small Cloud is relatively structureless, there is an extension, or "wing," in the direction of the Large Cloud, which is almost certainly a tidal protuberance produced by the gravitational attraction of the latter. . . .

Cepheids in the Small Magellanic Cloud

(*Sky and Telescope*, March 1967)

At the Harvard College Observatory the most extensive survey ever made of the variable stars in another galaxy has recently been completed by Cecilia Payne-Gaposchkin and Sergei Gaposchkin. Even in highly condensed form, the far-reaching results of this immense labor fill a 205-page book.[1]

Although our own Milky Way system contains about twenty thousand known variables, they are poorly suited for answering questions of a statistical character—questions, for example, about the spread in intrinsic brightness of Cepheids having a particular period. The Milky Way variables are scattered around the sky at unequal and uncertain distances from us, veiled and reddened by patchy interstellar dust. Any statistical investigation of them is bedeviled by observational selection.

All these difficulties are eased by studying the variables of a very nearby galaxy such as the Small Magellanic Cloud (SMC). It is a compact enough system to be photographed on a single plate with a large Schmidt camera, rich in variables that are all at nearly the same distance from us, and almost free from dust. It is no wonder that the period-luminosity relation for Cepheids was discovered from observations of the Small Magellanic Cloud. This was in 1912 by Henrietta Leavitt at Harvard, where almost all of more than 1500 known SMC variables have been discovered in the course of the last seventy years.

The Gaposchkins and their associates used about 560,000 brightness estimates of these stars, mainly on about five hundred photographs taken with Harvard's 24-inch wide-angle Bruce refractor between 1898 and 1950. These limits were widened to 1888 and 1962 by inclusion of some earlier Harvard and some later Boyden Observatory plates. The estimates could be converted into magnitudes by comparing them with brightness estimates for stars whose magnitudes were accurately measured photoelectrically by Halton C. Arp of Mount Wilson and Palomar Observatories.

In the SMC the great majority of variable stars are Cepheids, with periods ranging from about one day to over two hundred days, and with magnitudes of about 17 for those that pulsate fastest and about 12 for the slowest. The distribution of the periods was known to be very different from that in our own Galaxy, where five-day Cepheids are commonest; the peak for the SMC comes at three days.

Commenting on their study of period frequency for 1144 SMC Cepheids, the Gaposchkins note, "The well-known preponderance of short periods is enhanced by the results of our work, which has almost doubled the number of known Cepheids in the Small Cloud. More

[1] *Variable Stars in the Small Magellanic Cloud* by Cecilia Payne-Gaposchkin and Sergei Gaposchkin, Volume 9 of *Smithsonian Contributions to Astrophysics*. Available at $1.50 from Superintendent of Documents, U.S. Government Printing Office, Washington, D.C. 20402.

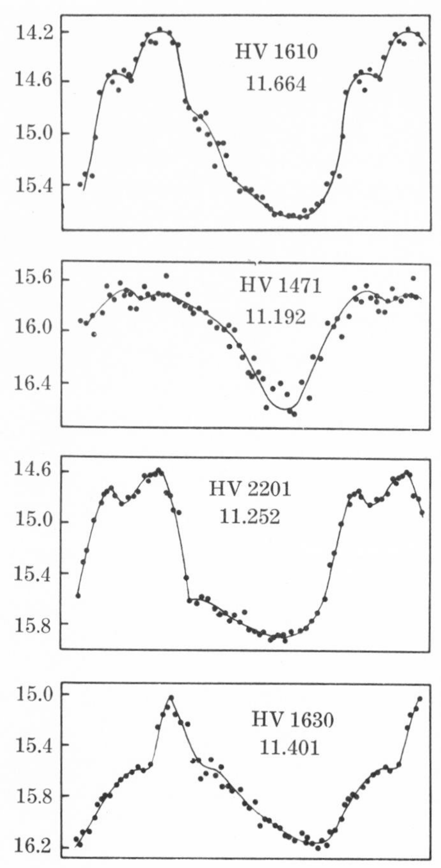

FIG. 15. Light curves (magnitude *vs* time) of four Cepheid variables in the Small Magellanic Cloud, as determined, from Harvard photographs, by Cecilia Payne-Gaposchkin and Sergei Gaposchkin. These stars are identified by their HV (Harvard Variable) numbers; their periods are between 11.192 and 11.644 days. All four stars have twin maxima, of which the heights differ. The first is higher than the second in HV 1471, but merely a shoulder on the left (rising) side of HV 1630. (From Volume 9, *Smithsonian Contributions to Astrophysics*, 1966)

Cepheids are in fact now known in that system than in any other galaxy, including our own."

In redetermining the period-luminosity relation, the Harvard astronomers point out that the occasional Cepheids with sinusoidal light curves are systematically brighter than the ordinary ones, with skew light curves, and hence should be excluded. A plot for the other stars shows an apparent median magnitude of 17.63 for a one-day Cepheid, and 15.50 for a ten-day one.

Since for many of the Cepheids studied, the plate material spans more than sixty years, changing periods are easy to recognize. But *bona fide* cases seem rather rarer than among galactic stars. The Gaposchkins say, "The sporadic occurrence of sensible changes of period among the stars investigated suggests that changes of period may be an evanescent phenomenon, may operate in either direction, and perhaps become progressively more probable the longer the period."

About a hundred pages of the Gaposchkin volume are occupied by diagrams of the light curves of individual variables, arranged in order of period. For each star, an IBM 7094 computer calculated the phases of the observations and averaged them into normal points, which it then plotted. The collection shows very clearly how the shape of the light curve depends upon the period of a Cepheid. For example, stars with periods near eleven days tend to have a double maximum, as shown in Figure 15.

The Gaposchkins estimate, on the basis of a formula first proposed by Andrew T. Young (University of Texas), that the ages of the Cepheids in the SMC range from about 10 million to 400 million years. Each star that becomes a Cepheid remains such for an interval equal to about 15 per cent of its previous lifetime. Thus a ten-day Cepheid would continue to pulsate for perhaps 11 million years, the amplitude finally dying away.

Cepheids of the Andromeda Galaxy

GEORGE S. MUMFORD

(*Sky and Telescope*, January 1966)

When Walter Baade died, in 1960, he had under way a very extensive investigation of variable stars in Messier 31, the great Andromeda galaxy, on photographs he had taken with the 200-inch telescope at Palomar Observatory. Since then the analysis of this plate material has been continued by Henrietta H. Swope and by Sergei Gaposchkin. The third and final paper presenting the results has now appeared in *The Astronomical Journal*, with Baade and Swope named as coauthors.

During this project 684 variable stars to as faint as magnitude 23 were studied in four areas at different distances from the center of M 31. Of these variables, 401 are Cepheids for which periods have been determined. In the latest paper Swope has given particular attention to the Cepheids and their statistical properties. Her main conclusion is that the Cepheids in M 31 resemble those in our own Galaxy but differ significantly from those in the Small Magellanic Cloud.

The evidence from the numbers of variables having different periods is clear-cut. Both the Milky Way system and M 31 contain about 280 Cepheids with periods longer than 5⅔ days. In both galaxies there is a shortage of periods near nine days. The discovery of variables with periods shorter than five days is less complete in M 31 than in our Galaxy, due to the intrinsic faintness of such stars, "but," says Miss Swope, "the distribution looks as if it would be the same as in the Galaxy and not like that of the Small Magellanic Cloud."

In the latter, Cepheids with periods longer than 5⅔ days are relatively rare, and there is no dip in the period-distribution curve at nine days. On the other hand, the Small Cloud is very rich in Cepheids of shorter periods, with the most frequent period about 2½ days.

These differences between the variable star populations of different galaxies have a far-reaching significance. A vital link in the cosmic distance scale consists of the distances to nearby galaxies as deduced from observations of the apparent magnitudes and periods of their Cepheids. This link is seriously weakened if significant differences exist between Cepheids in otherwise similar galaxies. Fortunately, this is now shown not to be the case for the Milky Way system and M 31.

The lengthy photographic studies by the Gaposchkins, Baade, and Swope allow greater confidence in distances estimated by the brightnesses (apparent magnitudes) of pulsating variable stars. The danger is that a Cepheid in one galaxy might have different intrinsic luminosity than does one with the same pulsation period in another galaxy, due to some local effect of chemical composition. Actually, such a difference was found by Walter Baade over fifteen years ago. As a consequence, the distances of galaxies had to be revised then.—TLP

The Distance Scale of the Universe

OTTO STRUVE

(*Sky and Telescope*, June and July 1953)

The year 1952 will go down in astronomical history as one in which astronomers accepted a tremendous change in their concept of the distance scale of the universe. Also, it will mark an astronomical milestone that we have reached directly as a result of observations with the 200-inch Hale Telescope—observations which no other instrument in the world is capable of making. The significance of this change is concisely described in the following paragraph quoted directly from the official minutes of Commission 28, Extragalactic Nebulae, of the International Astronomical Union, at its Rome meeting, September 4–13, 1952:

"Baade then went on to describe several results of great cosmological significance. He pointed out that, in the course of his work on the two stellar populations in M 31, it had become more and more clear that either the zero point of the classical Cepheids or the zero point of the cluster variables must be in error. Data obtained recently—Allan Sandage's color-magnitude diagram of the globular cluster M 3—supported the view that the error lay with the zero point of the classical Cepheids, not with the cluster variables. Moreover, the error must be such that our previous estimates of extragalactic distances—not distances within our own Galaxy—were too small by as much as a factor of 2. Many notable implications followed immediately from the corrected distances: the globular clusters in M 31 and in our own Galaxy now come out to have closely similar luminosities; and our Galaxy may now come out to be somewhat smaller than M 31. Above all, Hubble's characteristic time scale for the universe must now be increased from about 1.8×10^9 years to about 3.6×10^9 years."

In the final analysis, our knowledge of the distances of stars and galaxies rests upon the determination of the scale of the solar system; for several hundred years astronomers have been concerned with this first stage of the problem—until now we know solar-system distances with good accuracy. Then, using as a base line the astronomical unit (the distance between the earth and the sun), astronomers succeeded in 1837 in measuring the parallactic displacement and, hence, the distance of one of the nearest of the stars. The trigonometric parallax determinations which followed in the next 116 years form the second stage in the exploration of the universe.

The third stage resulted from Henrietta Leavitt's discovery, in 1912, of a close relation between the apparent brightnesses and the periods of Cepheid variable stars in the Small Magellanic Cloud. From a study of twenty-five variables, she made a diagram in which the apparent magnitude of each star at maximum and at minimum light was plotted against the logarithm of the period [Fig. 16]. A variable with a period of exactly ten days would have a logarithm of the period equal to 1.0 and would, according to the diagram, appear on a photographic plate with an apparent magnitude of 13.5 when at maximum light and about 14.7 when at minimum light. Fainter variables were found to have shorter periods; those of magnitudes between 15 at maximum and about 16.5 at minimum correspond to a period of about one day.

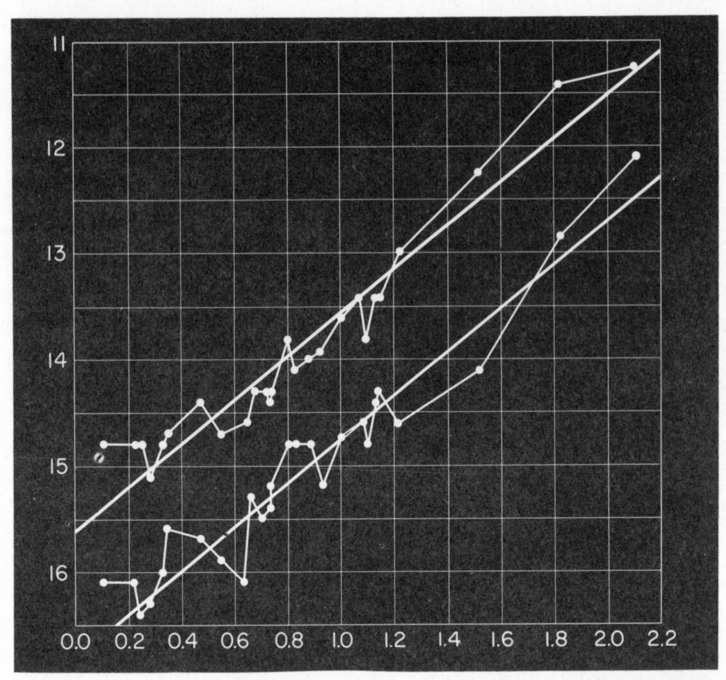

FIG. 16. The original period-luminosity curve, copied from Harvard Observatory *Circular* 173 (March 3, 1912), in which Henrietta Leavitt's discoveries are described. Photographic magnitude is plotted vertically and logarithm of period horizontally. The upper lines and set of points are for maxima of Small Magellanic Cloud variables, the lower for minima.

This diagram was the beginning of all our recent knowledge concerning the distances between our Galaxy and other galaxies, such as the Andromeda nebula, M 31, and the great spiral in Triangulum, M 33. Although at the time Miss Leavitt made this discovery, the importance of the period-luminosity relation for distance measurements was not appreciated, either by her or by E. C. Pickering, then the director of the Harvard Observatory, Miss Leavitt did remark:

"Since the variables are probably at nearly the same distance from the earth, their periods are apparently associated with their actual emission of light, as determined by their mass, density, and surface brightness. . . . Two fundamental questions upon which light may be thrown . . . are whether there are definite limits to the mass of variable stars of the cluster type, and if the spectra of such variables having long periods differ from those . . . whose periods are short."

It should be realized that Miss Leavitt's diagram as such did not directly provide the means for determining distances. It merely showed, for instance, that a Cepheid whose period is ten times longer than that

FIG. 17. The Small Magellanic Cloud, in which are contained the variable stars used by Miss Leavitt to discover the period-luminosity relation. (Harvard College Observatory photograph)

of another variable of this kind is approximately two magnitudes brighter than the latter, or that its intrinsic luminosity is 6¼ times brighter. Thus if we observe two variable stars of the same apparent magnitude at maximum light, the one having a period of four days, the other of forty days, we can conclude that the longer-period variable is 6¼ times as luminous but considerably farther away. As brightness varies inversely with the square of the distance, the actual relative distance would be the square root of 6¼, or 2½; the star of longer period must be 2½ times as distant as the one of shorter period.

But if we have not yet determined the distance of the nearer star, our observation provides only relative distances and does not give us any information regarding the absolute scale of distances. This latter and important aspect of the period-luminosity relation of the Cepheid variables is usually described as setting the zero point of the period-luminosity curve.

The extension of Miss Leavitt's work and its practical use as a yardstick for measuring the distances of globular clusters and extragalactic systems was made by Harlow Shapley in a long series of papers published in *The Astrophysical Journal* more than thirty years ago, while he was a member of the Mount Wilson Observatory staff. In some of his early work he determined the zero point, although this had also, in effect, been done by E. Hertzsprung in 1913 on the basis of the proper motions of typical Cepheids measured by L. Boss. If we could measure trigonometrically the distance of a single Cepheid variable star in our Milky Way system, and if we could estimate the amount of interstellar absorption between us and this variable star, we would have an immediate determination of the zero point.

Unfortunately, even the brightest of the Cepheid variables are so far away that ordinary trigonometric parallaxes are useless. Nor is there any indirect method available of the kind astronomers use when they take advantage of a special type of star occurring as a member of a visual binary system. Cepheid variables do not even occur among moving clusters of stars, such as the Hyades, or among other galactic clusters, like the Pleiades, for which the distances can be estimated with the help of the Hertzsprung-Russell diagram [see p. 318]. We are left with the difficult method of estimating statistically the distances of certain groups of galactic Cepheids with the help of their radial velocities and proper motions.

The method is, in principle, quite simple. Suppose you are located on a high building and you watch the movement of persons in the streets below. Suppose you know that the average speed of a person walking is one yard per second; you will also know that some persons are walking considerably faster than the average, and others much slower. Therefore you observe some of them to change their positions hardly at all, while others appear displaced about 20° in 10 seconds. Between these extremes there are individuals that appear to move with various speeds, and you conclude that their average apparent displacement is 10° in 10 seconds, or one degree in one second. You would further conclude that, on the average, a displacement of 1° per second corresponds to the known average linear velocity of one yard per second. In other words, at your distance above the street, one yard subtends an angle of 1°; from this you can easily compute that your elevation is about 60 yards.

The apparent angular motions of the bright Cepheid variables have been determined from accurate measures of their positions in the sky extending over hundreds of years. For the same stars, motions in the line of sight, or radial velocities, have been determined from measures of the

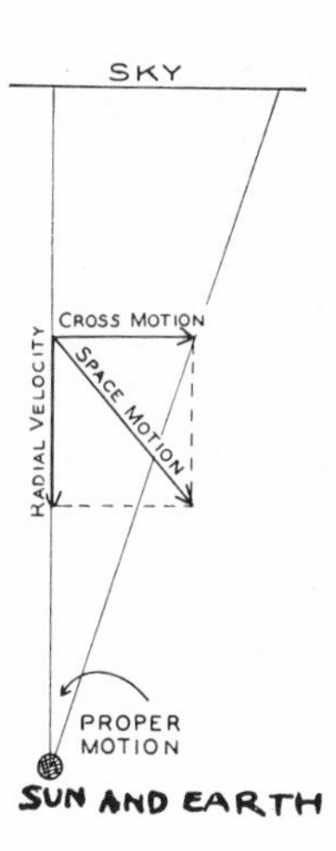

FIG. 18. Here is shown the relation among cross motion (tangential velocity), radial velocity, and the space motion of a star. Proper motion is the angular change of position produced by the cross motion, but its observed value (seen from the earth, very close to the sun) depends also on the distance of the star.

Doppler shifts of their spectral lines. An individual Cepheid, however, may be moving only at right angles to the line of vision, thus having large proper motion and no radial velocity with respect to the earth (see Fig. 18). Another star could be moving entirely along the line of sight, and would then have no detectable proper motion or tangential velocity. For an individual star we cannot determine whether a small proper motion results from an actual small velocity at right angles to the line of sight, or whether the motion seems small because the star is very far away. But if we consider the problem statistically, we may assume the motions to be at random, as large a component being radial as tangential, and we may roughly equate the radial and linear tangential velocities. On this basis, the measured radial velocity should be approximately a measure of the linear motion of the star, while the angular proper motion should be a measure of this same velocity as seen from the distance between the earth and the star. Thus if a star of large radial velocity has a small proper motion, its distance must be relatively great; stars of larger proper motion must be nearer.

The mathematics of this process is not difficult to understand, but it is unnecessary for our present purpose. The important point is to realize that if, over a large statistical sample of stellar data, we grant the similarity of the radial and tangential velocities, relatively reliable statistical parallaxes may be obtained.

In this manner, Shapley set the zero point of his famous period-luminosity curve [see Fig. 19], in which the abscissa has the same meaning as in the period-luminosity relation of Miss Leavitt, but the ordinate represents the intrinsic or absolute photographic magnitudes of the variable stars halfway between maximum and minimum light. This

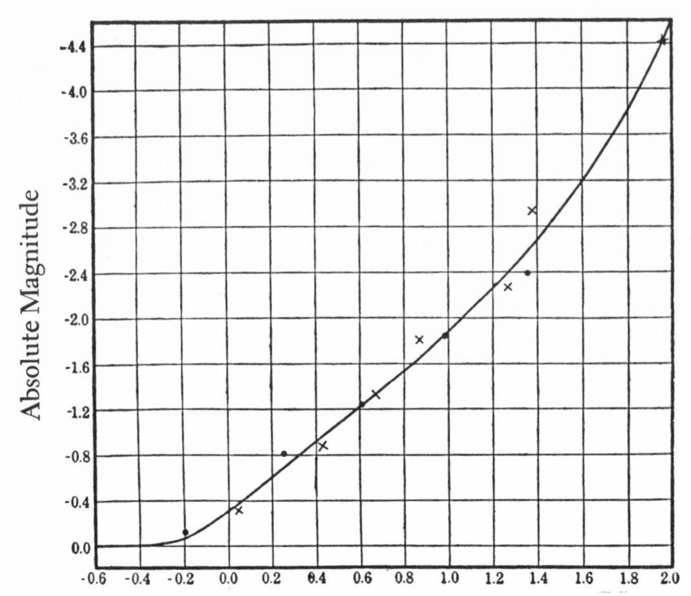

FIG. 19. The photographic period-luminosity relation adopted by Harlow Shapley in his book *Star Clusters* (McGraw-Hill, 1930). The dots represent means for intervals of logarithms of the periods; the crosses, for intervals of absolute magnitude.

diagram therefore gave us a means of determining the individual distances of the Cepheid variable stars once we observed their periods.

Let us suppose that in our Galaxy we have observed a variable star whose apparent magnitude, halfway between maximum and minimum light, is +8.2. After a series of observations we find that the period is ten days. We take the logarithm of this period, namely 1.0, and read the ordinate scale for that point on Shapley's curve, and find that the absolute magnitude of the variable is —1.8. The star thus appears ten magnitudes fainter when actually observed on a photographic plate than it would be if it were located at the standard absolute magnitude distance of 10 parsecs, or 33 light years. A difference of 10 in magnitude corresponds to a ratio in light of 10,000 [see p. 320]. Since light intensity diminishes with the square of the distance, the real distance of this variable is 100 times greater than 10 parsecs and is thus equal to 1000 parsecs.

Since we know that interstellar absorption tends to make the stars look fainter than they would otherwise, we have to introduce a correction to the observed apparent magnitude, allowing for intervening dust and gas, and thus reducing the distance somewhat.

This method is perfectly straightforward, and it was used successfully at Mount Wilson by E. P. Hubble and all of his followers in determining the distances of the nearer galaxies. In these systems many Cepheid variables were discovered, their periods were determined, and the rest of the procedure resembled the example we have worked out in the preceding paragraphs.

In this manner Hubble, in 1929, found that Cepheids of corresponding period in the Andromeda galaxy are about 4.6 magnitudes fainter than in the Small Magellanic Cloud. This would make the distance of the Andromeda galaxy 8½ times greater than the distance of the cloud. Using Shapley's value of 106,000 light years for the cloud, Hubble set the distance of the Andromeda galaxy as 900,000 light years.

Except for a relatively minor correction in this figure—bringing it down to 750,000 light years—which is due to the small amount of interstellar absorption in the direction of Andromeda, this result remained generally accepted until Baade's report at the Rome meeting. Not only that, but all other extragalactic distances were tied to the distance found by Hubble with the help of the Cepheid variables. Hence the framework of the metagalactic system rested entirely upon Shapley's zero point of the period-luminosity law. . . .

Since Shapley's first determination of the zero point of the period-luminosity curve, numerous attempts have been made to revise the original value. Many astronomers—B. P. Gerasimovich, J. H. Oort, K. Lundmark, Priscilla F. Bok, H. Mineur, and J. Schilt, among others—have indicated various corrections to Shapley's zero-point determination. The range covered by these corrections is about 1.5 magnitudes, indicating among other things how difficult it is to derive the zero point accurately.

Shapley himself, however, estimated at one time that the error of the adopted zero point would not exceed a quarter of a magnitude. This prediction, made in 1930, was seemingly confirmed by R. E. Wilson at the Mount Wilson Observatory in 1939, when he rederived the zero point on the basis of a large mass of new data on the radial velocities and proper motions of the Milky Way Cepheids. . . . In spite of these earlier results, it is now probable from the work of Baade and his collaborators that the zero point of the classical Cepheids was in error by 1.5 magnitudes, while that of the cluster-type Cepheids (RR-Lyrae variables) was about correct. In Shapley's period-luminosity diagram these two groups of variables form one continuous curve, the cluster-type variables with periods of one-half day being about two magnitudes fainter, intrinsically, than the classical Cepheids with ten-day periods. The latter, on direct photographs of M 31 with the 100-inch Mount Wilson telescope, have apparent magnitudes of 20. The former were too faint to be observed with the 100-inch, but were predicted to have apparent magnitudes of 22. They should therefore have been easily accessible to the 200-inch Hale reflector on Palomar Mountain.

Yet Baade stated at Rome, "The very first exposures on M 31 taken with the 200-inch telescope showed at once that something was wrong. Tests had shown that we reach with this instrument, using the $f/3.7$ correcting lens, stars of photographic magnitude 22.4 in an exposure of thirty minutes. Hence we should just reach in such an exposure the cluster-type variables in M 31, at least in their maximum phases. Actually we reach only the brightest stars in Population II in M 31 with such an exposure. Since, according to the latest color-magnitude diagrams of globular clusters, the brightest stars of Population II are photographically about 1.5 magnitudes brighter than the cluster-type variables, we must conclude that the latter are to be found in M 31 at photographic magnitude about 23.9, and not at 22.4, as predicted on the basis of our present zero points."

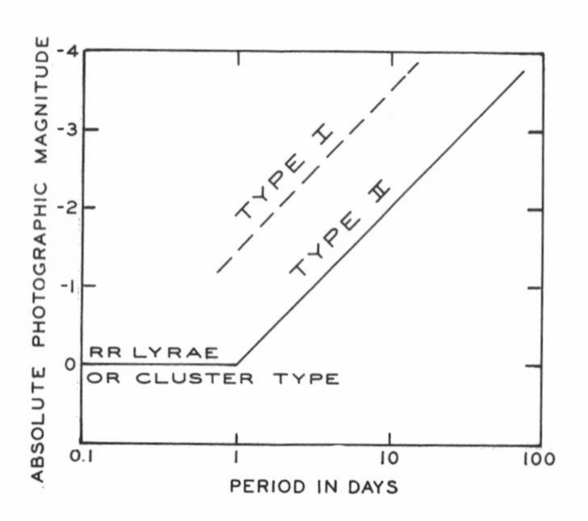

FIG. 20. There are now two period-luminosity curves instead of the single one formerly employed in estimating the distances of remote Cepheids. The classical, or type-I, Cepheids are about 1.5 magnitudes brighter than the type-II Cepheids.

Thus the Andromeda galaxy must be much farther away than we had heretofore considered it to be, for not only are we unable to observe its cluster-type variables at the limit of the 200-inch telescope, but at that limit we only begin to see the brightest stars of Population II. Baade pointed out that there is convincing proof that these brightest stars (red long-period variables) are properly identified because they emerge above the plate limit just when the globular clusters of the Andromeda system begin to be resolved into stars. (This resolution is another achievement of the 200-inch instrument.)

Obviously, this result does not of itself tell us whether the period-luminosity curve of the classical Cepheids or that of the cluster-type stars is in error by 1.5 magnitudes. This point was settled by Baade with the help of Allan Sandage's color-luminosity diagram of the globular

FIG. 21. The globular cluster Omega Centauri, shown here, is about 20,000 light years distant, an estimate based on the brightness of RR Lyrae variables in it and unaffected by the new period-luminosity diagram. (Royal Cape Observatory photograph)

cluster M 3, in which the dwarf branch was recorded. If we compare this dwarf branch with Harold Johnson's dwarf branch of the nearby stars (for which distances are very well determined), we find that Sandage's and Johnson's diagrams can be brought into coincidence when the cluster-type variables in M 3 are given an absolute magnitude of 0.0—in exact agreement with the earlier determinations. [That is, the distances of globular clusters in the Milky Way Galaxy are not changed; their variable stars are all of Type II.]

The error of 1.5 magnitudes, corresponding to an error in the resulting distances of a factor of two, in the sense that they must all be increased by this number, is thus attributable to the classical Cepheids. The curve representing their intrinsic brightness-period relation should apparently be raised 1.5 magnitudes above that for other long-period pulsating variables, such as the W Virginis stars.

Perhaps the best way to make the distinction between the Cepheids is to follow Baade in separating them into type I and type II, corresponding to his general division of all stellar objects into Populations I and II. The classical or Type-I Cepheids are those for which the period-luminosity curve must be raised—that is, they are intrinsically 1.5 magnitudes brighter than Type-II Cepheids of corresponding periods (including W Virginis stars). The cluster-type or RR Lyrae variables are of very short period; their luminosities remain unchanged, and they still form part of the original period-luminosity curve along which the Type-II Cepheids fall, as shown in Figure 20.

The Type-I Cepheids, in our Milky Way and in the distant galaxies, are therefore twice as far away as had been previously assumed. But the

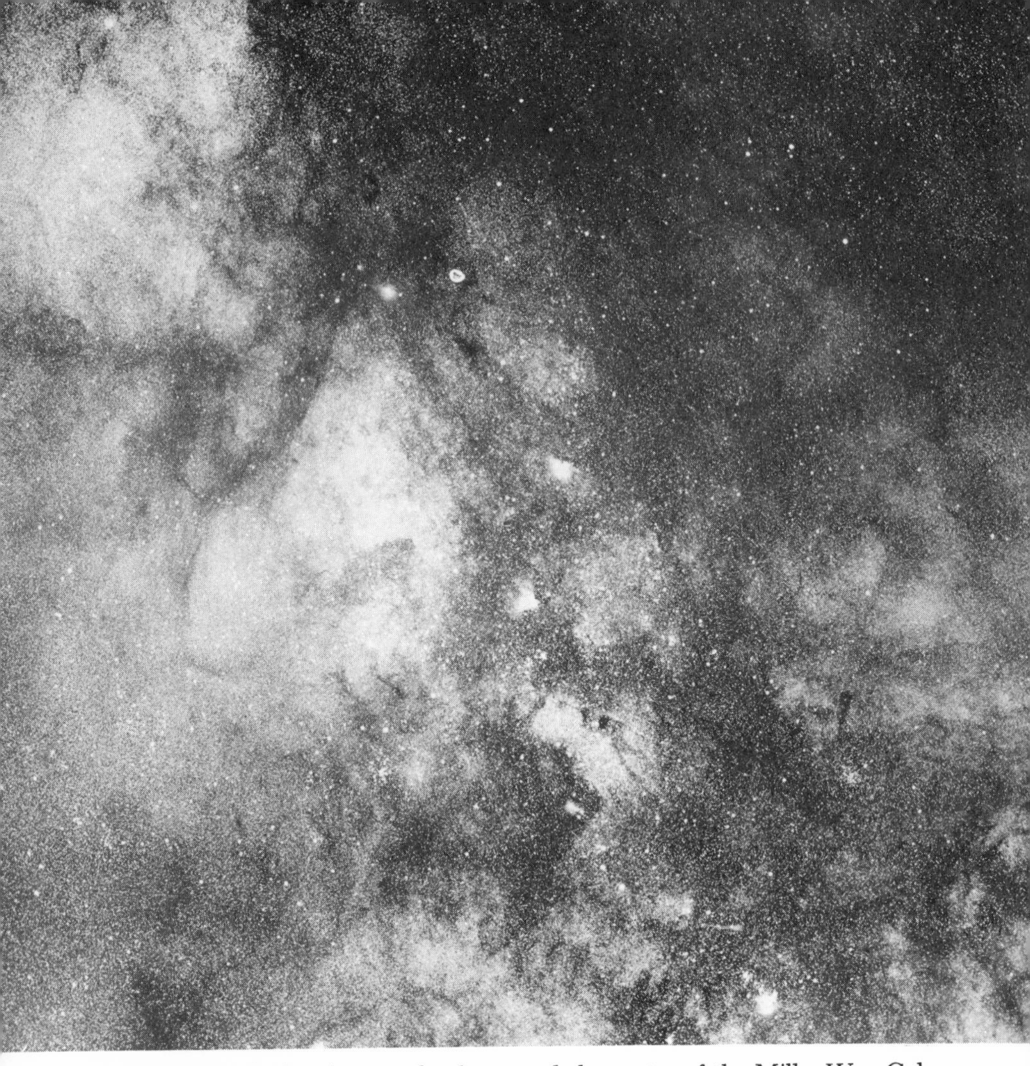

FIG. 22. Sagittarius star clouds, toward the center of the Milky Way Galaxy. Their distances, up to 30,000 light years, are unaffected by the new period-luminosity diagram. (Mount Wilson and Palomar Observatories photograph, courtesy of Yerkes Observatory)

distances of the globular clusters, and of the Milky Way star clouds in Sagittarius, which have been estimated with the help of the cluster-type variables, remain unaffected. Our Galaxy's size remains the same as before, but the Andromeda galaxy is now placed at a distance of 1.5 million light years (distance modulus [see p. 60] 23.9 magnitudes). Hence, it is also twice as large in diameter as was previously thought likely. It is, in fact, a little larger than our Milky Way system.

Baade's work has been confirmed by the independent results of A. D. Thackeray with the 74-inch Radcliffe Observatory reflector in South

Africa. At the Rome meeting he announced that the first cluster-type variables found in the Small Magellanic Cloud have magnitudes of the order of 19.0. The old period-luminosity relation had predicted a magnitude of 17.4. Again the classical Cepheids, with periods of between one and forty days, are at fault. The Magellanic Clouds are twice as distant as we had thought previously.

Also in line with Baade's results are some recent observations of colors and magnitudes of star clusters in the two Magellanic Clouds, by S. C. B. Gascoigne and G. E. Kron, on the basis of work at the Mount Stromlo Observatory in Australia. Some of their clusters are probably globular in kind—yet they have found these clusters about 1.5 magnitudes fainter, intrinsically, than the globular clusters of the Milky Way. Since the globular clusters are probably all rather similar to one another, the suspicion arose, as in the Mount Wilson work, that the distances of the Magellanic Clouds had been underestimated. Work of a similar nature on the Magellanic Clouds has been recently carried on at Harvard's southern station, and Shapley sets the revised distance to the clouds at 150,000 light years, making their respective diameters 30,000 and 20,000 light years. This result was announced by Shapley at the December 1952 meeting of the American Astronomical Society at Amherst. Thus it would appear that there is complete unanimity among the active workers in this field that the new zero point of the period-luminosity relation has now been fairly accurately determined.

The new distance scale helps to reconcile a number of contradictions that had resulted from the previous scale. The observable part of the universe is now so large that a time interval of 3.6 billion years is required to account for the expansional velocities of the galaxies—that is, the expansion began 3.6 billion years ago. This new time scale agrees with the radioactive determination of the age of the earth and with the evolutionary time scale of the stars.[1] . . .

Still unanswered is the question, Why was the zero point of the classical Cepheids in error by as much as 1.5 magnitudes? In Shapley's early work it was not yet realized that the cluster-type variables belong to Population II (old stars), while the classical Cepheids are nearly all members of Population I (young stars). Hence in the earlier discussions there was a tendency to force the period-luminosity relation to form a continuous curve from the shortest periods of two hours, to the longest periods, of forty or fifty days.

[1] More recent research has increased all of these age estimates; the age of the universe from the Hubble constant (see p. 25) is now estimated at 10^{10} (10 billion) years, the age of our sun at 5×10^9 years, that of the earth at more than 4×10^9 years, and that of some star clusters at more than 10^{10} years. Of course, the universe must be older than any of its contents.—TLP

We now know that the chemical composition and structure of the stars of Populations I and II are not the same and that the zero points for the two groups of variables are independent quantities. . . .

Struve goes on to discuss the physical mechanism of pulsating variables, a topic covered in Volume 6 of this series, The Evolution of Stars. *The scale of extragalactic distances, with which we are here concerned, was in much better order after Baade's recognition of Type-I Cepheids. Together with red-giant stars, blue giants, globular clusters, and novae, they helped determine the distances of many nearby galaxies from the ratio of observed brightness to intrinsic luminosity of each type of star. This ratio is usually expressed as the difference between apparent magnitude (m) and absolute magnitude (M), the "distance modulus," which is related to the distance (D) by the formula*

$$m - M = 5 \log D - 5,$$

assuming that there are no obscuring dust clouds along the distance D, so that brightness decreases with the "inverse-square" ($1/D^2$).

From distances measured in this way, together with total magnitudes of whole galaxies, it is possible to get absolute magnitudes or intrinsic luminosities[1] *of galaxies, and they are found to range from* $M = -12$ *to* -20 *or* $L = 4 \times 10^6$ *to* 10^{10} *times the luminosity of the sun. More distant galaxies, in which individual stars or clusters cannot be resolved, are assumed to have absolute magnitude about* -16, *and this is used to estimate distances from total apparent magnitudes. In fact, the Hubble law is often plotted as logarithm of the red shift (z) versus total apparent magnitude (m) rather than v versus D. Because of differences in M, points representing single galaxies on such a plot are scattered around the straight line* $\log z = 0.2m + k$ *(a constant). The scatter is greatly reduced when the plot is made for clusters of galaxies. Then z is the average red shift for several galaxies in a cluster, and m is the magnitude of the brightest galaxy in the cluster, showing that the brightest galaxies in clusters have more nearly the same M (or luminosity).*

In this way, astronomers have pushed distance estimates to almost a billion light years. The last step, as shown in Chapter 6, is to use the measured red shift as a distance indicator, assuming that the Hubble law is correct. For a single galaxy so distant that its individual stars cannot be resolved, this method undoubtedly yields the best distance estimate, since the Hubble constant, on which it depends, has been so carefully determined. There are possibilities of refining the magnitude method by using some other observable characteristic to distinguish

[1] If luminosity is measured in "suns," $M = 5 - 2.5 \log L$.

between galaxies of high luminosity and less luminous ones. It may also be possible to use angular diameter as a distance indicator for certain types of galaxies, although there is difficulty in measuring diameter because the brightness of a galaxy fades off toward the border, without any sharp edge.

Not only do these distance estimates set the "scale of the universe," they also directly affect estimates of luminosities, sizes, and masses of galaxies, as the next three chapters will show.—TLP

2

Structure and Content of Galaxies

Many attempts have been made to classify the patterns or structures of galaxies. The fascination of this activity derives from the regularities that are not quite perfect, allowing each viewer to pick the variation, or combination of patterns, he prefers. The most obvious pattern led to the designation "spiral nebulae," and before 1940 that term (a misnomer —since "nebula" means gas cloud) was used for the broad class of all galaxies. Actually, only a fraction of all galaxies are spirals. Other patterns and striking features, all of which are illustrated in one or more figures in this book, are: elliptical outlines, bars, rings, disks, halos, nuclei, dark lanes, spikes, jets, tails, mottled structures, and smooth structures.

The most significant feature of galaxies in general is the disk or saucerlike shape, with a diameter four or five times its thickness. The plane of such a disk, corresponding to the galactic equator or the plane of the Milky Way, may be viewed by us edge-on (as in Fig. 23), or square-on, or at some angle of tilt between 0° and 90°. If the disk were thin and perfectly circular, the tilt angle (*i*) could be calculated from the major and minor axes of the elliptical projection: $\sin i = B/A$, where A is the largest dimension, and B the smallest (perpendicular to A). However, the spiral arms in actual galaxies distort the circular disks

FIG. 23. Edge-on spiral, NGC 4565 in Coma Berenices, photographed with the Lick 120-inch reflector. (Lick Observatory photograph)

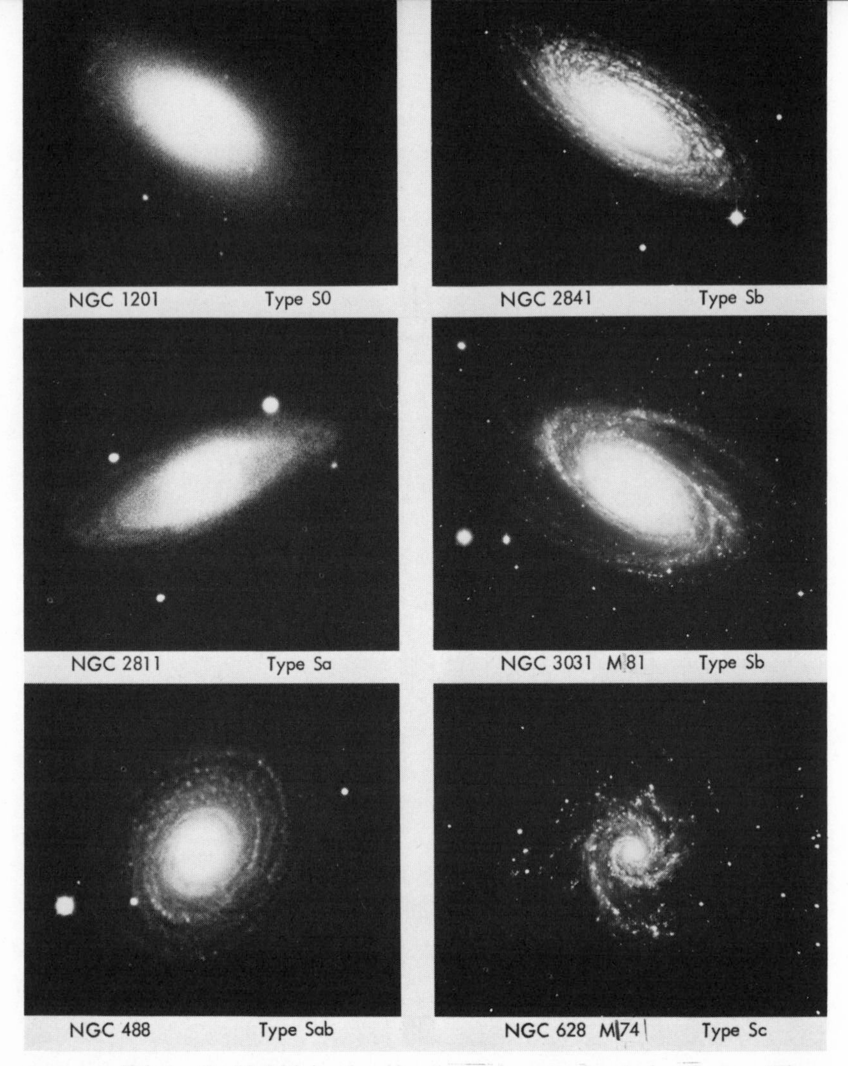

FIG. 24. Edwin P. Hubble's classification of normal spiral galaxies. These six systems exhibit a continuous gradation of forms, from type SO (flattened systems with no trace of spiral structure) through the tightly wound Sa galaxies and the looser-armed Sb class to the very open Sc galaxies. Note particularly how in Sa the arms are smooth and featureless, in Sb somewhat structured, and in Sc broken up into fine detail. (Mount Wilson and Palomar Observatories photographs)

and prevent accurate measurement of A, B, and i. Moreover, galaxy disks are thicker near the center, and some of the smooth, circular galaxies (Fig. 26) are spherical, rather than disk shaped.

Edwin Hubble's classification of galaxies, illustrated in Figures 24, 25, and 26, identified four major types: normal spirals (S), barred spirals (SB), ellipticals (E), and irregulars (Irr). In the first two he allowed for tilt (i); in effect, Hubble's "morphological type" is based on how each

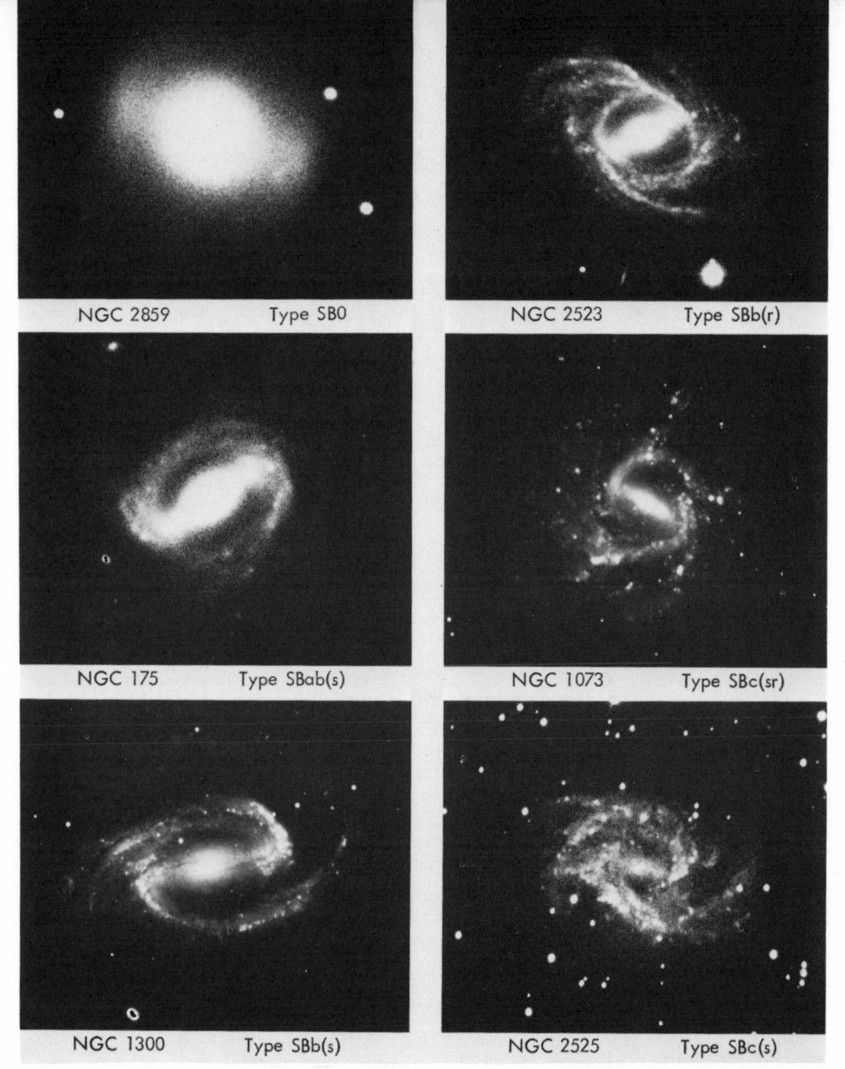

FIG. 25. The name "barred spiral" was coined by Hubble for such galaxies as these, which form an SB sequence parallel to the S sequence in Figure 24. The SB sequence starts with SBo, transitional between elliptical galaxies and the true barred ones. Note the progressive strengthening of arms relative to bar, as the texture of the arms becomes more and more broken. The added symbol "(r)" indicates an external ring. Barred spirals do not differ from normal spirals in intrinsic luminosity, dimensions, spectral characteristics, or distribution over the sky. (Mount Wilson and Palomar Observatories photographs)

galaxy would appear if viewed square-on. The sequences Sa-Sb-Sc and SBa-SBb-SBc run continuously from "tightly wound" spiral arms to "loosely wound." An Sc galaxy is more mottled than an Sa; in side view it is flatter and usually shows a dust lane (Fig. 23).

The E galaxies are all smooth, and form a continuous sequence from

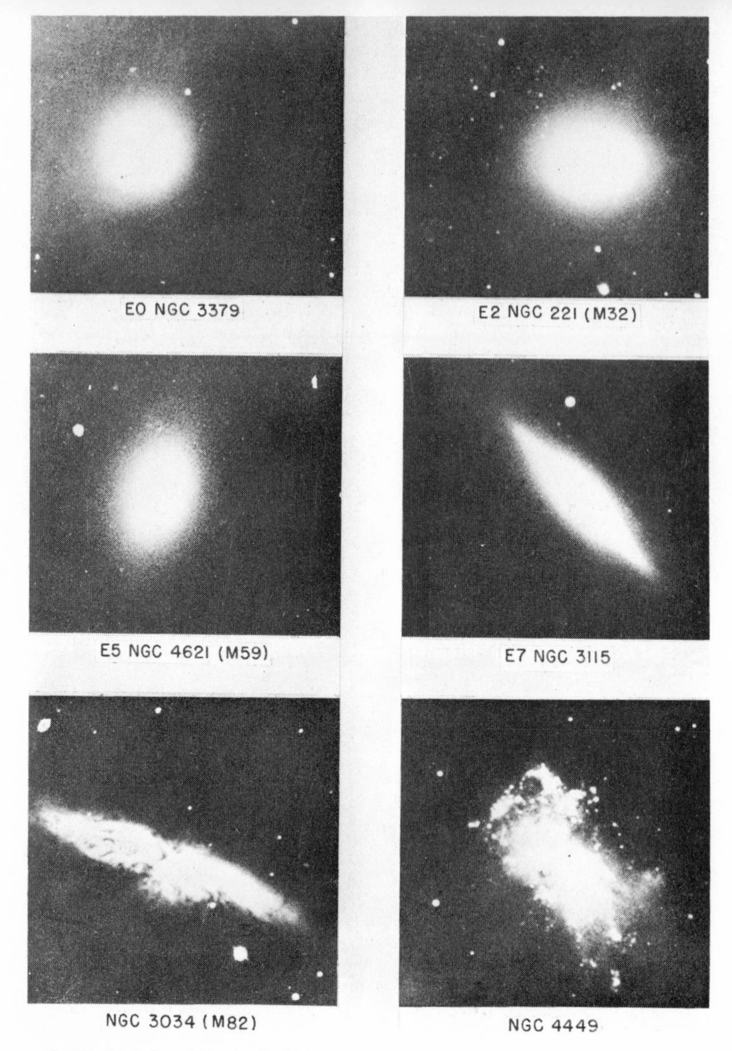

FIG. 26. Hubble's elliptical (E) and irregular (Irr) galaxies. The E types have ellipticities from 0 to 0.7. (From *The Realm of the Nebulae* by Edwin Hubble, Yale University Press, 1936)

E0 (circular) to E7 (ellipticity, 1 − B/A = 0.7), without regard for tilt. That is, an E0 galaxy may be spherical, or it may be a saucer-shaped galaxy viewed square-on. The Irr galaxies are mottled and show no obvious pattern. In 1930, when this classification was first published, Hubble connected the four types in an overall "branched sequence," illustrated by the "tuning-fork diagram" in Figure 27, running from E0 on the left through E7, then splitting into the sequences ending in Sc or SBc on the right. He considered that the irregular (Irr) galaxies belong near

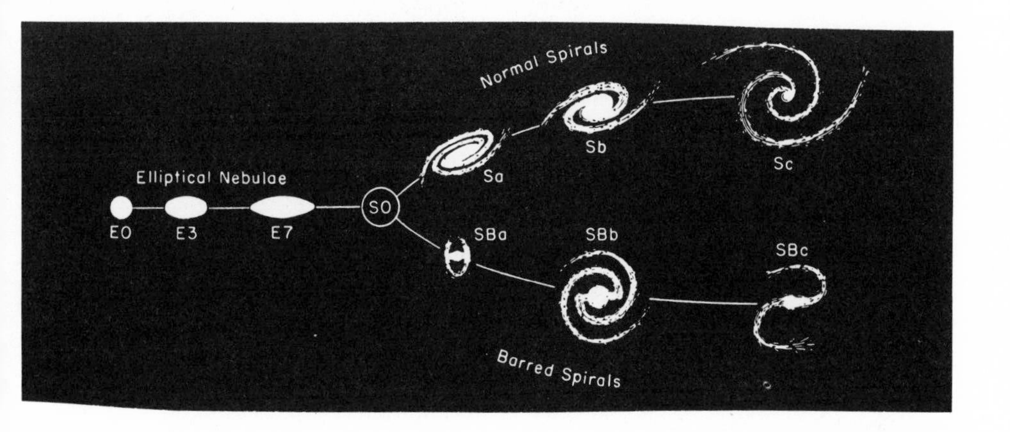

FIG. 27. Hubble's sequence of galaxy types. (From *The Realm of the Nebulae* by Edwin Hubble, Yale University Press, 1936)

the Sc and SBc types in this diagram because of their mottled appearance. In a later refinement, Hubble introduced the "transition classes" of So and SBo galaxies in the sequence between E7 and Sa or SBa. More recently, the irregulars have been divided into two classes: a "loose" Irr-I type, which is illustrated in Figure 27, and a compact Irr-II type, which seems to be more closely related to So than to Sc.

Hubble's classification scheme works remarkably well in relating pattern (structure) to other characteristics. For instance, the overall color of E-type galaxies is red, of Sa and SBa types, yellow, and of Sc, SBc, and Irr types, blue. Also, the dust content runs from near zero in E galaxies to large amounts in Sc, SBc, and Irr galaxies. Of course, no two galaxies are exactly the same, and some galaxies are so peculiar that they cannot be fitted into Hubble's classes, and are classed E(pec) or S(pec). In some cases the peculiarity is obviously due to gravitational forces from another nearby galaxy (Fig. 67); in other cases there seems to have been a violent explosion that distorted a normal galaxy (Fig. 70). Catalogues of "interacting" galaxies and "peculiar" galaxies, containing photographs of hundreds of such galaxies, have been published by B. A. Vorontsov-Velyaminov in Moscow, U.S.S.R., and by H. C. Arp at Mount Wilson and Palomar Observatories.

Other investigators have proposed more detailed classification schemes, dividing the range of spirals into four or more types, such as Sa, Sab, Sb, Sbc, Sc, and Sd. Some add the letter "r" to indicate a faint ring *structure around the spiral arms. Of course, Hubble's original types do include several subtypes. Obvious examples are right-handed and left-handed spirals (which seem to be rotating in opposite directions as in Figs. 42 and 44), a difference which may be significant in pairs and small groups of galaxies. There is also good evidence of dwarf and giant E galaxies, differing by large amounts in both mass and luminosity, and the class of Irr galaxies is divided into Irr 1 and Irr 2, according to dust content as well as concentration. Also, the strength of radio emission differs widely among galaxies of the same Hubble type. The next article describes one of the most recent efforts to refine the morphological classification of galaxies, making use of spectra.*—TLP

Classification of Galaxies

(*Sky and Telescope*, September 1961)

The world's largest refracting telescope, the 40-inch at Yerkes Observatory, has been used effectively for photography of galaxies as part of a classification program being carried on by W. W. Morgan, who gave the Henry Norris Russell lecture on this subject [at the June 1961 meeting of the American Astronomical Society in Nantucket, Massachusetts]. Noted for his work on the spectral classification of stars, he has recently been applying similar techniques to exterior stellar systems.

First he noted the long-continued usefulness of the famous Hubble sequence, set up a quarter century ago. In this simple scheme, elliptical systems of increasing apparent flatness blend into two series of spirals, normal and barred, the most open spirals being at the opposite end of the sequence from the roundest ellipticals.

Some years ago Harlow Shapley suggested that the actual evolutionary progression might be in the reverse direction, and in Morgan's scheme, sketched in Figure 28, the spirals provide the great bulk of objects between loosely knit irregulars on the one hand and highly concentrated ellipticals on the other. Central condensation is expressed as 1 for the least concentration to 7 for the strongest.

The labels indicate major population differences among the types, based on each system's integrated light, which gives a composite spectrum of a multitude of stars of different classes. Those contributing the most light dominate the spectrum, even if they do not constitute most

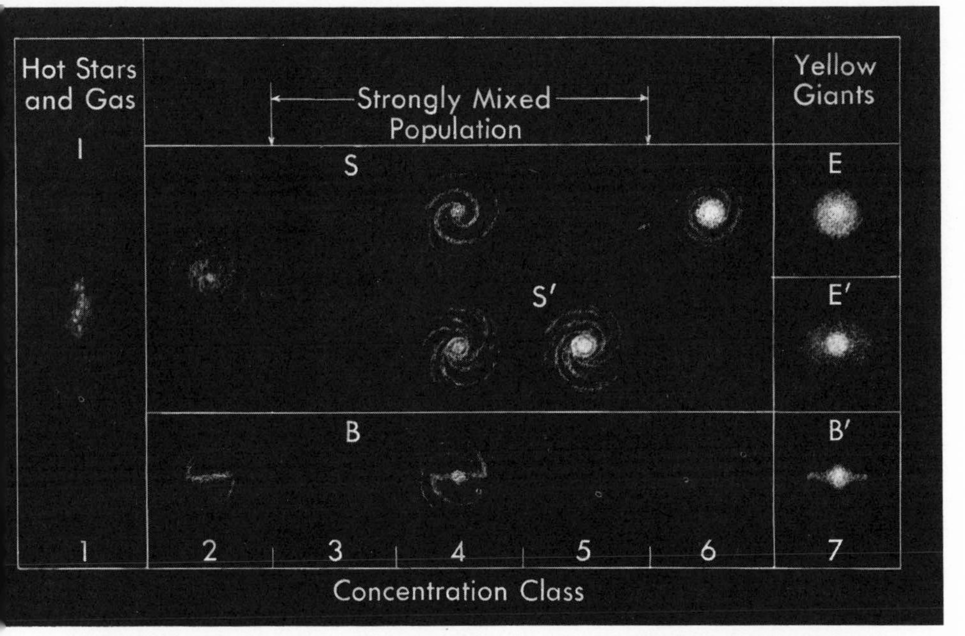

FIG. 28. Irregular galaxies (I, at the left) contain mostly hot stars and gas, but some have *A*-type spectra. The least concentrated normal (S) and barred (B) spirals also have early spectra, but as their central condensations become more prominent the composite spectra are of later type. The latest spirals and the ellipticals have central regions containing many *K*-type yellow giants. (From *The Astrophysical Journal*)

of a system's mass. Can these stars be identified unambiguously and thus used to place galaxies in the classification system?

Morgan attacked this problem with the aid of the now universally adopted Morgan-Keenan-Kellman classification of stellar spectra in which the Draper classes, *O, B, A, F, G, K, M* [see p. 323], are broken down into luminosity groups, "I" indicating a supergiant star of greatest luminosity and "V" a dwarf, faintest in its spectral class. Detailed criteria, such as relative line strengths, permit astronomers working with modest equipment to assign spectral classifications to stars they are studying.

Galaxies are more difficult to observe spectroscopically, only the nearest ones being resolvable into stars and groups of stars. The classification system must apply to the generally more distant objects for which only composite spectra can be obtained. Morgan has found criteria for establishing a gradual sequence from irregular galaxies, which contain hot young stars and gas, through spirals with mixed populations, to ellipticals, which have late-type spectra.

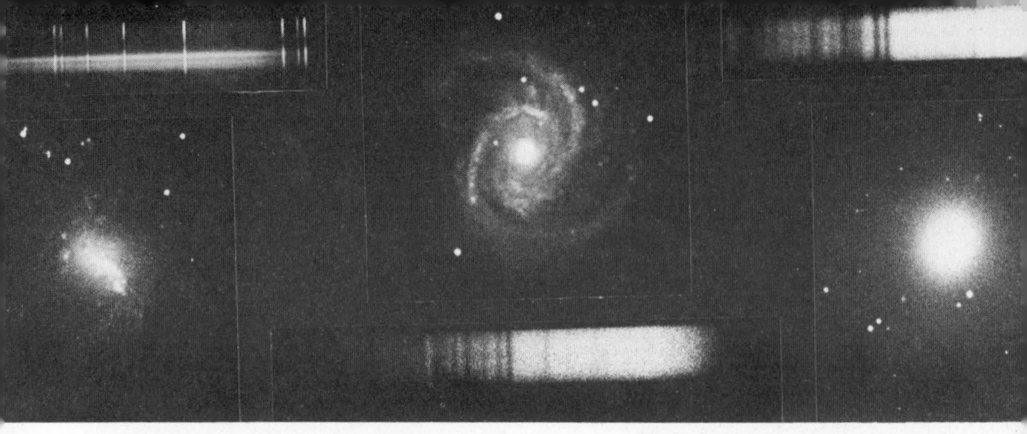

FIG. 29. These galaxies and their spectra represent types in the Morgan classification scheme. At the left, NGC 4214 is an irregular of extremely early type, showing hydrogen, helium, and ionized oxygen emission lines. In the center is NGC 4321, having moderate central concentration and a highly composite spectrum. At the right is a strongly concentrated elliptical system, NGC 4374, with a *K*-type spectrum dominated by yellow-giant stars. The direct photographs were taken with the Yerkes 40-inch refractor by Robert Garrison; the spectra were obtained with the McDonald 82-inch reflector. (Courtesy Yerkes Observatory)

Among the earliest irregular systems is NGC 4214 (on the left in Fig. 29), which has a spectrum resembling that of the Orion-nebula star cluster. The continuous spectrum is strong far into the ultraviolet, and the nebula emission lines of ionized oxygen are also observed. Other irregulars of this kind are NGC 3991 and 6052, all rich in young stars of type *B*. Somewhat later spectral types are exemplified by NGC 4490, an irregular whose spectrum resembles the Pleiades cluster taken as a whole, average type *A*.

The spirals have varying characteristics, but are generally a mixture of types. Such objects as M 100 (NGC 4321, center in Fig. 29) possess spectroscopic compositeness to an extreme degree, seeming to consist of both young and old populations of stars.

Some spirals, such as M 31 in Andromeda, have central regions dominated by yellow *K*-type giants, resembling in this respect many globular clusters in our own Milky Way Galaxy. The giant elliptical systems, including NGC 4374 in Virgo (M 84; on the right in Fig. 29), have rather pure *K*-type spectra, in which the absence of stars of the early spectral types (*B*, *A*, *F*) is pronounced.

The 40-inch refractor with focal ratio *f*/19 is far too slow for photography of most galaxies. However, a reducing camera was constructed by A. B. Meinel at the University of Arizona whereby the long-focus telescope can be made as fast as *f*/3. A field lens is placed at the focal plane of the 40-inch objective so that all the light from a large field will pass through a "collimator" lens that is followed closely by a filter and a "camera" lens. The size of the photographed image depends on the ratio of the collimator- and camera-lens focal lengths, and the overall speed of the telescope is increased by this same factor. The camera lens can be changed, a 2-inch focal length being used to make the 40-inch instrument *f*/3, a 3-inch for *f*/4.5, and a 4-inch for *f*/6. . . .

The second part of the lecture concerned the relative compositions of clusters of galaxies, particularly the Ursa Major and Virgo clusters. In general, the latter has more member galaxies of high concentration (late type) than the former, and this characteristic has been found for some other clusters. Do these differences indicate evolutionary changes in galaxies?

Morgan concluded by saying, suggestively, "Is our present epoch unique for observing these systems? If we were to look at the rich clusters of galaxies at some time in the distant past, there is evidence that they would have a systematically different aspect from the way they appear now."

Evolutionary change in a galaxy is one of the current research problems reserved for the next chapter. The remainder of the present chapter is devoted to detailed studies of the Milky Way and of nearby galaxies of various types. The Milky Way, our own Galaxy, is definitely spiral, of type Sb, although our view from inside it is sometimes confusing.—TLP

Symposium on the Galaxy

(*Sky and Telescope*, August 1950)

The dedication of a new Schmidt camera [p. 322] and a symposium on the structure of the Galaxy brought astronomers from all parts of the country to Ann Arbor, Michigan, June 22–24 [1950]. The 24-36-inch Schmidt, now operating at the University of Michigan's new Portage Lake Observatory, was dedicated to the memory of the late Heber D. Curtis.

More than thirty years ago, Curtis was among the first to adhere to the belief that the nebulous "island universes" investigated by Sir William Herschel lay outside our own Galaxy, although the majority of astronomers pictured a larger, more inclusive Milky Way system. It was therefore appropriate for the symposium to discuss questions of galactic structure.

Inasmuch as opaque dust clouds interfere with our view of the center of the Milky Way system, the 200-inch telescope has begun the investigation of other galaxies to give us greater understanding of our own system. Walter Baade of Mount Wilson and Palomar Observatories said that the neighboring spirals Messier 31 in Andromeda and Messier 33 in Triangulum are believed to be similar to our own, but the analogies are not all simple. With the 200-inch instrument and blue-sensitive plates, he expected to resolve into stars the nucleus of the Andromeda galaxy, as had already been done for this and its smaller companions with red-sensitive plates and the 100-inch Mount Wilson reflector. The brightest stars of the nuclei of galaxies are red giants; the brightest blue stars in M 31 were expected to be of apparent magnitude 22.2, but recent long-exposure 200-inch photographs reaching 22.7 failed to show any resolution of the nucleus of M 31, nor were its companions, M 32 and NGC 205, resolved.

On the positive side, however, with 2-hour exposures Baade for the first time has resolved into stars the outer portions of three globular clusters associated with M 31. The present estimated distance of up to 800,000 light years for this galaxy is based on observations of classical Cepheid variable stars in the arms of the system. Now it should be possible to determine the distance by means of the cluster-type variables in the globular clusters, as is done in our own Milky Way system.

The brightest Andromeda globulars seem to be about a magnitude fainter in total luminosity than Milky Way globulars, —8 absolute magnitude instead of —9. But there is a recognized discrepancy in the zero point of the period-luminosity curve of classical Cepheids as compared with cluster-type variables, and observations of the Andromeda globulars may clear up this problem. [See p. 56.]

N. U. Mayall, [then] of Lick Observatory, University of California, proposed further analogies after studying the rotational motions of M 31 and M 33. His observations of radial velocities of bright-line emission objects in M 31 show that maximum rotational velocity occurs about 70 minutes of arc from the nucleus, at each end of the spiral. As in M 33, indications are that farther out the velocities are less (as with the planets of the solar system); several outlying bright-line objects show periods from 90 to 200 million years.

In what Mayall called a "first crude attempt" to locate the analogous region of maximum rotation in the Milky Way, he has analyzed radial velocities of some 150 Cepheid variable stars determined by A. H. Joy at Mount Wilson. The maximum rotational velocity probably occurs at a distance from the center of the order of 9/10 that of the sun; this maximum velocity appears to be only a few per cent greater than that at the sun's distance. The speed of the sun itself may be considered as being between 180 and 280 kilometers per second, depending on the values used for its distance from the center—6000 to 9000 parsecs.

A systematic search of the Milky Way system, reported by R. Minkowski of Mount Wilson and Palomar Observatories, has revealed 212 new planetary nebulae, bringing the known total to 371. Of these, 295 are in the strip between galactic latitudes $+10°$ and $-10°$, indicating a moderate concentration toward the galactic plane. There is a high concentration toward the galactic center, and in this direction the mean apparent size of the planetaries is the least, 5.9 seconds of arc. This supports the conclusion that the bulk of the planetaries observed in this direction is actually a distant group at the center of the Galaxy. . . .

J. J. Nassau, [then] at Warner and Swasey Observatory, Case Institute of Technology, presented a report jointly with W. W. Morgan on a survey for *OB* (blue) stars. . . . Combining their results on nine hundred *OB* stars with other facts about our Galaxy and other galaxies, these astronomers suggested that the sun is located near the outer border of a spiral arm. The arm extends roughly from the constellation Carina to Cygnus. The fact that many faint and hence distant *OB* stars are found toward Cygnus indicates that we are observing the stars in the extension of this arm beyond the clustering in that constellation—that is, beyond 3000 light years.

The part of the spiral arm near our sun contains a large cloud, or groups of small clouds, of interstellar dust and gas which obscures the distant stars and divides the Milky Way into two branches, easily visible to the unaided eye. This obscuring cloud, or rift, is in the shape of a slightly bent cigar and is over 3500 light years long. At one end of it is the southern Coalsack and at the other the brilliant group of *OB* stars of the Northern Cross. . . .

The remainder of this article summarizes surveys of other classes of stars and includes an estimate of the absolute magnitude of our Galaxy, as seen from outside. Harlow Shapley puts this as −18 magnitudes, photographic, to be compared with −16.2 and −14.5 for the Large and Small Magellanic Clouds. That is, the LMC is 18 per cent as luminous as the Galaxy; the SMC, 4 per cent.

In 1950 no mention was made of the 21-cm radio emission of hydrogen (H-I) clouds in the Galaxy. More recently, radio measurements have shown that hydrogen outlines the spiral arms fairly accurately.—TLP

The Galaxy's Spiral Shape

GEORGE S. MUMFORD

(*Sky and Telescope*, May 1965)

Ever since the first evidence of spiral structure in our Galaxy was found, many astronomers have attempted to trace its detailed pattern. The work is complicated by the fact that the distribution of interstellar gas as deduced from 21-cm observations of neutral hydrogen does not show clear-cut spiral arms. In addition, observational errors, ambiguities in determining distances, and large irregularities in the arrangement of the gas may obscure the sought-for structure.

But the Russian astronomers N. S. Kardashev, T. A. Lozinskaya, and N. F. Sleptsova write, "It appears that the picture of the distribution of hydrogen clouds in the Galaxy available at the present time does, on the whole, reflect the true state of affairs, namely, that even if there is spiral structure, the angle of torsion of the spiral cannot be very different from 90°."

They have examined all currently available profiles of the 21-cm line obtained near the galactic equator. In addition, they have compiled data on discrete radio sources emitting at the hydrogen-line frequency, and optical observations. They conclude that the Milky Way system can be represented by a double-branched logarithmic spiral, its angle of torsion varying smoothly from 83° in the central parts to 85° in the outer regions of the galactic disk. (By torsion angle is meant the angle between the radius vector to a point on a spiral arm and the tangent to the arm at that point.) A value of 83° for the Galaxy had been suggested by the Australian radio astronomer B. Y. Mills in 1959.

More than three complete turns of each spiral arm can be traced, according to the report by Kardashev and his associates in the *Soviet Astronomical Journal.*

Current Studies of the Magellanic Clouds

D. J. FAULKNER

(*Sky and Telescope*, August 1963)

As our two nearest neighbor galaxies, at only a tenth the distance of the Andromeda nebula, the Magellanic Clouds occupy a unique place in extragalactic research. Their structure and dynamics as galaxies, the evolutionary processes at work, their usefulness in settling questions about distance measurement—these are but a few of the intriguing aspects of Magellanic Cloud research.

The southern declination of the Magellanic Clouds puts them out of reach of the large American and European reflectors, and it is only in the last ten or fifteen years that instrumental advances in the Southern Hemisphere have enabled astronomers to give them the attention they deserve. Hence it was very appropriate that the Magellanic Clouds, together with our own Galaxy, formed the subject of the International Astronomical Union's symposium in Australia this March [1963].

The Magellanic Clouds are classified as irregular galaxies, or sometimes as irregular barred spirals. Both the Large and Small Clouds are so near to us that their contents can be studied effectively with medium-size telescopes. For example, it has been pointed out that the Mount Stromlo 30-inch reflector shows more detail in the Clouds than the 200-inch Palomar telescope does in the Andromeda galaxy. Among the stars, clusters, and nebulosities in the Clouds we can recognize representatives of Baade's Populations I and II, making possible some very instructive comparisons of their distributions.

Detailed radio charts of the two Magellanic Clouds have been prepared showing the intensities in both 21-cm radiation and the spectral continuum. Comparing these charts with optical data has yielded two particularly interesting results. First, the two Clouds are enclosed within a common envelope of neutral hydrogen gas, which has very sharp outer extremities. Second, the Large Cloud's centroid of radio emission is displaced a full degree from its optical center.

G. de Vaucouleurs, University of Texas, told the symposium how a comparison of the Magellanic Clouds with other galaxies can help interpret their forms. Observations of barred spirals have given convincing evidence of large-scale streaming of gas outward along the bar on either side of the nucleus. He proposed a similar model for the inner portions of the Large Cloud.

New evidence that the Large Cloud is a flattened, rotating system was provided by M. W. Feast of Radcliffe Observatory in South Africa. With the coudé spectrograph of the 74-inch reflector there, he measured the radial velocities of forty emission nebulosities. His results fall on a clearly defined rotation curve, centered on the centroid of the radio rotation curve. But while it is thus becoming increasingly clear that the Large Cloud is a flattened system, there is still considerable uncertainty as to its orientation in space.

The 210-foot steerable radio telescope completed in 1961 at the Australian National Radio Astronomy Observatory, in Parkes, New South Wales, was used by R. X. McGee and J. V. Hindman to observe the Clouds for the 21-cm hydrogen line. They achieved a resolution of 7 kilometers per second in radial velocity, and 14 minutes of arc in angle.

The Large Cloud, they find, has many individual complexes of neutral hydrogen standing out from a background of much lower intensity. Some of these complexes are as large as 500 parsecs or more in diameter, with masses of 100,000 to 10 million suns. Many of them are correlated with optically observed H-II regions (ionized hydrogen near hot stars) and star associations, and with radio continuum sources.

On the other hand, the Small Magellanic Cloud presents quite a different picture. Here the distribution of neutral hydrogen is smooth, with only a few individual complexes. The most startling finding is that the 21-cm line has a double-peaked profile over most of the Small Cloud area. The two peaks are separated by an almost constant gap that corresponds to a velocity difference of 30 to 40 kilometers per second, suggesting the presence of two substantially distinct bodies of gas. Considerable discussion was provoked by these results, but no single explanation was immediately obvious.

Radio continuum observations of the Clouds at 1410 and 408 megacycles per second were presented by D. S. Mathewson and J. R. Healey of Sydney. Superimposed on a large-scale structure that has roughly the same limits as the visible Clouds, there are about 240 discrete sources (Fig. 30), many of them identifiable with optically observed hydrogen patches. No radio coronas, like the one around M 31, are observed about the Clouds, and they have a significantly lower ratio of radio to optical emission than do spiral galaxies of types Sb and Sc. A survey of other normal galaxies indicated, however, that these last two properties are common among other systems of the Magellanic type. . . .

Some very interesting surveys of the Large Cloud have been carried out with the 20-26-inch Schmidt telescope of the Uppsala southern

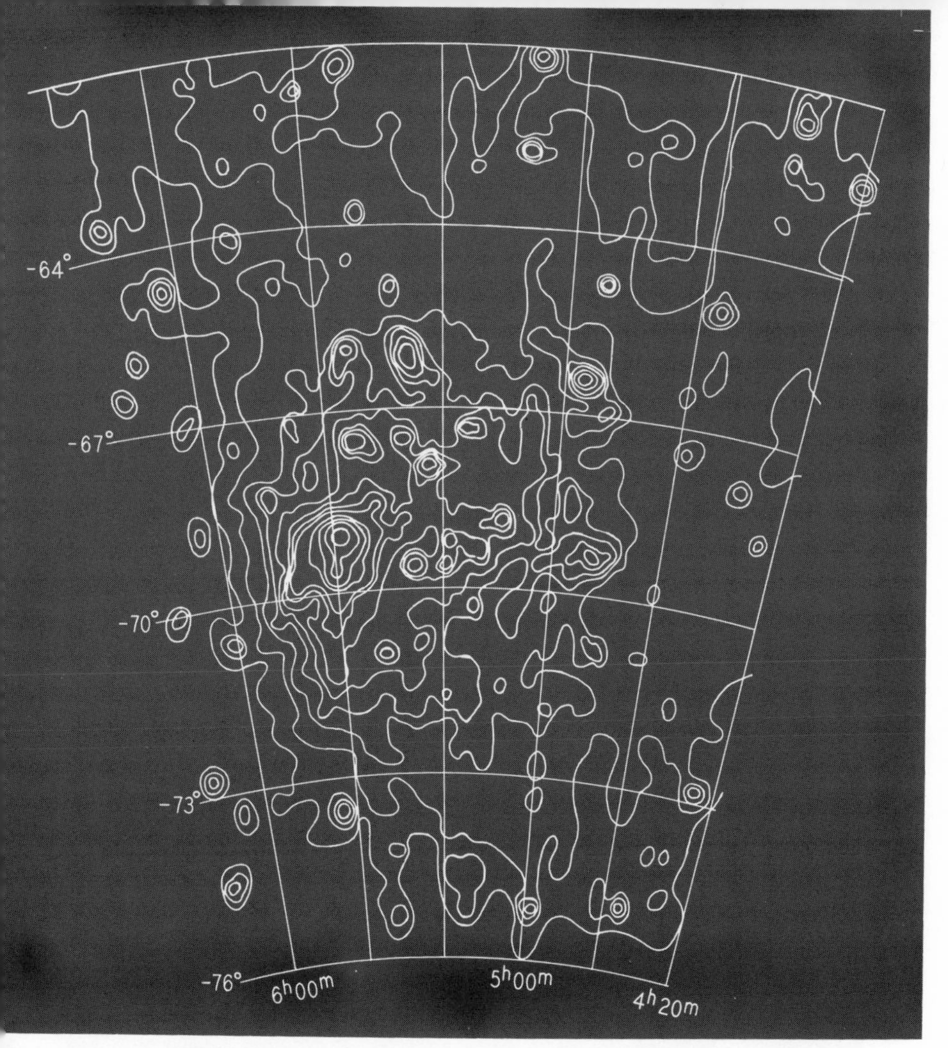

FIG. 30. Radio-continuum intensities in the Large Magellanic Cloud as observed by D. S. Mathewson and J. R. Healey with the 210-foot telescope at Parkes, New South Wales. The LMC "bar" (shown in Fig. 31) is centered at declination −69°, right ascension 5^h25^m, and runs northwest-southeast. The small round contours represent "peaks," about 70 of which appear here. Over twice as many other smaller ones are omitted in this contour plot. (Chart from Commonwealth Scientific and Industrial Research Organization)

station on Mount Stromlo. B. E. Westerlund reported that 480 star clusters . . . and 400 carbon stars were found. They appear to define a spiral structure for the Large Cloud in agreement with what de Vaucouleurs found from star counts.

Westerlund also presented evidence for population differences within the Small Cloud. From a photoelectric study of clusters, field stars, and

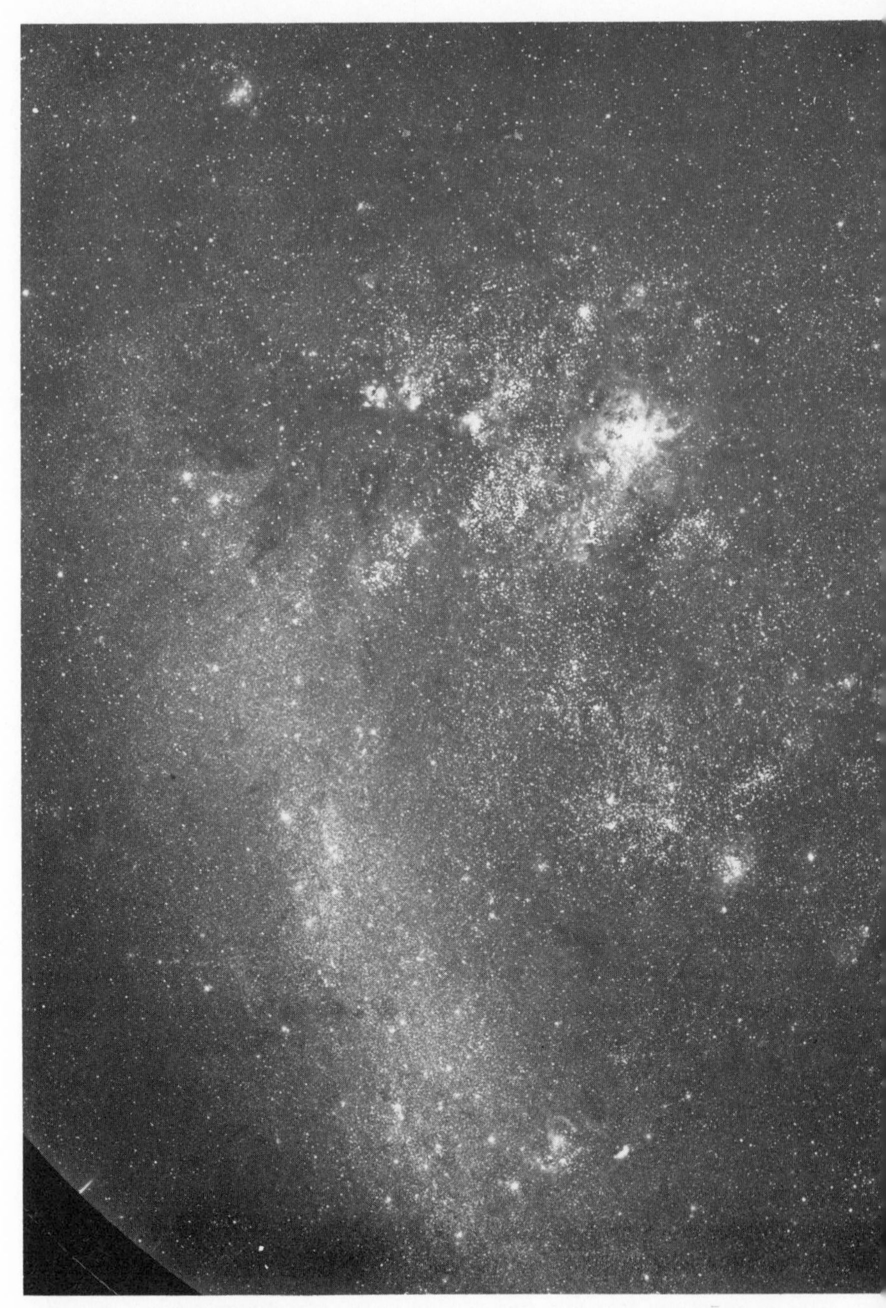

FIG. 31. A photograph showing details of the main "bar" of the Large Magellanic Cloud, which extends along the left part of the field. At the top, right of center, is the very bright nebula 30 Doradus. (Harvard College Observatory photograph)

emission regions, he ascertained that the wing of the Cloud has a predominantly younger population than Halton C. Arp had found for the main body. . . .

Some of the most intriguing questions in Magellanic Cloud research relate to the evolution of these systems. Such questions are not easily answered; only the gradual accumulation of data on many objects at all stages of evolutionary development can build up a complete picture of what has happened and is still happening.

Star clusters and associations provide a particularly fruitful line of attack, since the members of a cluster may be assumed to have a common origin and very nearly the same age. There are about thirty red globularlike clusters in the Clouds, which are brighter than visual absolute magnitude −7 and resemble Milky Way globulars. But there are also other globularlike objects that are much younger, and appear more like Milky Way open clusters, except for their much larger number of stars, and for the anomalous location of giant stars in their color-magnitude diagrams.

The oldest of the red globular clusters (for example, NGC 121 in the Small Cloud) appear quite similar to globulars in our own Galaxy. They are metal poor and have estimated ages of about 10 million years. On the other hand, the youngest globularlike clusters differ appreciably from their nearest counterparts in our system. The observations suggest a marked difference between the evolutionary patterns in the Clouds and in the Galaxy.

We may postulate that star formation began about the same time in the Galaxy and in the Clouds, and that the initial chemical compositions were similar, but it seems the evolutionary processes have been more rapid in the Galaxy. Thus, while our globular clusters are old objects, the Cloud ones are still seen in all stages of development. . . .

The Great Nebula in Andromeda

(*Sky and Telescope*, June 1948)

Three photographs by John C. Duncan, [then] at Whitin Observatory, Wellesley College, made in August 1933 with the 100-inch Hooker reflector at Mount Wilson Observatory are combined in Figure 32. As

FIG. 32. A composite of three photographs of the galaxy in Andromeda, M 31, taken by John C. Duncan on August 19–20, 1933. (Mount Wilson and Palomar Observatories photograph)

Duncan points out in his book *Astronomy* (Harper, 4th ed., 1946), this is the only extragalactic nebula distinctly visible to the naked eye. There are only five galaxies (excluding the Magellanic Clouds) brighter than the eighth photographic magnitude. The Andromeda nebula was just a hazy spot in the heavens until Simon Marius, in Germany, first observed it in 1612 with a telescope and compared it to a candle shining through a plate of horn. With the advent of the reflecting telescope and greater light-gathering power, the nebula was found to resemble an indistinct lens with a sharp center and fading-off at the edge. Through the large reflecting telescopes of the Herschels and of Lord Rosse in Ireland, suggestions of dark rifts could be seen visually. It remained for photography to record the individual characteristics of the numerous objects that go to make up this brightest galaxy.

Star clouds, globular clusters, open clusters, and clouds of nebulosity have been found in this great stellar system. In only twenty years, more than a hundred novae have been observed, chiefly in the region of its nucleus. Forty Cepheid variable stars are known and watched by observatories with large telescopes. The famous supernova of 1885, S Andromedae, reached a maximum absolute magnitude of about −14.5 (100 million suns), which is brighter than many whole galaxies. Telescopes of today cannot record individual stars much fainter than the twenty-second magnitude. If an "Andromedean" were to try to observe our sun from the Andromeda system, using our present-day equipment, he would be unsuccessful; old Sol would be more than a hundred times fainter than this limiting brightness.

Figure 32 shows the center of the galaxy unresolved so far as individual stars are concerned. In 1943, however, Walter Baade, using the 100-inch instrument on nights of good seeing and minimum temperature change (to avoid change of focus during long exposure), succeeded in resolving the nucleus of M 31 into great numbers of twenty-first–magnitude stars. He employed red-sensitive plates and a red filter to cut down fogging by the light of the sky.

In Duncan's pictures, the resolution of the spiral arms into individual stars, star clusters, and clouds of interstellar gas is quite evident. NGC 206 is the outstanding star cloud visible in the upper portion of Figure 32. The large white image, above the center, is the elliptical nebula M 32, a companion of M 31. Another and more distant companion, NGC 205, is included on the original negative. . . .

At the Newtonian focus of the 100-inch telescope the image of the main body of this galaxy is about two feet in length. Most authorities agree that this portion of the system measures 160 minutes of arc (2⅔ degrees) long and 40 minutes wide on ordinary photographs with large reflectors. Microdensitometer tracings show much greater extent than this. By means of the magnitudes and periods of its Cepheid variables, the distance of the Andromeda system is now placed near 805,000 light years.[1] That far away, an angle of one degree corresponds to 14,000 light years.

Fortunately for observers on the earth, the great Andromeda spiral is seen with its plane tipped about 15° to the line of sight. This makes it possible to observe the spiral structure and to study to better advantage this most important of our neighbors.

[1] More recently estimated at 2 million light years (see p. 11).—TLP

Rectified Photographs of M 31

(*Sky and Telescope*, April 1964)

The great Andromeda spiral galaxy, Messier 31, is the largest and most beautiful object of its kind in the sky. However, studies of its form are hampered by the large tilt (75°) of its plane to the plane of the sky. If this great system could be viewed directly from above—in orthogonal projection—its structure might become more evident.

It is not easy to remove the foreshortening by rectifying a photograph. The tilt angle and the position angle of the major axis (40°—about NE–SW) must be known with considerable precision; also, the large angular extent of the galaxy complicates matters, for the telescopic image is not merely a parallel projection of the original. The techniques needed are those used by photogrammetrists in rectifying aerial photo-

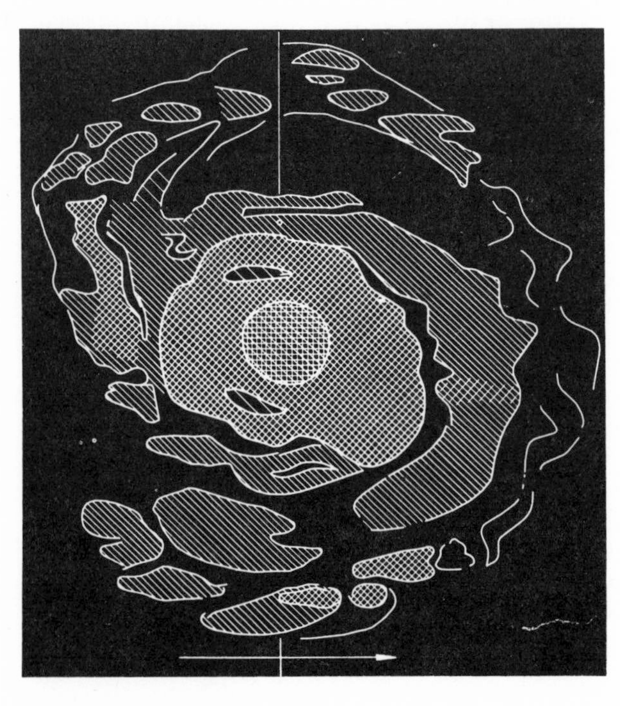

FIG. 33. If the Great Nebula in Andromeda (M 31) could be viewed square-on, its pattern would resemble this. The vertical line shows the orientation of the long axis of the galaxy as we see it, and the tiny circle on this line (above center) represents a bright star (BD +40°151) in M 31. (From *Jena Review*)

graphs to construct topographic maps, but with the added difficulty of a 75° projection angle instead of the usual few degrees.

At the Karl Schwarzschild Observatory in Tautenburg, East Germany, N. B. Richter and O. Weibrecht have obtained rectified photographs of M 31 in this way, starting with negatives taken by the 79-inch (2-meter) telescope.

From these pictures, shown schematically in Figure 33, they conclude that the Andromeda galaxy should be classified as type SA(rs) in the system proposed by G. de Vaucouleurs (University of Texas). The original definition of this class is: "SA(rs) has a fairly small, fairly sharp, bright and round nucleus in the center of a diffuse lens, or bulge, out of which two main arms and two or more additional, weaker arms emerge tangentially; the two main arms simulate an incomplete ring around the lens."

Nuclear Region of M 31

(*Sky and Telescope*, June 1964)

In the central parts of the great Andromeda galaxy, Messier 31, multitudes of very faint stars are so closely packed together that they cannot be studied individually with any telescope. The integrated spectrum indicates light of *K* stars. But is this caused by a minority of *K* supergiants, or by a predominance of dwarfs?

A new means of answering this difficult question was reported by W. W. Morgan and John S. Neff of Yerkes Observatory. They began by making photoelectric measurements at four wavelengths of the brightnesses of *K* stars in our own Galaxy. For each star, one color index was obtained from the difference between the magnitudes at 3300 and 3700 angstroms, a second index from 4700 and 5500. The relation between these two indexes for a particular star was found to be a good indication of whether it was a giant or a dwarf.

The Yerkes astronomers applied this test to the unresolved nuclear region of the Andromeda galaxy, making their observations with the 36-inch reflector at McDonald Observatory. The results indicate that in the ultraviolet spectral region the light of the central parts of M 31 is due predominantly to *K* giant stars.

This conclusion agrees with some earlier spectroscopic observations of M 31 in blue light by N. U. Mayall.

The Nucleus of the Andromeda Nebula

JOSEPH ASHBROOK

(*Sky and Telescope*, February 1968)

Long-exposure photographs of Messier 31 such as Figure 32 are intended to display the spiral arms. Hence the bright central regions are strongly overexposed, concealing a very small, almost starlike core at the precise center.

This innermost feature was concisely described in 1929 by Edwin P. Hubble, from his observations with the Mount Wilson 100-inch reflector: "The nucleus itself, on the shortest exposures, is sensibly round, with a diameter of about 3″ . . . and with an apparent photographic magnitude about 14.0. . . ." This is a good description still, but photoelectric data show that the nucleus is about a magnitude brighter (13.0).

A large, long-focus refractor like the 30-inch, *f*/18.5 Thaw telescope at Allegheny Observatory is effective in recording the tiny heart of the Andromeda system, as Figure 34, a photograph by Walter Feibelman, demonstrates. He writes, "The plate scale is 14.6 seconds of arc per millimeter. Measured on an enlarged print of the original negative, the nucleus comes out to be 4.5 seconds in diameter."

This feature should be carefully distinguished from the central condensation, or general brightening of the Andromeda nebula toward its midpoint. The condensation is a gross detail visible in small telescopes. In fact, the total light from an area 42 seconds in diameter is equivalent to a star of photographic magnitude 9.5, according to J. Stebbins and A. Whitford's photoelectric measurements at the Lick Observatory.

Thus, the inner part of the Andromeda nebula can be described in the same language that we use for a comet head: a coma containing a faint starlike nucleus. The problems of specifying the magnitude of a comet are closely paralleled here. A comet may look five magnitudes brighter in a small telescope than on a photograph taken with only enough exposure to record its nucleus; similarly, the central condensation of M 31 may appear approximately eighth magnitude in an amateur's 3-inch refractor and thirteenth magnitude on a short-exposure plate with the 100-inch telescope.

The published visual reports of the M 31 nucleus are extensive,

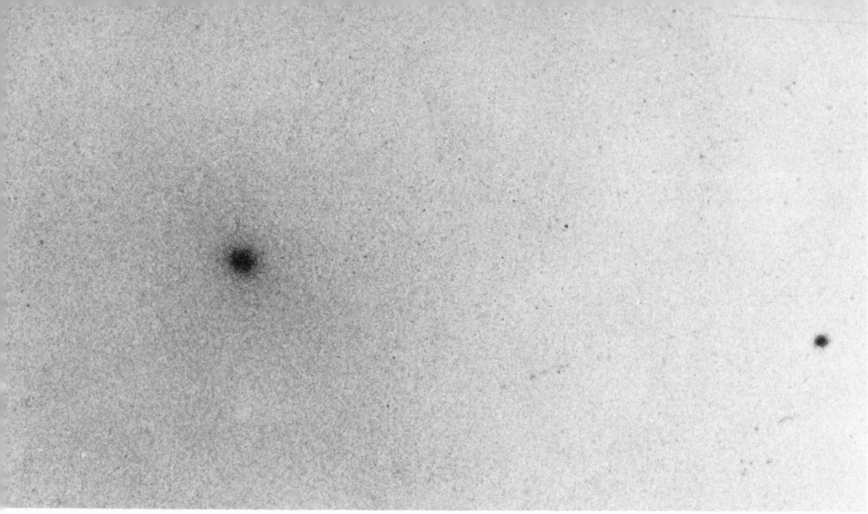

FIG. 34. The nucleus of the Andromeda galaxy (M 31), photographed with the 30-inch Thaw refractor by W. A. Feibelman. This is a negative, enlarged from an Eastman 103a-O plate exposed 10 minutes. On the same scale, the 3° length of M 31 that is ordinarily seen would stretch 23 feet! North is up, and Lamont's star is near the right edge, about 125 seconds of arc from the nucleus. In 1885 the supernova S Andromedae appeared 0.4 inch to the right of the nucleus and 0.1 inch below. Spectroscopic observations reveal that the nucleus is rotating rapidly in a period of about 520,000 years. Physically, it resembles an abnormally massive and luminous globular cluster. (Allegheny Observatory photograph)

though scattered, and go back many years. On the very first nights (July 15 and 20, 1847) that the 15-inch refractor at Harvard Observatory was used, William C. Bond noted that the great nebula in Andromeda had a nucleus resembling a star. "I do not recollect to have seen any notice of this appearance," he stated in a letter to President Edward Everett of Harvard.

Bond was not the discoverer, for on October 13 and 14, 1836, F. Lamont had seen the nucleus with the 10½-inch refractor of Munich Observatory. With a filar micrometer, Lamont determined its diameter as 6.9 seconds and measured its coordinates relative to a nearby twelfth-magnitude star (shown also in Fig. 34).

Especially important among the visual observations of the nucleus are E. E. Barnard's with the 40-inch Yerkes refractor, which he reported in *The Astronomical Journal* in 1917. He had long before thought to answer the question of whether the Andromeda nebula was another galaxy or lay inside our Milky Way by determining whether it had an appreciable proper motion. For this purpose, Barnard measured the position of the nucleus relative to three faint stars on many nights from 1898 to 1916. . . .

"The nucleus of the nebula," wrote Barnard, "is about 2″ to 3″ in diameter, but it is so strongly condensed that under good conditions it can be bisected with almost the same accuracy as the comparison stars. From its nature, the brightness of the nucleus varies greatly with the size of the telescope and the magnifying power used. With the 40-inch telescope it is of about thirteenth to fourteenth magnitude. . . ." [Barnard found no measurable proper motion.]

In the April 1960 *Publications* of the Astronomical Society of the Pacific, A. Lallemand, M. Duchesne, and Merle F. Walker reported how the nucleus of the Andromeda galaxy looked at the coudé focus of the 120-inch Lick reflector, where the scale is 1.9 seconds per millimeter. "Even with such a large scale," they noted, "the visual appearance of the nucleus on the slit was nearly stellar, with only a slight haze visible around it; the inner parts of the nebula, so bright on ordinary photographs, were completely invisible." The photographic diameter of the nucleus is 4.4 seconds, according to these investigators.

The central part of Messier 31 received searching attention from a host of observers upon the appearance of the supernova S Andromedae in August 1885 at a point only 16 seconds of arc from the nucleus, in the direction of Lamont's star. At maximum light the supernova was approximately sixth magnitude, and its gradual fading could be followed in large telescopes until early 1890.

At first, when S Andromedae was still very bright, its glare obliterated its immediate surroundings, and for a short while it was believed that an outburst of the nucleus itself had occurred. But as the supernova began to fade, the "old nucleus" became increasingly visible, and many observers measured the relative coordinates of the two with micrometers.

E. Hartwig, a young German astronomer then at Dorpat Observatory in Russia, suggested that the nucleus was variable in brightness. He noted that E. Schönfeld at Mannheim Observatory (presumably using a 6-inch refractor) had reported it about tenth magnitude, and that H. L. d'Arrest had assigned a magnitude of 9 or 10. On the other hand, Hartwig himself in late 1885 found "only a very faint star, certainly fainter than 12."

The correct interpretation of these statements, I suspect, is not that the nucleus is variable, but simply that Schönfeld and d'Arrest were describing the central condensation and Hartwig the true nucleus. . . .

These articles show some of the difficulties of observing galaxies—difficulties due to tightly packed stars, to tilt, which produces foreshortening, and to superposition of one part on another—and some of the similarities between M 31 and our Milky Way Galaxy, both spirals. An elliptical galaxy, like the nucleus of a spiral, is also difficult to resolve.—TLP

Resolution of NGC 205

GEORGE S. MUMFORD

(*Sky and Telescope*, March 1967)

The technique of superimposing several negatives of the same star field to clarify the images and enhance faint objects is well known. Last year it was applied at the David Dunlap Observatory of the University of Toronto by René Racine, who used plates from the 48-inch Palomar Schmidt telescope and from the 52-inch Schmidt of Karl Schwarzschild Observatory, Tautenburg, East Germany.

Superposition of many plates tends to cancel out granularity of the photographic images, thus smoothing the background while strengthening faint stars. By combining six photographs of the same region, each showing stars a little fainter than twenty-first magnitude, images down to nearly twenty-third magnitude could be recognized. This work was described in the *Journal* of the Royal Astronomical Society of Canada for April 1966.

Racine and Toronto astronomer Sidney van den Bergh, who had taken the Tautenburg plates, have made a composite photograph for NGC 205, one of the two elliptical companions of Messier 31 in Andromeda. Figure 35 was made from five 52-inch and four 48-inch plates, in visual (yellow) light. Comparison of this picture with a well-resolved 200-inch photograph of NGC 205 (Plate 3, *Hubble Atlas of Galaxies*, Carnegie Institution, Washington, D.C., 1961) reveals that some of the bright clumps on the composite are agglomerations of only two or three crowded stars. In regions where the stars are farther apart, the brightest individual members of NGC 205 can occasionally be seen.

A photoelectric brightness sequence for faint stars near the Andromeda nebula has been prepared by H. C. Arp, of Mount Wilson and Palomar Observatories. It turns out that the composite of NGC 205 has a limiting photovisual magnitude of 21.8, and van den Bergh concludes that this galaxy's stars can be resolved only on photographs that show stars at least this faint.

This result is at variance with work by the late Walter Baade, who during World War II used the 100-inch reflector at Mount Wilson Observatory to resolve the nuclear region of M 31, as well as of its companions M 32 and NGC 205. Afterward, he strongly adhered to 21.2

FIG. 35. This picture of NGC 205 is a superposition of nine plates taken with *two* Schmidt telescopes. South is up, east to the right. Of course, there are many foreground stars on the photograph—members of the Milky Way Galaxy. (David Dunlap Observatory photograph)

as the magnitude at which M 31's brightest stars and its globular clusters could be resolved. If his value is changed to 21.8, as van den Bergh and Racine believe necessary, the estimated distance to the Andromeda group of galaxies would be substantially increased.

Assuming that the brightest stars in NGC 205 are similar to the brightest ones in globular clusters, the Canadian astronomers suggest 24.4 as the apparent distance modulus[1] of M 31, M 32, and NGC 205, with an uncertainty of about 0.3. This value is in general agreement with recent evidence from studies of Cepheid variables, novae, and W Virginis stars in M 31. The modulus corresponds to a distance to that group of galaxies of about 630,000 parsecs, or 2 million light years, if it is assumed that space absorption has dimmed these galaxies by 0.4 magnitude.

It may appear from these professional studies that the Andromeda galaxies are only to be observed with large telescopes. As already noted, M 31 is visible without a telescope on a clear, dark night, although its structure cannot be seen. With a modest telescope, much more becomes visible, and photographs that record the overall shape, size, and pattern can be made by amateur astronomers.—TLP

A Messier Album[2]

JOHN H. MALLAS AND EVERED KREIMER

(*Sky and Telescope*, May and October 1967)

It has been over 200 years since the noted French comet hunter Charles Messier started his famous list of nebulae and clusters. He wrote:

"What caused me to undertake the catalogue was the [Crab] nebula I discovered above the southern horn of Taurus on September 12, 1758, while observing the comet of that year. . . . This nebula had such a resemblance to a comet, in its form and brightness, that I endeavored to find others, so that astronomers would not confuse these same nebulae with comets just beginning to shine."

Messier was one of the most active and successful comet discoverers of all time, having found twenty-one of them according to his own count.

[1] Distance modulus is the difference between apparent magnitude and absolute magnitude, $m - M = 5 \log D - 5$. See p. 60.—TLP

[2] The full Messier Album is to be published as a separate book by the Sky Publishing Corporation and will include finder charts, photographs, and drawings of all Messier objects. See p. 30 for a list of Messier's galaxies.—TLP

His searches were made at the Marine Observatory in Paris, where he was at first an assistant to J. N. Delisle; in 1759 he became chief astronomer. This observatory was an eight-sided tower forming part of the Hotel de Cluny, now an art museum.

Actually, Messier compiled three lists of objects rather than a single catalogue. The first, containing positions of the objects now called M 1 through M 45, was published in 1774 in the *Mémoires* of the Paris Academy of Sciences. Twenty-three more (up to M 68) were announced in the 1781 volume of the French almanac *Connaissance des Temps*, and the 1784 volume added M 69 to M 103. Another six, M 104 through M 109, are other objects known to Messier but catalogued by his colleague Pierre Méchain in the *Berliner Astronomisches Jahrbuch* for 1786.

Not all of these objects were discovered by Messier himself. M 45, for example, is the bright Pleiades Cluster, conspicuous to the naked eye. More than two dozen were reported to Messier by Méchain, who was also an active comet hunter. . . .

The telescopic equipment used by Messier was modest. His favorite instrument was a 7½-inch Gregorian reflector with a length of 32 inches and a magnifying power of 104. Because of the lower reflectivity of its speculum-metal mirrors, especially when tarnished, the light-grasp of this telescope was greatly inferior to that of a modern aluminum-on-glass reflector of similar size. He also made use of achromatic refractors of about 3-inch aperture.

With such limited equipment, many of the sky objects originally classified by Messier as nebulae are actually star clusters that he could not resolve. Moreover, the distinction between galaxies and nebulae was unknown in his time, and so both were called nebulae.

Because of its origin, Messier's catalogue is a nearly complete listing of the most spectacular deep-sky objects visible from north temperate latitudes. It has had permanent value to astronomers in providing handy and distinctive designations for the more important galaxies, nebulae, and clusters. And for the amateur observer it forms a convenient checklist of about 100 of the most rewarding objects in the sky.[1]

[1] Some *Sky and Telescope* references:

O. Gingerich, "Messier and His Catalogue," August 1953, p. 255; September 1953, p. 288.

O. Gingerich, "Observing the Messier Catalogue," March 1954, p. 157.

O. Gingerich, "The Missing Messier Objects," October 1960, p. 196.

J. H. Mallas, letter about M 40, August 1966, p. 83.

K. G. Jones, "Some New Notes on Messier's Catalogue," March 1967, p. 156.

"The Messier Catalogue." A card giving a detailed list and a sky map for all Messier objects. Available from Sky Publishing Corporation, 49-50-51 Bay State Road, Cambridge, Massachusetts 02138, for 35 cents postpaid.

We present here a photograph, a drawing, a finder chart, and a description of the visual appearance of each Messier object [only four of the galaxies are reproduced here], all from our own observations. The only parts of "A Messier Album" that are borrowed from published sources are the paragraph or two of basic data for each subject, and its description in the *New General Catalogue* by J. L. E. Dreyer (1888).

Between 1958 and 1962, in the course of a more extensive sky survey, one of us [Mallas] examined all the Messier objects. This work was done at Covina, California, with a 4-inch *f*/15 Unitron refractor that has an equatorial mount and clock drive. So far as light-gathering power and resolution are concerned, this telescope is no doubt superior to the various small eighteenth-century instruments employed by Messier.

At Covina, the sky was dark in the east and reasonably so to the north and south. But to the west, even after midnight, there was a bothersome glow on the horizon caused by Los Angeles city lights. When a temperature inversion prevailed, smog existed until after midnight. By necessity, most clusters, nebulae, and galaxies were observed in the second half of the night. Observing was done only when conditions were rated as good or excellent for the 4-inch.

The Messier list contains such a variety of objects that no one aperture or magnification can give good views of them all. Hence published descriptions may mean little unless the size of telescope used is specified. Of course, there is no substitute for the personal experience of viewing the Messier objects.

For example, the Pleiades cluster (M 45) is best seen with a 2-inch at about 15X, but is disappointing in the 4-inch. However, the nebulosity associated with the Pleiades requires a 4-inch or larger. Similarly, while the globular cluster M 13 in Hercules needs a 6-inch telescope to be well resolved, a 4-inch will suffice for M 4 and M 22 if the observer's location is favorable for viewing these southerly globulars. . . .

Of the sixty-odd drawings made, most are of globulars and galaxies. The sketches were made on vellum-type drafting paper with a soft pencil, using finger smudging and erasing until the desired effects were achieved. . . . While we had previous knowledge of how *types* of objects appeared photographically, we had not seen a photograph of the individual object beforehand. It was only after the survey was complete that we compared all the drawings with photographs.

It became apparent that the visual impressions of *galaxies* can be correlated with their photographic properties, and in fact give clues to their Hubble types [see p. 64]. The most useful criterion is the size of the bright central region. Visually, an Sa spiral can often be recognized by its large nuclear region, whereas an Sc spiral has a small, almost starlike core. In poor seeing, this difference is not apparent. The experienced eye under favorable observing conditions can judge the texture of an object as fine grained, granular, fluffy, or mottled. Long practice counts greatly in this type of observing. . . .

The distribution of Messier objects on the sky is far from uniform; clusters and diffuse nebulae are bunched in the summer Milky Way, while galaxies abound in the region of Virgo, Coma, and Leo (Fig. 36). The constellations containing the largest numbers of Messier objects are Sagittarius with fifteen and Virgo with eleven.

All the photographs were taken by Kreimer with a 12½-inch *f*/7 Cave reflector with big setting circles and push-button electric slow motions in both coordinates. . . .

Comfort during a photographic session is necessary for good guiding on picture after picture. An adjustable observing chair helps bring the observer's eye to the level of the guiding eyepiece, while a box attached to the chair keeps filmholders and tools within handy reach. It is not difficult to average eight pictures during a four-hour session. . . .

When the Virgo group of galaxies was being photographed, bright sky conditions did not permit going much below magnitude 16. So, for three or four of these individual galaxies, two negatives were superimposed during printing. . . . In almost all cases the negatives were printed on No. 5 paper, which helped bring out faint details, but at the cost of leaving heavily exposed areas a nearly featureless white. . . .

Of course, no one telescope can be best for photographing as varied a set of objects as those in the Messier catalogue. Many kinds of darkroom techniques are needed to bring out the desired details in all of them. . . .

M 81 NGC 3031 $9^h51^m.5$ $+69°18'$
Sb Galaxy in Ursa Major

Messier 81 has a total magnitude of about 7.9 visually and 8.4 photographically. Its extreme dimensions on photographs are 21′ by 10′ (minutes of arc). There is a fine photograph taken with the 200-inch telescope of this splendid Sb spiral in the *Hubble Atlas of Galaxies*. It is about 7 million light years distant, according to Allan Sandage. As is to be expected for such a nearby galaxy, the red shift is small: radial velocity 88 kilometers per second.

NGC description. A remarkable object, extremely bright and extremely large; extended in position angle 156°. It increases in brightness inward, first gradually and then suddenly to a very much brighter center. Bright nucleus.

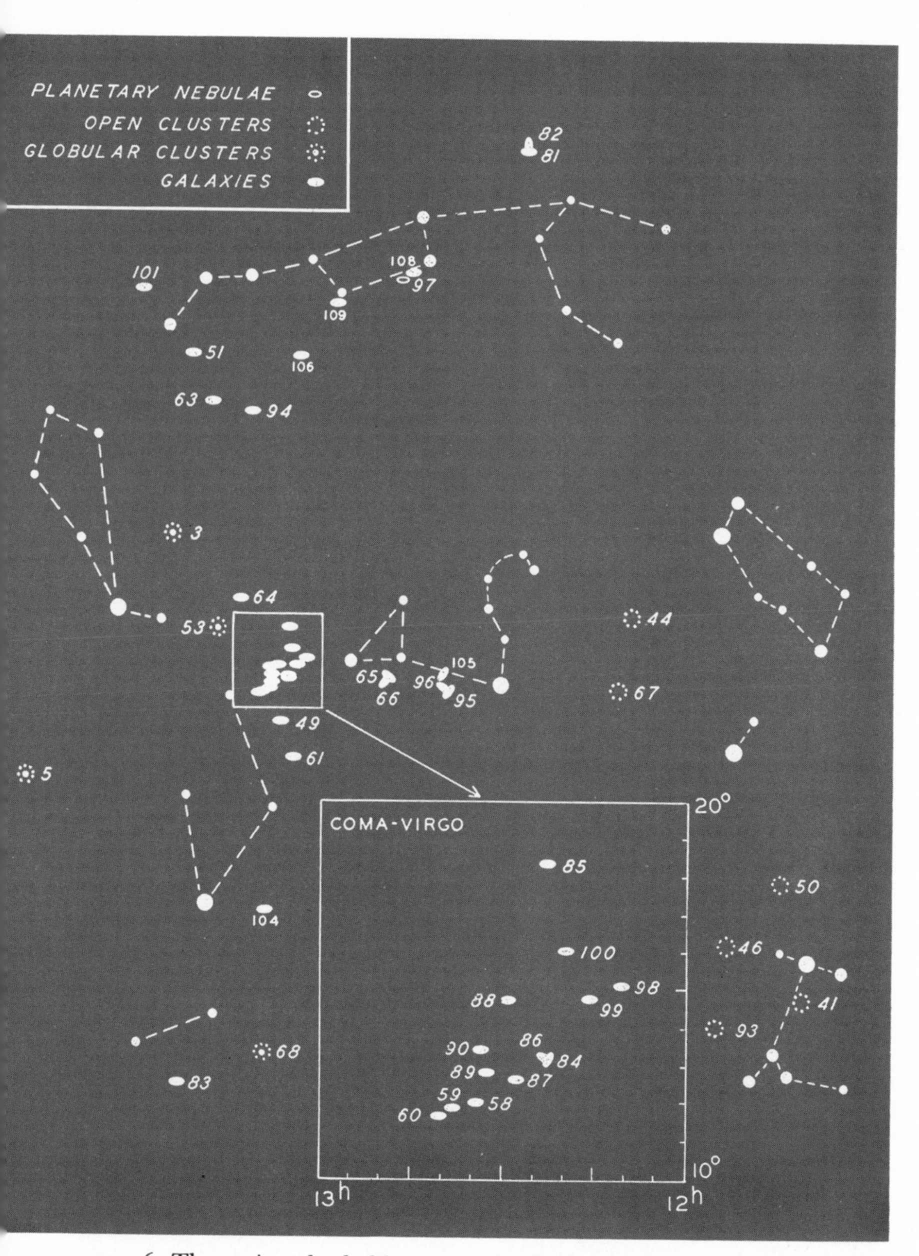

FIG. 36. The spring sky holds many Messier objects, including more than a dozen galaxies in the Coma-Virgo region (enlarged, bottom center). This schematic map, centered on Leo, extends from about declination +75° (top) to −45°, and from right ascension 6^h (right) to $15^h.5$. Northwest of the bowl of the Big Dipper are the galaxies M 81 and M 82. (From a complete plot of the Messier objects prepared by Owen Gingerich, Harvard College Observatory)

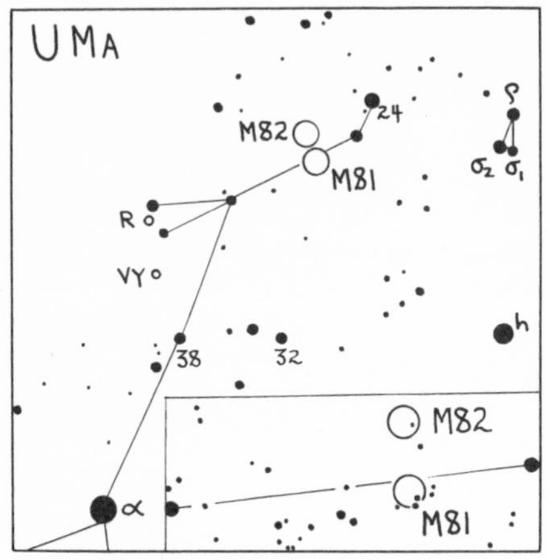

FIG. 37. Finder chart for the famous pair of galaxies M 81 and M 82 in Ursa Major, with north at the top and a scale of about 5.5° per inch. At lower left is the northern "Pointer," Alpha Ursae Majoris. The inset is of the galaxies' immediate surroundings.

Visual appearance. As seen in a 4-inch telescope, M 81, a beautiful object, has just about the most strongly granular central region of any galaxy. The outer parts are mottled and uneven in brightness and texture. Two other visual characteristics are the fairly sharp outer edge and the bright arcs at the ends of the major axis. Comparison of the photograph in Figure 38 with the drawing, in which the fainter details are exaggerated intentionally, indicates that these arcs are portions of spiral arms. M 81 rates as one of the finest galaxies for small apertures.

FIG. 38. At the left is Evered Kreimer's photograph of M 81, taken December 5, 1965, with 20 minutes' exposure in the low-temperature camera on his 12½-inch telescope and enlarged six times to a scale of about 7′ per inch. At the right is John Mallas' drawing at about the same scale, made from visual observations with his 4-inch refractor.

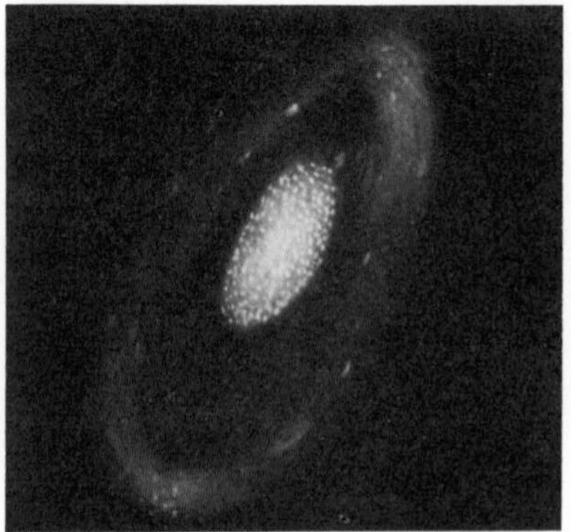

M 82 NGC 3034 $9^h51^m.9$ $+69°56'$ Irr Galaxy in Ursa Major

This irregular galaxy is about 9′ by 4′ in angular extent, and of magnitude 8.8 visually and 9.4 photographically. As pictured in the *Hubble Atlas of Galaxies,* faint slender filaments extend from its boundaries, and it is streaked with dark lanes and patches of dust. This galaxy is remarkable for the great explosion now in progress in its central regions. About the same distance from us as its neighbor M 81, it also has a small spectral red shift, 322 kilometers per second.

NGC description. Very bright and very large. An extended ray.

Visual appearance. A gem! In a low-power field, it forms a beautiful pair with M 81. In shape and color M 82 appears as a silver sliver, with its brightest part off center, as in Figure 39. The dark absorption band seen in the photograph (Fig. 39) was not detected in the 4-inch. This telescope showed the galaxy to be highly uneven in brightness, but with little or none of the grainy texture seen in M 81. The drawing does not accurately represent the irregularities in surface brightness of this unusual object.

FIG. 39. At left is Kreimer's photograph of M 82, taken on December 4, 1965, and enlarged six times to a scale of about 7′ per inch. At the right is Mallas' drawing at the same scale.

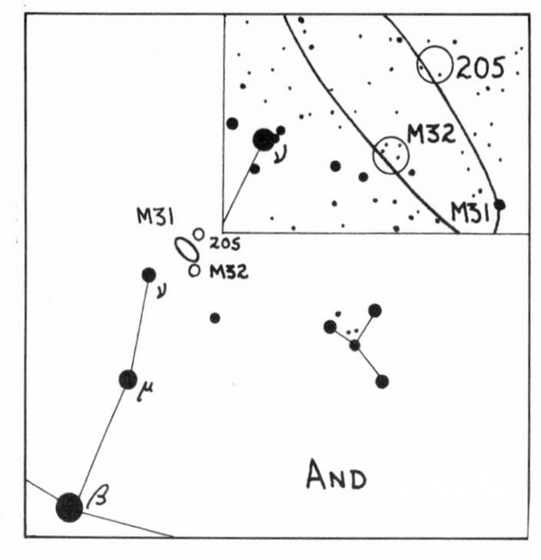

FIG. 40. Finder chart for M 31 and its two bright companions. The visual extent of M 31 is indicated in the small-scale map; the inset shows photographic dimensions. North is up.

M 31 NGC 224 $0^h40^m.0$ $+41°00'$ Sb Galaxy in Andromeda

M 31 has a total visual magnitude of about 4. On the best photographs taken with Schmidt telescopes it appears some 3° long. Yet careful binocular estimates by the French astronomer Robert Jonckheere in 1952–1953 gave a length of 5°10′ and a width of 1°05′. (His 2-inch binoculars had all optical surfaces coated, with the oculars centered so that both pupils were fully used. Other precautions included ten minutes of dark adaptation, and exclusion of all light except from Andromeda.)

Photographically, this is an Sb galaxy (a type marked by quite tightly wound arms and a nucleus of intermediate size) which is approaching us at a velocity of 68 kilometers per second. M 31 is a member of the Local Group of about seventeen galaxies, to which our Milky Way belongs. One of the latest determinations of the distance to M 31 is 2.2 million light years, by Henrietta Swope at Mount Wilson and Palomar Observatories.

NGC description. Magnificent object, extremely bright, extremely large, very much extended.

Visual appearance. The Andromeda nebula is impressive in small telescopes, although the beginner may be disappointed that he cannot see the details that are photographed. In his 4-inch refractor, Mallas was not able to see as large an extent of M 31 as is photographed. The best way to detect the faint extensions shown in the drawing (Fig. 42) is to let the galaxy drift through the field of the telescope.

The brighter portions of the arms form a flattened diamond-shaped figure. The central condensation appears very intense and starlike at low power. The 4-inch at a magnification of 25 does not reveal a grainy

FIG. 41. This mosaic of M 31 was prepared by Evered Kreimer from three 1-hour exposures taken in December 1963 with his 12½-inch *f*/7 reflector at Prescott, Arizona. The spiral structure was not seen visually. Lying south (above and left) of M 31's core is M 32. The scale is about 0.25° per inch.

FIG. 42. At left is John H. Mallas' visual impression of the central hub and lens of M 31. At right, M 32 appeared grainy but otherwise featureless to him in his 4-inch Unitron refractor. Mallas made his observations from Covina, California.

central region, yet this visual characteristic is prominent to Mallas in other galaxies of the Messier catalogue.

The photograph (Fig. 41) reveals the spiral arms. These are separated by lanes of obscuring matter, of which one or two have been detected visually with large apertures. Also well shown in the photograph is M 31's small, bright core, which is usually obliterated by overexposure on photographs with large instruments.

M 32 NGC 221 $0^h40^m.0$ $+40°36'$ E2 Galaxy in Andromeda

M 32 has a visual magnitude of 8.7, and is located 24′ south of M 31's core. It is one of a pair of bright companions to the Andromeda nebula, the other being ninth-magnitude NGC 205. (In addition, there are two twelfth-magnitude companions, NGC 147 and NGC 185, that lie about 7° to the north.) M 32 is an elliptical galaxy, with an apparent diameter of about 3′, which corresponds to a linear extent of 2000 light years.

NGC description. Remarkable; very bright, large round; suddenly much brighter in the middle, toward the nucleus.

Visual appearance. This is a beautiful galaxy, one of the best examples of an elliptical. Visually it has an oval form, and bears magnification well. In the 4-inch, the brightness of M 32 is pretty uniform nearly to the edge, then fades rapidly into the sky background. In the photograph of M 32 (Fig. 41), note how the faint outlying reaches of the Andromeda nebula appear to blend into the image of M 32.

Spiral Structure in Galaxies

JOSEPH ASHBROOK

(*Sky and Telescope*, June 1968)

In all of Sir William and Sir John Herschel's many observational descriptions of the forms of nebulae, the word *spiral* does not seem to occur. The term *spiral nebula* dates back only to 1845, when the Earl of Rosse introduced it to describe his discovery of this characteristic. By the start of the twentieth century photographic surveys showed that the faint nebulae existed by the hundred thousands, and that a large fraction of them were spirals. Today, explaining why so much of the material in the universe has assumed such a shape is among the central problems of astronomy. The beginnings of this epic are worth some review.

William Parsons (1800–67) was one of the very few amateur astronomers who dreamed of building the largest telescope in the world, then did so. He enjoyed the advantages of being son to a wealthy Irish peer and of being a highly talented engineer. At Birr Castle, he began making small speculum-metal mirrors in 1827, and completed a 36-inch altazimuth reflector by 1839 (the same year Sir William Herschel's 48-inch was finally dismantled). Parsons was well advanced with the construction of a 72-inch reflector when, upon his father's death, he became the third Earl of Rosse.

This instrument was tested in February 1845 and demonstrated that its light grasp and definition were excellent for a survey of the Herschel nebulae. With a focal length of 53 feet, the 72-inch leviathan was slung from chains between two stonework walls 56 feet high, the tube sup-

FIG. 43. This photograph of Lord Rosse's 72-inch telescope, near the present town of Birr, County Offaly, Eire, may have been taken around 1880, to judge from the growth of ivy. The instrument had two primary mirrors of speculum metal, each 6 inches thick, weighing 3½ and 4 tons, respectively. While one mirror was being repolished, the other was used in the telescope. Above the western (left) wall, note the movable gallery upon which the observer stood to reach the Newtonian focus, when the telescope was pointed near the zenith. The dark horizontal bar below the mouth of the tube is the screw that drove the telescope in hour angle. This was turned by an assistant, giving a motion smooth enough to permit rough micrometer measurements at the eyepiece. (Science Museum photograph, Crown copyright reserved)

ported at its lower end by an enormous universal joint.[1] Thus the telescope was restricted to a strip of sky along the meridian, roughly one hour wide, within which visual observations could be made (see Fig. 43).

"The spiral arrangement of Messier 51 was detected in the spring of 1845," Lord Rosse told the Royal Society in 1850.[2] Presumably he made the discovery personally, although J. P. Nichol of Glasgow made the drawing [Fig. 44] of M 51 that Lord Rosse displayed at the 1845 meeting of the British Association for the Advancement of Science.

[1] See M. A. Ellison, "The Third Earl of Rosse and his Great Telescopes," *Journal* of the British Astronomical Association, **52**, 267, 1942.

[2] *Philosophical Transactions* of the Royal Astronomical Society, 1850, p. 505.

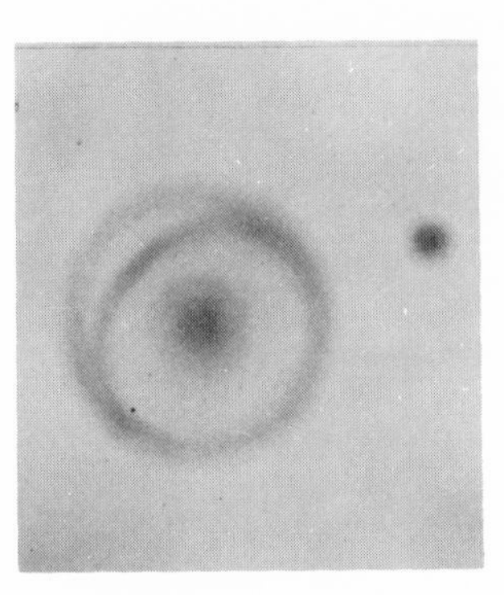

FIG. 44. To the left is Sir John Herschel's drawing of M 51, made more than a decade before Lord Rosse's discovery. It shows a split ring surrounding the central condensation, closely resembling Herschel's idea of the Milky Way's structure. To the right is a picture published in 1850, of the Whirlpool nebula as seen in the large Birr Castle reflector. Lord Rosse later concluded that the lesser nucleus (NGC 5195) is spiral; photographs, however, show it as irregular. Both sketches show bright positions in black.

A second spiral, M 99 in Coma Berenices, was recognized during the spring of 1846. Lord Rosse wrote in June 1850: "The other spiral nebulae discovered up to the present time are comparatively difficult to be seen, and the full power of the instrument is required, at least in our climate, to bring out the details. It should be observed that we are in the habit of calling all objects spirals in which we have detected a curvilinear arrangement not consisting of regular re-entering curves; it is convenient to class them under a common name, though we have not the means of proving that they are similar systems. They at present amount to fourteen. . . ." This list included M 33 in Triangulum, NGC 2903 in Leo, and NGC 7479 in Pegasus.

Lord Rosse's announcement concerning M 51, the Whirlpool nebula in Canes Venatici, raised much excitement at the time because of its bearing on current cosmological thinking. M 51 had been carefully studied by Sir John Herschel with an 18¾-inch reflector. He described it in 1833 as consisting of a bright round nebula surrounded at some distance by a ring that was double in its southwestern part. "Supposing it [the ring] to consist of stars, the appearance it would present to a spectator . . . near the central mass would be exactly similar to that of our Milky Way. . . . Can it be, then, that we have here a brother-system bearing a real physical resemblance . . . to our own [Milky Way system]?"[3]

[3] *Philosophical Transactions* of the Royal Astronomical Society, 1833, p. 496.

The spiral arrangement in M 51 and similar objects immediately suggested internal motions, governed by dynamical laws that Lord Rosse in 1850 thought "almost within our grasp." This lead was followed up by an American scientist, Stephen Alexander (1806–83), in a remarkable series of papers in *The Astronomical Journal* for 1852.

Alexander argued that "the Milky Way and the stars within it constitute a spiral with several (it may be *four*) branches, and a central (probably spheroidal) cluster. . . ."[4] He even attempted to trace the course of these spiral arms across the sky, on the basis of Sir John Herschel's maps and description of the naked-eye Milky Way.

Bravely, he sketched in qualitative terms the evolutionary history of a typical galaxy. The initial state was a huge rotating mass in dynamical equilibrium. As it cooled it contracted and, becoming unstable as the rotation accelerated, began to shed its matter in the form of spiral arms.

So far as I know, Alexander was the first to suggest seriously that our Galaxy has a spiral form. He was professor of mathematics and astronomy at the College of New Jersey, now Princeton University. . . .

Curiously, the great Andromeda galaxy M 31 was not recognized to be spiral until very late. However, the dark "canals" discovered in 1847 by G. P. Bond with the 15-inch Harvard refractor are the interarm spaces. The true form was not revealed until long-exposure photographs were taken in 1888 by Isaac Roberts with a 20-inch reflector and in 1890 by E. E. Barnard with a 6-inch *f*/5 refractor.[5] . . . Neither Roberts nor Barnard used the word "spiral" in describing their pictures, but spoke of "rings." . . .

Developments in astronomical photography over the past century have made it possible to photograph more detail than can be seen at the eyepiece of a telescope. Part of this photographic superiority depends on the use of filters to select one color of light that shows significant detail (such as the red emission of interstellar hydrogen). After such use was made of filters twenty years ago, astronomers tried "subtracting" one photograph from another to bring out differences in sharp contrast.—TLP

[4] *The Astronomical Journal*, **2**, 101, 1852.

[5] *Monthly Notices* of the Royal Astronomical Society, **49**, 65, 1888; **49**, 121, 1889; **50**, 310, 1890.

Composite Photographs of the Whirlpool Nebula

DORRIT HOFFLEIT

(*Sky and Telescope*, March 1956)

At Mount Wilson and Palomar Observatories, Fritz Zwicky has been applying what he calls analytical photography to studies of the distribution of stars and nebulosity in our own and other galaxies. He now has obtained excellent results in the case of the famous Whirlpool nebula, Messier 51, and its companion system, NGC 5195, and has published a discussion of these in the *Publications* of the Astronomical Society of the Pacific.

Two composite photographs are reproduced in Figure 45. They were made from a pair of plates, one with an ordinary blue-sensitive emulsion, the other with an orthochromatic emulsion exposed through a yellow filter. Their exposure times have been adjusted so that corresponding images of *O*- and *B*-type (blue-white) stars are of the same size on both pictures. The images of orange and red stars are then larger on the yellow plate than on the blue one.

Using the proper exposure time, a positive transparency of the yellow plate was then made, the resulting white images of very blue stars being identical in size with their black counterparts on the blue negative. The left-hand picture in Figure 45 shows the effect of superimposing this yellow positive on the blue negative. When the procedure is reversed, with a blue positive superimposed on the yellow negative, the right-hand picture in Figure 45 is obtained. Very blue stars appear as black spots on the left, but as white spots on the right. Very red stars are white spots on the left, and black on the right. Stars of intermediate colors give black spots surrounded by white rings, the relative diameters being a measure of blueness or redness.

A glance shows that these composites are very different. Much more detail is seen in the blue-negative yellow-positive combination. Bright blue stars of Population I are distributed unevenly along the galaxy's spiral arms. The irregular companion galaxy, across which one of the Whirlpool's spiral arms extends, is smaller in blue light than in yellow.

On the blue-positive yellow-negative composite, on the other hand, the spiral arms shown by the black images of unresolved red stars are very smooth and well defined, suggesting a higher degree of organization of the fainter Population-I red stars than of the blue supergiants in the

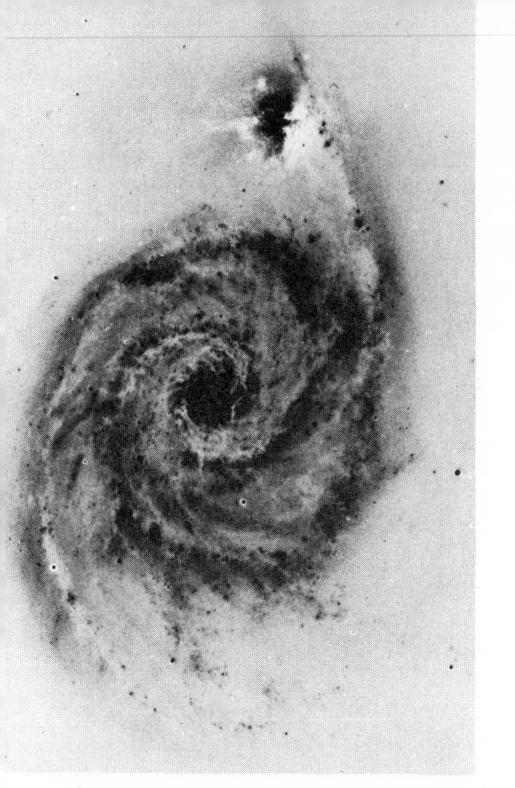

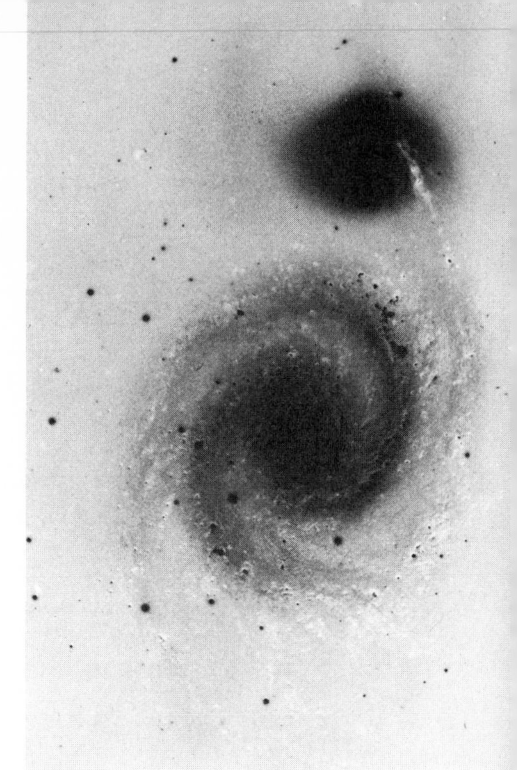

FIG. 45. These composite photographs of the galaxy Messier 51 in Canes Venatici were obtained by Fritz Zwicky with the 200-inch Hale Telescope. The scale is 4.9 seconds of arc per millimeter. The left-hand picture accentuates blue detail; the other the structure in yellow light. (Mount Wilson and Palomar Observatories photograph)

same regions of the Whirlpool galaxy. The companion appears large in this combination and shows none of the structural detail seen in the left-hand picture. Zwicky points out, however, that different galaxies have different relative distributions, and that some which look almost identical on single direct photographs may appear quite different by analytical photography.

Photographs like those in Figure 45 are clear evidence of the spiral structure of galaxies seen less well in many other photographs in this book. Apparently, the material (stars and interstellar gas) is concentrated in two spiral arms winding outward from opposite sides of the nucleus, but there is also material between the arms. The whole structure "looks as if it were rotating," and Doppler shifts in the spectra of tilted spirals like M 31 indicate the rotational motions. For many years there was a controversy as to the direction of rotation, but the evidence is now fairly definite that the arms trail; in Figure 45 the rotation is clockwise.

The reason for the rotation of a galaxy is simple; unless the material is in orbit around the nucleus it would be pulled in by the gravitational attraction of the rest of the galaxy. In fact, Newton's laws of gravitation and of motion under a force can be used in a simple calculation of the galaxy's mass from the rotational speed of the outer parts—just as the sun's mass is calculated from the speed and orbit size of a planet moving around it.) As shown in Volume 7 of this series, Stars and Clouds of the Milky Way, *rotation of our Milky Way Galaxy was detected in 1924. However, the explanation of the spiral patterns of our Galaxy, and others, was not easy. In fact, the first theoretical attempts, by the late Bertil Lindblad of Stockholm, predicted spiral arms leading (not trailing).*—TLP

Spiral Patterns in Galaxies

BEVERLY T. LYNDS

(*Sky and Telescope*, June and July 1967)

Astronomical observers have been familiar with spiral galaxies for a long time. As early as 1845, Lord Rosse saw the whirlpool pattern of Messier 51 in Canes Venatici, and in 1888 a photograph by another Englishman, Isaac Roberts, revealed the spiral form of the great Andromeda nebula. In 1900 James Keeler announced that most of the faint nebulae visible by the thousands on Lick Observatory photographs were spirals. During the present century the patterns in our own and other galaxies have been mapped by many radio and optical observers, who have gathered a rich harvest of empirical facts.

Astronomical theoreticians have been much slower in finding acceptable explanations of why such a spiral pattern should exist in galaxies. One of the most recent theories is that developed at Massachusetts Institute of Technology by C. C. Lin and his collaborators, especially F. H. Shu (now at Harvard Observatory).

Fundamentally, a typical spiral galaxy is a concentration of stars and gas into a relatively flat disk. The entire disk rotates about its center in much the same way as the planets go around the sun: the outer parts move with smaller speeds than do the regions closer to the central mass of the galaxy. The problem presented by the spiral structure of such a system was clearly defined by J. H. Oort in 1961, when he addressed a conference held at Princeton on interstellar matter in galaxies:

"In systems with strong differential rotation, such as is found in all nonbarred spirals, spiral features are quite natural. Every structural irregularity is likely to be drawn out into a part of a spiral. But this is not the phenomenon we must consider. We must consider spiral structure extending over the whole galaxy, from the nucleus to its outermost parts, and consisting of two arms starting from diametrically opposite points. Although this structure is often hopelessly irregular and broken up, the general form of the large-scale phenomenon can be recognized in many nebulae."

Oort's remarks and those of his colleagues at the Princeton meeting stimulated Lin to seek a theoretical model that would develop the type of spiral pattern just described. Like the pioneering work by the late Swedish astronomer B. Lindblad, his theory is based upon the gravitational effects operating in a galaxy. Lin met with some eighty observational and theoretical astronomers during a symposium on the spiral structure of our Galaxy, held at Tucson, Arizona, on March 16 and 17, 1967. During the two-day session at the University of Arizona's Steward Observatory, the details of Lin's theory were presented.

In any galaxy, every star or bit of interstellar matter will exert a gravitational force on every other one, accelerating it. Thus the mass distribution within a galaxy will essentially control the motions within it, as a consequence of the combined effect of all these gravitational attractions.

There are other forces acting in our Galaxy; in particular, a relatively weak magnetic field may exist. Lin has shown that the accelerations produced by magnetic forces are probably small compared to the gravitational accelerations, over distances like the size of a galaxy. However, if we are interested in the structural details of a spiral arm, then the effect of a magnetic field should probably be considered, since distances within the arm are very small compared to the entire system, and the strength of the magnetic field may change considerably from place to place. . . .

Even though a galaxy as a whole rotates about its center of mass, we know that in our own system not all stars are moving in circular orbits; this is analogous to the elliptical orbits of planets around the sun, and the deviations from perfectly circular motion are called dispersion speeds. . . . As described by Lin, "The theory based purely on gravitation would say that the spiral structure is perhaps primarily determined by all the stars whose dispersion speeds are not too large—very young and moderately young stars. But those stars whose dispersion speeds are extremely large would not show the effects of the spiral gravitational field that might be present. The essential idea is that if we observe a spiral pattern of gas and young stars, then underlying this there is a gravitational potential, low in the arms and relatively high in between the arms. The gravitational structure is maintained by gas and by stars of small dispersion speed."

How does one go about solving the complex set of differential equations describing these motions? Lin has used what he calls the "hypothesis of quasi-stationary spiral structure." Suppose we start with a galaxy that is symmetrical about its axis of rotation. Suppose, also, that a small perturbation in the gravitational field occurs, changing it by only a few per cent, in such a manner that the points of lowest field strength trace out a spiral pattern through the galaxy.

What will be the response of the gas and stars to such a spiral field? Will matter tend to concentrate in regions where the potential is low? If so, then this concentration of matter would sustain the initially assumed pattern of the gravitational field. Lin describes such a situation as a "self-sustained field." Once the field exists, the matter of the galaxy will move so as to maintain that field.

The hypothesis of quasi-stationary spiral structure assumed that there are solutions representing a *permanent* spiral pattern, and the mathematical analysis has justified this assumption. It is next necessary to see if these solutions agree with observation. Lin has shown that in a rotating flattened system, like a galaxy, small density fluctuations may be unstable and tend to grow. These density "waves" produce the perturbation in the gravitational field.

One of the most interesting of Lin's results is a demonstration that the perturbations will grow and lead to a periodic spiral pattern if the wave is trailing, but the perturbations are suppressed if the wave is leading. As a result, the theory predicts spiral arms that trail as a galaxy rotates, and forbids galaxies with leading arms—in most satisfactory accord with observation!

Lin has deduced an approximate theoretical relation between pitch angle and arm spacing:

$$\tan i = \pi^{-1} \ln (d_2/d_1),$$

where i is the pitch angle (angular deviation of an arm from the circular direction), and d_1 and d_2 are the distances of the first and second arms from the center of the galaxy. (Here π is 3.1416, and ln denotes natural logarithm.) Thus if we assume that arms exist 5000 and 7000 parsecs from the center, the pitch angle should be about 6°. Further, the next arm (if the same pitch angle holds) should be at 10,000 parsecs. These numerical values are in fact roughly those for the arms of our own

FIG. 46. This full-face view of M 74, in Pisces (type Sc), shows two spiral arms, originating from opposite ends of the central mass. (Lick Observatory photograph)

FIG. 47. The four-armed spiral NGC 6946 is a nearby galaxy on the Cygnus-Cepheus border. (Mount Wilson and Palomar Observatories photograph)

Galaxy as determined from 21-cm observations; the gaseous arms also seem nearly circular, so the 6° pitch angle is reasonable.

It should be emphasized that the spiral patterns calculated by Lin represent the form of the gravitational field. The arms are density waves through which flow the individual atoms and stars of the galaxy. The pattern velocity need not be the same as the velocity of a gas atom or star. . . .

Lin has proved that certain conditions are required for a quasi-stationary spiral structure to be possible. In general, such structure can be established over an entire galaxy only if its pattern is two-armed; a pattern of more than two arms would result only in pieces of spiral arms. [See, however, Figure 47, where two arms in NGC 6946 apparently split into four.]

Using a two-arm model for our own Galaxy, Lin deduces that a spiral pattern can exist only between 4000 and 13,000 or 14,000 parsecs from the center. These values are again consistent with the observations of radio astronomers. Lin also emphasizes our lack of clear understanding of the Galaxy's central region, from both theoretical and observational points of view. . . .

As is well known, radio maps of the Milky Way Galaxy can be constructed from observations of the 21-cm emission line of neutral hydrogen. These impressive maps are based on an adopted *rotation curve* of the Galaxy, which can be prepared on the assumption that the hydrogen gas is moving in circular orbits around the galactic center, and that it forms a continuous sheet in the galactic plane.

Then, if we look in any direction within about 90° of the center, the gas having the highest observed velocity of approach or recession is

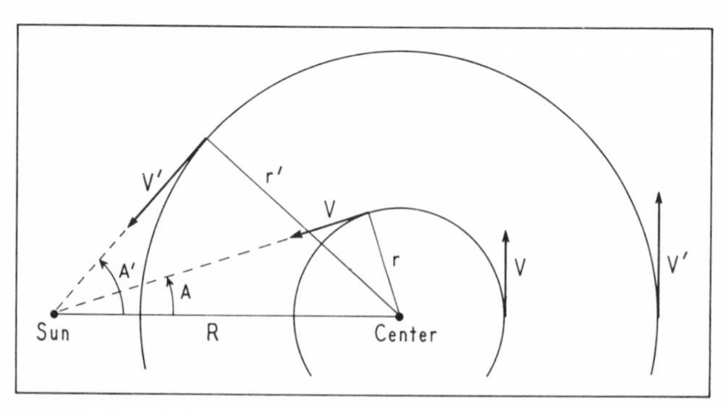

FIG. 48. Along the directions indicated by angles A and A′, the most rapidly approaching gas will have the orbital velocities V and V′. Since the distance R is known, the gas-to-center distances r and r' can be found and two points established on the galactic-rotation curve.

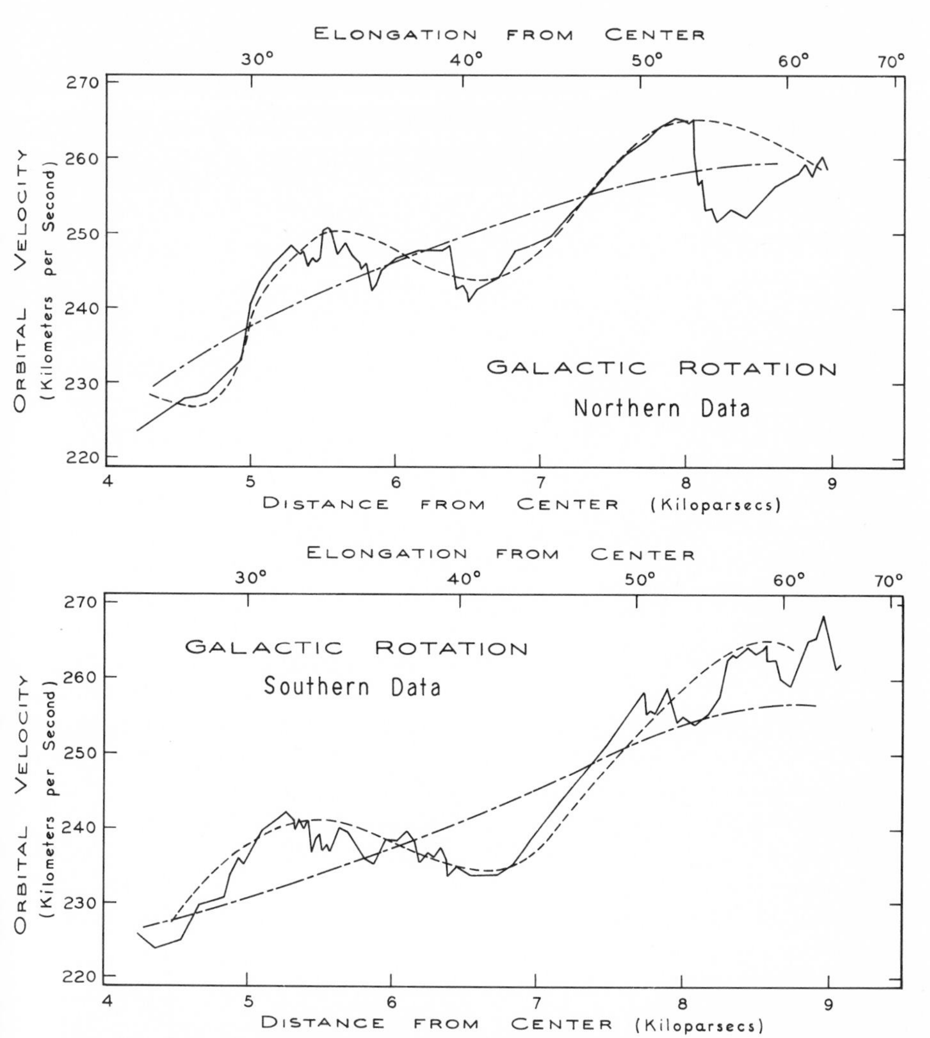

FIG. 49. Galactic-rotation curves prepared by C. Yuan. Above, galactic longitudes 0° to 70° are represented, the solid observed curve matching fairly well the wavy dashed lines of C. C. Lin's theory. Below are Southern Hemisphere observations for galactic longitudes 360° to 290°, and the match is similar. (Massachusetts Institute of Technology diagrams)

that whose orbital motion is directly toward or away from us (as shown in Fig. 48). The line of sight to this gas is tangent to its orbit. The point of tangency, the galactic center, and the sun form a right triangle, and we can solve the triangle to find the distance (r) of the gas from the galactic center. Repeating this process for different directions of observation permits building up the rotation curve point by point.

In the two charts of Figure 49, prepared by C. Yuan at MIT from observations and from theory, we see that the actual rotation curves of the Galaxy based on radio observations are not smooth, but have bumps. In the past, the generally accepted explanation of such bumps was that part of the gas is missing; along some lines of sight there is no hydrogen at the tangent point and the gas producing the highest radial velocities has been incorrectly placed too close to the galactic center. . . .

Lin interprets the systematic gas motions (ups and downs in the rotation curve by 8 kilometers per second) as flow produced by the gravitational field of the spiral structure. His theory predicts the bumps in the rotation curves. The gas concentrations required to produce these bumps are not excessive, the hydrogen density in the spiral arms being only three or four times more than between them. . . .

The two-day March gathering of experts on galactic structure raised more questions than it answered. For example, if Lin's theory is correct, then the neutral-hydrogen maps of our Galaxy must be redetermined by using the rotation curve appropriate to a galaxy having density waves in it. . . .

In summarizing the conference, W. W. Morgan of Yerkes Observatory suggested questions and answers that may be paraphrased this way:

Which components of a spiral galaxy are most fundamental? Dynamically, the stars may well control the gravitational field producing the spiral pattern. But the resulting mass concentrations are more clearly shown by the obvious H-II regions, neutral hydrogen, and dust lanes.

What is the time sequence? If, as Lin says, the spiral pattern is a quasi-permanent feature of a galaxy, how do the arms themselves change with time? The work of the Danish astronomer Bengt Strömgren on the older *B* stars and their birthplaces may provide a partial answer to this.

What differential effects are associated with the various kinds of arm components? An example is the action of the magnetic field, which influences the gas strongly, but is ignored by the stars.

Today an astronomer interested in galactic structure is almost like the proverbial donkey between two haystacks. He can find satisfaction from the observational haystack, or he can move to the theoretical one of spiral patterns and their ramifications. Or, as Bart Bok suggested at the Tucson meeting, it is also quite legitimate for the donkey to nibble for a while from one haystack, and then the other, growing wiser all the time.

3

Evolution

Can the galaxies maintain their good looks? Will they look the same a billion years hence as they do now? Three hundred years of observation, the last eighty aided by photography, have shown no signs of change (except for a supernova outburst here and there). Yet astronomers can say with full confidence that galaxies are aging. In 100 million years they will certainly be different.

It is more difficult to say just how each one will change. This problem is like predicting the future change in a city from a set of aerial photographs taken in an interval of two minutes. (In fact, the three-hundred-year age of a city is in about the same ratio to two minutes as the age of a galaxy is to our eighty years of photographic observations.) Motions can be observed in the aerial photos, and differences can be recognized between new buildings and old ones. With high magnification, the differences between children and adult people might also be seen. Similarly, in a nearby galaxy, astronomers can recognize old and new structures and distinguish between young, Population-I stars and old, Population-II stars (see p. 321).

Three types of change can be expected to take place in a galaxy over long intervals. (1) Motions of the stars and interstellar material may change the overall structure (and thus account for the different classes of E types and spirals). (2) Formation of new stars from interstellar clouds of gas and dust may take place, "cleaning up" the dust lanes and mottled regions in spirals. (3) Evolution of the stars will change their colors and luminosities and hence the overall color and luminosity of the galaxy.

The energy output of each star implies of itself a change in the star—primarily a conversion of hydrogen to helium in its core, where nuclear reactions convert a small fraction of its mass to energy. Details of this process are discussed in Volume 6 of this series, The Evolution of Stars. *Most of the observations used to build and check out the theory of stellar evolution are of nearby stars in the Milky Way Galaxy, but we can expect similar evolution in other galaxies.*—TLP

Ages of the Stars

OTTO STRUVE

(*Sky and Telescope*, September 1960)

The stars that make up our Milky Way system differ greatly in age. Very luminous hot blue objects, like Vega, are comparatively young; their enormous outflow of energy cannot have been continuing for more than a relatively few million years. Because such stars are observed in large numbers, the supply of them must be replenished, and some are being formed at the present time.

At the other extreme, the oldest stars are now believed to have ages as large as 20 billion years, four or five times greater than the best estimates of astronomers a few years back. This far-reaching finding is based on studies by many scientists, notably Fred Hoyle in England and Martin Schwarzschild and Allan R. Sandage in the United States.

The breakthrough was made possible by a much improved knowledge of the thermonuclear reactions that are the sources of stellar energy and by a more refined understanding of the internal structure of stars. It became feasible to trace by calculation the manner in which a star evolves, as well as the time scale involved. Using a large electronic computer, today's theoretical astrophysicist can follow, step by step into the remote future or past, the changes of a model star in radius, luminosity, mass, and chemical composition.

If these theoretical properties are compared with the observed characteristics of actual stars, much can be learned about stellar ages. Particularly suitable for this comparison are members of clusters—stars of similar ages, which have evolved at different rates, primarily because of differences in mass from star to star.

In 1957 Sandage presented a paper on the color-magnitude diagrams

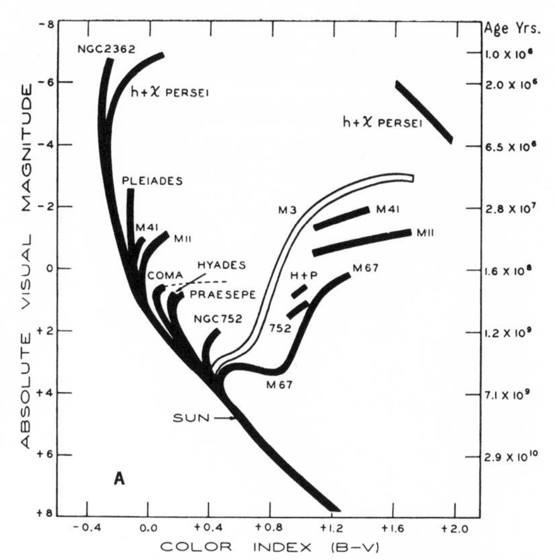

FIG. 50. In this 1957 color-magnitude diagram by Allan R. Sandage, the ages of stars in clusters may be read on the scale at the right, valid for the upper ("turn-off") end of each main sequence. The globular cluster M 3 and the sun are included for comparison. (Courtesy Mount Wilson and Palomar Observatories)

of star clusters, at the Vatican symposium on stellar populations. His basic observational material is summarized schematically in Figure 50, which gives for each of ten galactic clusters [in the Milky Way] the relation between the absolute magnitudes and colors of its stars. [Each star should be plotted as a point; Sandage uses the black lines to show schematically the strips of points plotted for each cluster.] The lower parts of the main sequences coincide, but for each cluster the upper end of the sequence turns to the right. The most luminous stars on these "stubs" are of absolute magnitude −7 in the clusters NGC 2362 and h and *x* Persei, but only about +2 in NGC 752. . . .

Sandage told the astronomers gathered at the Vatican symposium, "Current ideas of star formation and subsequent evolution . . . require that stars are formed from the interstellar medium. As they contract, their points [plotted in Fig. 50] move toward the main sequence [the black strip running upward to the left]. . . . The central temperature rises during contraction, until, at a certain critical value, thermonuclear reactions begin, contraction stops, and a stable star is born. The luminosity of the stable star depends upon the mass of the initial condensation. Because different size masses start condensing there are stars spread continuously along the main sequence at the time of stellar birth.

"The first result of the nuclear reactions is to convert hydrogen into helium in the central regions of the star. This causes a readjustment of the stellar structure so as to compensate for the increase in the mean molecular weight. Detailed computations of this structural change were first made by Schönberg and Chandrasekhar and later by many other authors. The general result is that the evolving star remains close to the main sequence until a critical fraction, q_c, of its mass has been exhausted of hydrogen, at which time the star rapidly expands and moves redward (rightward) in the color-magnitude diagram into the region of the yellow giants.

"These theoretical considerations find direct support in Figure 50. When a cluster has been in existence for a time, T, all stars brighter than a certain luminosity will have exhausted the critical mass q_c and will have left the main sequence. Stars only slightly fainter than this limit will have exhausted a smaller fraction and will have evolved only slightly from their initial stellar structure on the main sequence. The details of the theoretical evolution explain rather well the observed change of slope of each galactic cluster main sequence near its termination point.

"The age of each cluster follows immediately if we identify the main sequence termination point with the stage when a star has exhausted q_c of its mass of hydrogen."

In this way Sandage obtained the cluster ages listed in Table 3. They

TABLE 3. AGES OF GALACTIC STAR CLUSTERS

Cluster	*Constellation*	M_t	*Age in millions of years*
NGC 2362	Canis Major	<-7.0	<1
Double Cluster	Perseus	-7.0	1
Pleiades	Taurus	-2.5	20
M 41	Canis Major	-1.5	60
M 11	Scutum	-1.3	60
Coma	Coma Berenices	$+0.5$	300
Hyades	Taurus	$+0.8$	400
Praesepe	Cancer	$+0.8$	400
NGC 752	Andromeda	$+1.9$	1000
M 67	Cancer	$+3.5$	5000

can be read off directly on the scale at the right side of Figure 50, at the level of the absolute magnitude, M_t, where the upper end of a cluster's main sequence terminates. For M 67 there is no gap between the main-sequence stars and giants, but $+3.5$ can be taken as the brighter limit for the former.

Thus the galactic cluster NGC 2362 in Canis Major (Fig. 51) consists of very young stars, less than a million years old, while M 67 in Cancer has an age of five billion years, according to Sandage's 1957 work.

During the three years since, several important advances have been made. The time scale of stellar evolution has been more accurately

FIG. 51. The very young cluster NGC 2362 surrounds fourth-magnitude star Tau Canis Majoris. This exposure of 5-minute duration was taken with Harvard's 60-inch Boyden Station reflector. (Harvard College Observatory photograph)

FIG. 52. The cluster NGC 188, from a 90-minute blue-light exposure with the Harvard 61-inch reflector, taken by Ivan King on July 10, 1948; the photograph is here enlarged to six minutes of arc per inch. The member stars are unusually faint for so large a cluster. (Harvard College Observatory photograph)

calculated by several astrophysicists, most recently by Hoyle and his associates in England. These computations enabled him, in 1959, to revise the age of M 67 upward to 9.2 billion years. . . . In 1948 Ivan King had called attention to the odd open cluster NGC 188 (Fig. 52), . . . and he and Sidney van den Bergh, at David Dunlap Observatory, suggested to Sandage that NGC 188 might be as old as or older than M 67.

Sandage undertook a detailed investigation of the cluster, measuring the brightness of its stars in two colors on photographs taken with the Mount Wilson 60-inch reflector, in terms of sequences obtained photoelectrically with the 200-inch Palomar telescope.

Figure 53 is a color-magnitude diagram for NGC 188, from Sandage's own observations. Each point represents the apparent magnitude and the color index of a measured star. [When apparent magnitudes are converted to absolute magnitudes, the main-sequence break falls below the one for M 67 on Figure 50.]

There is a remarkable similarity in the color-magnitude patterns for M 67 and NGC 188, but clearly the latter cluster is considerably older, because its turn-off point from the main sequence is at absolute magnitude +5 instead of +4. Using Hoyle's dating process that gave an age of nearly 10 billion years for M 67, Sandage finds that NGC 188 is perhaps 24 billion years old.

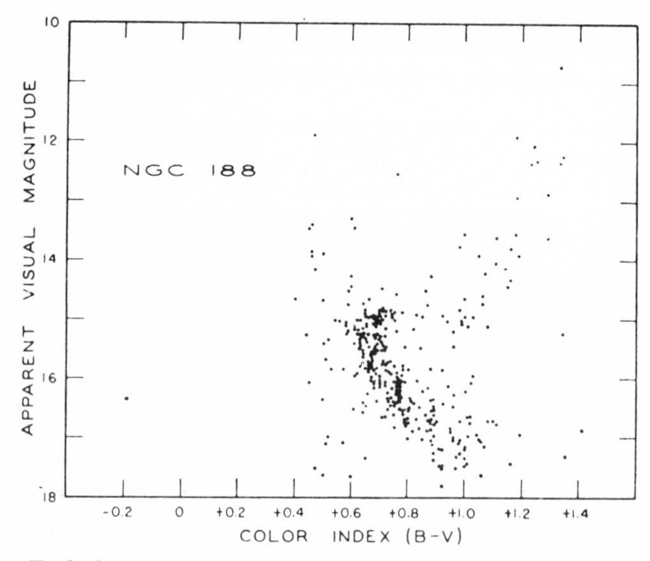

FIG. 53. Each dot represents a star in the cluster NGC 188, plotted by Allan Sandage in accordance with his measurements of its brightness and color. (Mount Wilson and Palomar Observatories diagram)

Sandage cautiously remarks in a letter, "I personally do not much believe this large age and hope other theoretical people will compute additional models." But there is hardly any doubt that the *relative* ages of the galactic clusters are reasonably correct.

These new estimates by Sandage and Hoyle, of the ages of the oldest stars, run up to five times greater than the 5 billion years suggested not long ago for the entire Galaxy. This longer time scale is consistent with some other kinds of astronomical data. For example, the age of the earth is fairly reliably known to be between 4 and 5 billion years; it is not unreasonable that the Galaxy may be many times older. Furthermore, recent calculations by Hoyle and W. A. Fowler at Cal Tech on the formation of chemical elements in the Milky Way system give an age of between 12 and 20 billion years.

Another argument can be based on the red shift of the galaxies. Present data indicate that the velocity of recession increases by 75 kilometers per second for each million parsecs of distance, although this value of the Hubble constant is highly uncertain. On simplified assumptions, its reciprocal—11 billion years—represents the lapse of time since the expansion started.

On the other hand, the very long time scale of the oldest stars contradicts some other results. In a discussion of Sandage's paper at the Vatican symposium, the Dutch astronomer J. H. Oort raised the following question: Can any galactic cluster resist, for longer than 5 billion years, the disruptive action of tidal forces exerted by other stars, cosmic clouds of dust and gas, and the Galaxy as a whole? Assuming that a cluster is dissolved only by gravitational forces, Chandrasekhar found that the Pleiades will have broken up after 3 billion years. . . .

Other evidence against extremely great stellar ages comes from a recent study of the masses of stars within 10 parsecs of the sun by Su-Shu Huang [then at the Goddard Space Flight Center]. There is a striking lack of stars with masses less than 1/20 of the sun's in this sample, although current ideas about star formation suggest that they should be very numerous. Huang proposes that we see so few because they are not yet shining by thermonuclear reactions, but are still in the contracting stage of their evolution and too faint to be detected.

This would suggest that the age of the Milky Way is roughly equal to the time interval required for a star of about 0.08 solar mass to have contracted from a globule to a main-sequence *M* star—6 to 8 billion years. . . .

It seems likely that the age estimates of M 67 and NGC 188 given above are too large. Although fairly accurate nuclear data were used by

Hoyle in his calculations, there are still many uncertainties about conditions at the core of a star. Since 1960 the estimate of the Hubble constant has increased to 100 km/sec per megaparsec, implying a somewhat smaller age of the universe. Considering the uncertainties, it is remarkable that the estimated ages are all so near 10 billion years.

The formation of stars from interstellar material is an essential step in all evolutionary theories. The evidence for interstellar hydrogen gas in "H-II regions" in other galaxies has been known for many years. Note that the H-II regions also imply the presence of hot young stars, whose short lifetimes further imply continual star formation in old galaxies.—TLP

Interstellar Gas in Galaxies

(*Sky and Telescope*, September 1950)

Spectra of fifty galaxies on red-sensitive film, taken this spring by Thornton Page with the 82-inch McDonald Observatory reflector, show hydrogen-alpha emission in twenty-seven cases. It is believed that very hot stars in these galaxies emit invisible ultraviolet light which causes the interstellar gas, consisting of hydrogen, oxygen, nitrogen, neon, and other common elements, to shine with characteristic visible wavelengths, as in galactic H-II regions.

Page also finds fainter emission lines in some of these spectra for these other elements as well as hydrogen, and by two independent methods, is able to estimate the temperatures of the hot stars emitting the exciting ultraviolet radiation as being from 13,000° to 25,000° absolute, corresponding to early *A*- and *B*-type stars. He points out that galaxies showing no emission lines must be practically devoid of early-type stars, or of interstellar hydrogen, or of both. If the galaxies have actually two kinds of stellar population, his method of observing their spectra should aid in classifying them.

Interstellar Gas

(*Sky and Telescope*, November 1957)

Sidney van den Bergh, [then] of Perkins Observatory, calls attention to a surprising property shared by both the Milky Way Galaxy and the Andromeda nebula: the relatively small amount of interstellar gas in the spiral arms. While ejection of matter from stars provides some such gas, a larger quantity is consumed in the process of star formation. Were no external supply available, the gas in the solar vicinity would be exhausted in about 7×10^8 years.

Furthermore, the amount of gas now in the galactic nucleus is only about 1/250 or 1/300 that which should have been ejected from the stars there during the 5-billion-year history of our Galaxy.

Van den Bergh suggests that both these points can be explained by a continuing escape of gas from the nucleus, the gas moving outward along the spiral arms. The present rate of gas loss from the central part of the Galaxy would match the rate at which the interstellar gas is being removed from the spiral arms by star formation.

Hydrogen in Spiral-Galaxy Halos

GEORGE S. MUMFORD

(*Sky and Telescope*, March 1967)

Only for regions in or near the plane of the Milky Way Galaxy is the average density of interstellar neutral hydrogen known. However, some 98 per cent of the volume of the Galaxy is contained in a nearly spherical halo, about 50,000 light years in radius, that is defined by the system of globular clusters.

According to Morton S. Roberts, National Radio Astronomy Observatory, we need to know the hydrogen density in this halo to evaluate the motions of cosmic rays, as well as to determine whether our Galaxy has a radio halo or not.

Unfortunately, if the density of hydrogen in the halo is low, any 21-cm radiation from it will be largely masked by that from hydrogen in the vicinity of the sun. A promising alternative is to observe nearby spiral galaxies—systems similar to our own—that happen to appear edge-on, so that their disks and halos are not confused with each other.

In *Physical Review Letters* for December 5, 1966, Roberts reports such observations for NGC 4244 in Canes Venatici and NGC 7640 in Andromeda. Using the 300-foot telescope at Green Bank, West Virginia, he found no measurable radiation that could be attributed to neutral hydrogen in their halos. This set an upper limit of 0.001 neutral hydrogen atom per cubic centimeter at a distance of 20,000 light years above the central plane in each of these objects.

By analogy, this same limit appears reasonable for the halo of the Milky Way system.

It may well be that the Milky Way's interstellar gas is being used up, and that there was much more during the early history of our Galaxy. In fact, most astronomers consider that every galaxy was formed from a condensing gas cloud.—TLP

Collapse of the Galaxy

(*Sky and Telescope*, May 1963)

About 10 billion years ago, the Milky Way Galaxy underwent a rapid collapse, shrinking to about a tenth of its former diameter and about a twenty-fifth of its thickness. This idea is proposed by three astronomers at Mount Wilson and Palomar Observatories: O. J. Eggen, D. Lynden-Bell, and A. R. Sandage. Their evidence consists of the observed motions and colors of 221 well-studied dwarf stars. They have analyzed the motion of each of these stars around the galactic center, computing its orbital eccentricity and angular momentum.

The California astronomers find that the stars with abnormally strong ultraviolet radiation (indicating low abundance of heavy elements) all move in highly elliptical orbits. On the other hand, stars with little or no ultraviolet excess (indicating more heavy-element content) follow nearly circular paths. In addition, the former stars tend to travel farther

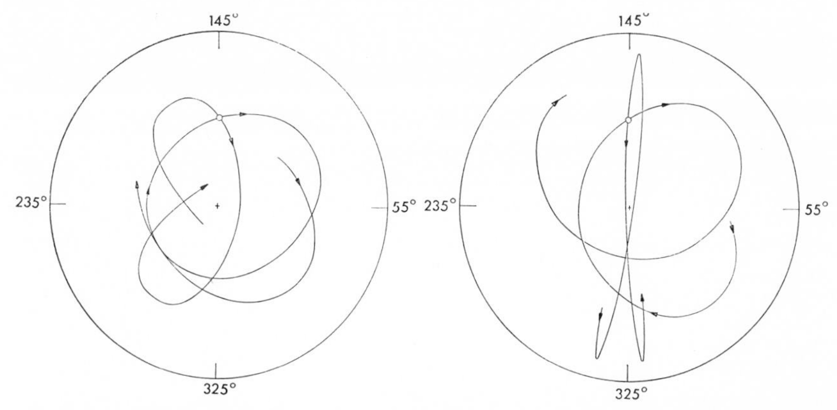

FIG. 54. Orbits of four stars now near the sun (small circle), 10,000 parsecs from the center of the Galaxy (small cross). In each diagram the outer circle is 20,000 parsecs from the center, and the young-star orbit is a series of wide loops around the center. The two older stars (low metal abundance) follow long, narrow loops. (From *The Astrophysical Journal*)

above and below the galactic plane, and they also have smaller angular momenta.

Such patterns of motion can persist over time intervals that are long compared with the present age of the Galaxy, since the stars interact exceedingly slowly to exchange their energies and momenta. Hence a knowledge of the present motions of individual stars can tell something about the conditions when they were formed.

It is now known that metal-poor stars are older than metal-rich ones. Hence the older stars are those with highly elliptic, strongly inclined galactic orbits, while the younger ones move nearly in the galactic plane in circular orbits.

To account for these facts, Eggen, Lynden-Bell, and Sandage suggest that the following sequence of events may have taken place during the history of our Galaxy. About 10^{10} years ago the protogalaxy started to fall together out of intergalactic material. It was either already rotating, or acquired angular momentum from nearby, similar systems. In the protogalaxy, condensations formed that later became globular clusters and stars like those found in such clusters.

After a few hundred million years, say the California astronomers, the inward collapse was balanced by the rotation. However, material continued to fall toward the galactic plane from both sides, giving rise to the present thin disk of the Galaxy. As the gas density increased, the rate of star formation was stepped up.

Through collisions, gas streams lost their motion in the direction of the galactic center, and settled into nearly circular orbits. They still produced stars, which move in the circular paths that the gases had settled into. On the other hand, the older stars continue in highly eccentric orbits produced in the original collapse.

The Heart of the Milky Way

GEORGE S. MUMFORD

(*Sky and Telescope*, May 1965)

The ages and chemical compositions of stars in the central regions of a galaxy should provide very important clues as to how galaxies form and evolve. But even in such a nearby system as M 31, the stars in the central part are too closely packed for individual study. Our only hope lies in probing the center of our own Galaxy, despite its almost complete obscuration by interstellar dust clouds.

Fortunately this obscuration is very uneven, and some sections of the Milky Way are relatively free from it. One of these windows lies in the direction of the distant globular cluster NGC 6522 in Sagittarius, only about 4° from the galactic center.

This region has been intensively studied by Halton C. Arp of Mount Wilson and Palomar Observatories, whose results are given in the January [1965] issue of *The Astrophysical Journal*. His analysis included photoelectric measurements to determine the distance of NGC 6522 from the magnitudes and colors of its stars and a photographic study of about 1300 stars in its vicinity on plates obtained with the 200-inch reflector.

The globular cluster itself is 7500 parsecs (about 24,000 light years) distant from us. This is some 2400 parsecs nearer than the galactic center, where stars are dimmed about 1.4 magnitudes by interstellar dust, as observed in yellow light.

The Milky Way system should be classed as an Sc spiral because of its relatively small nuclear region, concludes Arp. On photographs, the central bulge of M 31, an Sb spiral, appears as a flattened spheroid about 4000 parsecs in diameter. While the nuclear region of our Galaxy would have a similar appearance if viewed from outside, its diameter is probably only about 2000 parsecs. "From the Sun's position in the Milky Way the nucleus should outline a bulge about 11° in extent in longitude and somewhat smaller in latitude," notes the California astronomer.

The light of the Milky Way's nuclear region comes mainly from giant stars, of absolute magnitudes between +2 and −2. These stars are relatively rich in metals, resembling in this respect the members of a metal-rich globular cluster such as 47 Tucanae. Their color-magnitude diagram shows that the nuclear stars are generally older than 10^9 (1 billion) years, and probably the bulk of them are as old as stars anywhere in the Galaxy.

As more evidence is collected, astronomers are beginning to realize that conditions in the nucleus of a galaxy differ widely from those in the arms (where the sun is located in the Milky Way Galaxy). The "closely packed" stars near the center of each spiral or elliptical galaxy certainly dominate motions of the other stars by their gravitational attraction. Their age indicates that these central stars have existed in some galaxies since the "beginning of time." The outward flow of interstellar gas, and more dramatic signs of "explosions" in the nuclei of some galaxies, raise doubts about the stability, or equilibrium, of a nucleus.

Since gravitational forces play an important role in holding galaxies together, the total mass—and its distribution throughout a galaxy—must certainly affect the evolution of that galaxy.—TLP

Inclinations of Spectrum Lines in Spirals

N. U. MAYALL

(*Sky and Telescope*, November 1948)

Thirty-five years ago [at Lowell Observatory in 1913] V. M. Slipher placed the slit of a spectrograph on a spiral nebula and obtained a spectrogram that showed inclined lines. From this result he concluded that the spiral was rotating about an axis—that on one side of its center the material of the galaxy was approaching us (relative to the center) and on the other side receding. His pioneer observation and its interpretation not only antedated by more than ten years the discovery of the rotation of our own Milky Way Galaxy, but it opened a new

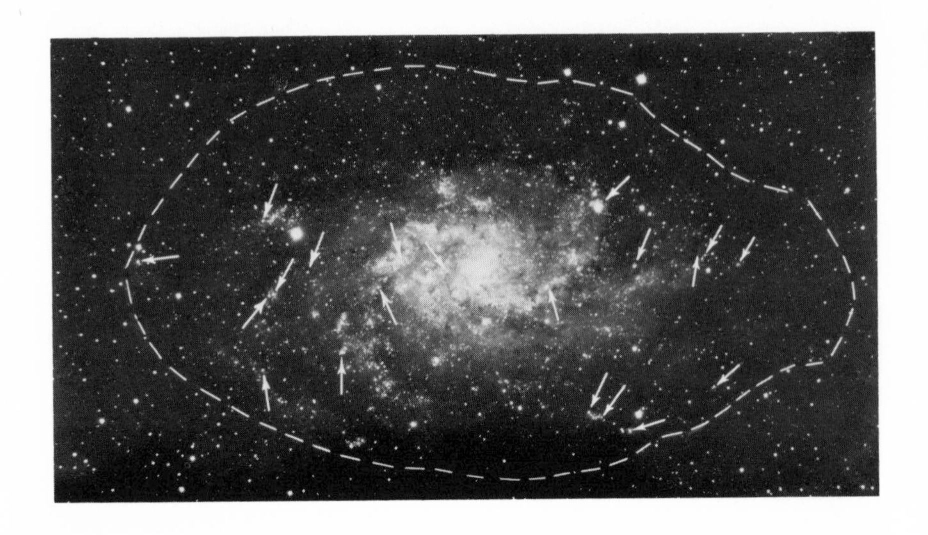

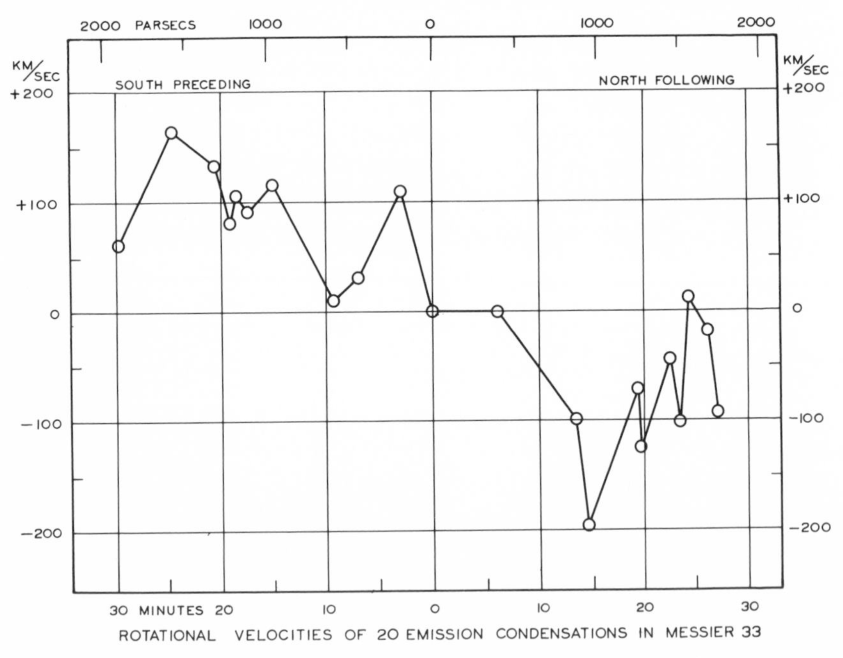

FIG. 55. A. The arrows on this photograph of M 33 indicate the emission patches for which rotational velocities derived from the observed Doppler shifts are plotted in chart B. Positive motion is away from the observer. The horizontal scales of B give linear (upper) and angular (lower) distance from the center of the spiral. Note that the rotational speed increases with distance from the center throughout the main body, and then decreases in the outer parts. (Lick Observatory photograph and diagram)

field of investigation: the study of dynamics in the largest known stellar systems. The field, however, has proved difficult to cultivate, and even now our factual knowledge in it is in the broad exploratory stage.

At the outset it should be stated that the interpretation of spectrum-line inclinations as rotations is fraught with many difficulties. When the slit of a spectrograph is placed on a spiral the resulting spectrogram represents a large-scale integration over many motions and radiations throughout a large portion of the system. This condition is especially true for the edge-on spirals and the large amorphous nuclear regions of early- and intermediate-type spirals. To be reasonably certain of what motion corresponds to what point in a spiral, it is desirable to observe individual objects, and to know where they are located with respect to the fundamental plane. Unfortunately, with present techniques, this ideal approach can be made in only a few of the nearest spirals; in fact, it has been done only for the two nearest of them, M 31 and M 33. Observations of their brighter emission patches, which probably correspond to the larger galactic nebulosities, indicate that, on the grand scale, the main bodies of these spirals rotate nearly as solid bodies (see Fig. 55), while their outer parts show a planetary or Keplerian type of motion (slowest farthest from the center).

For M 31 and several other spirals observed at Mount Wilson, initially by F. G. Pease and later more extensively by M. L. Humason, it was found that the spectrum lines are conspicuously inclined, and sensibly straight, in the vicinity of the nuclei (see Figs 56 and 57). Since the light from stars near a nucleus comes to us through absorbing gas clouds, the interpretation of the observations in terms of simple, solid-body rotation requires some justification. This seems to have been provided in a theoretical paper by E. Holmberg of Lund, Sweden.

In 1939 Holmberg investigated the ratio between observed radial velocity and circular orbital velocity as a function of nuclear distance for several assumed distributions of mass, light emission, and light absorption within spherical nuclei. The inclusion of a factor for absorption probably is not necessary, since there is increasing evidence that most of the absorbing material in a spiral is confined to a quite thin, patchy stratum in the principal plane. Nevertheless, Holmberg found that the ratio varies almost linearly with distance from the nucleus. . . . In short, the interpretation of straight inclined lines in nuclear regions as an indication of solid-body rotation seems to be acceptable as a working hypothesis. On this basis, the nuclear regions of M 31, three other spirals, and one elliptical appear to be rotating faster than their respective main bodies.

In a first crude attempt to expand this meager but hard-won body

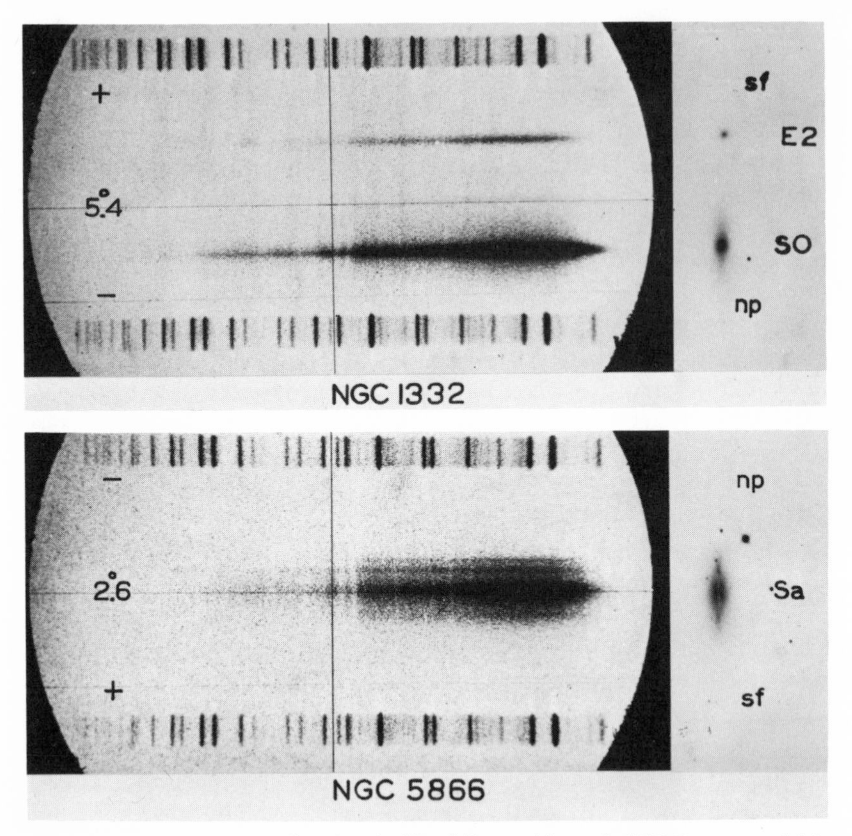

FIG. 56. Typical spectra showing inclined lines. Above is NGC 1332, which has a period of about 6 million years for the rotation of its nuclear region. It has a fainter companion elliptical galaxy. NGC 5866, below, rotates in about 6.5 million years. Spectra are enlarged about ten times from the original negatives. The direction of the spectrograph slit is indicated by the letters "sf" (southeast) and "np" (northwest). (Lick Observatory spectra)

of data, the present preliminary discussion is based on thirty low-dispersion spectra that show inclined lines. All these spectra were obtained with the nebular spectrograph of the 36-inch Crossley reflector at Lick Observatory. The linear dispersion varies from 220 angstroms per millimeter at wavelength 3727 to 430 at the hydrogen-gamma line. Some of the spectra date back to 1935, and were taken mainly for radial-velocity determinations, so that they are by no means the best that could be obtained in a well-planned study of nebular rotations. At first, when emulsions were slower, only the nuclear regions were recorded. After the war, when the faster and finer-grained Eastman IIa-O emulsion became available, it was possible to obtain spectra of most of the

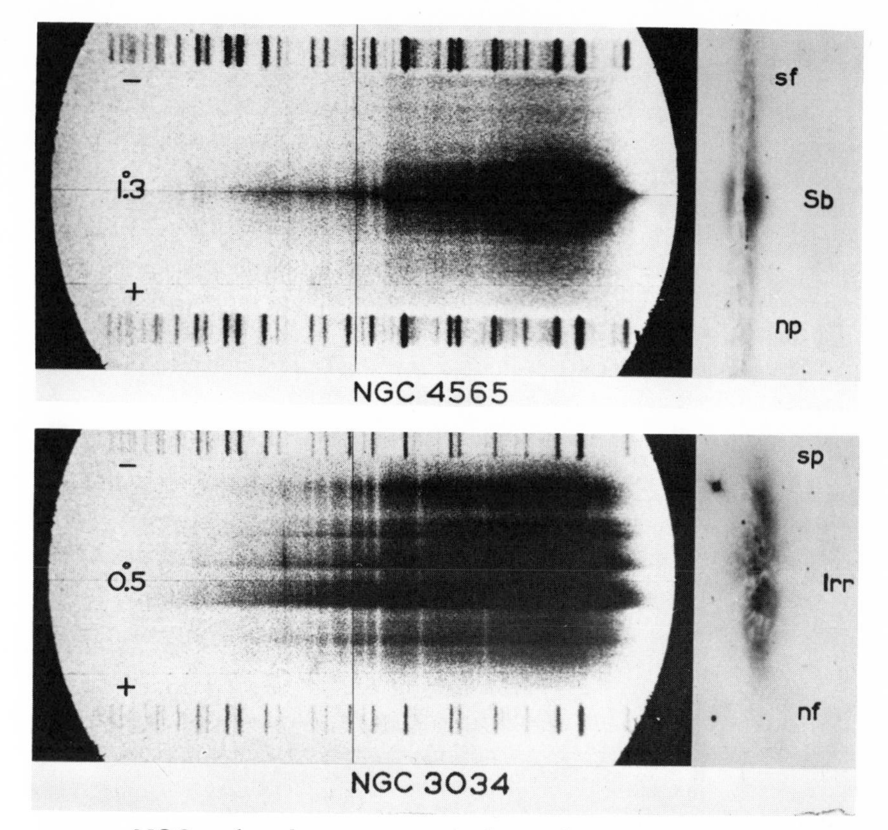

FIG. 57. NGC 4565, whose spectrum is shown above, is an edgewise spiral of Sb type. The inclined lines indicate a period of revolution of about 16 million years. Below is the spectrum of the irregular galaxy NGC 3034, showing very little rotation. Spectra enlarged thirteen times from the original negatives. (Lick Observatory spectra)

main bodies of a number of spirals. Wherever possible, the slit was placed along the major axis [longest dimension] of each system, and to check that the line inclination originated in the spiral, rather than from instrumental causes, several duplicate exposures were made with the slit rotated 180° between exposures.

In analyzing the spectra, plates of six late-type spirals oriented nearly edge-on were measured in the usual manner with a micrometer set on the spectrum lines at a number of different points along each line. For all six spirals an essentially straight-line relation was obtained between nuclear distance and Doppler-shift velocity. This result suggested that the laborious micrometer settings along each line might be replaced by measurement of the inclination of each spectrum line as a whole, by rotation of the spectrogram.

This procedure proved to be rapid and the results agreed with those of the usual method. A small Gaertner measuring engine was modified so that its plate-carrying table can be rotated and read to 0°.01 degree. The measurement error depended strongly upon the lengths and visibility of the lines. Under the most favorable circumstances, it was 0°.05 (plus or minus). For the narrowest measurable spectra, 0.11 millimeters wide, showing few lines of poor visibility, probable errors up to 0°.5 were obtained. It appeared that inclinations of the order of 0°.1 were the smallest that could be detected in spectra of average characteristics. As a rule, the ionized-oxygen line at 3727 angstroms, and the H and K lines of ionized calcium were best suited for measurement. . . .

Some idea of the problems of measurement may be gained from Figures 56 and 57. These were photographed through the eyepiece of the measuring engine, and the pictures therefore show how the spectra appeared during measurement. To the right of each spectrum is a direct (negative) photograph of the nebula, enlarged to the same scale as the direction perpendicular to dispersion.

The tilt in the early-type (S0) system NGC 1332 (shown in Fig. 56), 5°.4 at the H and K lines, is the largest measured. The spectrum is also the narrowest, so the inclination is uncertain by at least 0°.5. NGC 5866, also an early-type galaxy, has inclination of lines 2°.6.

NGC 4565 (see Fig. 57) is considered a typical edge-on spiral with a prominent dark lane silhouetted against the nuclear region. Although the exposure was five hours on IIa-O emulsion, the spectrum shows only the brighter nuclear region in measurable strength. A longer exposure could be used, but the spectrum of the outer parts of the spiral would be confused with night sky radiations. The inclination for NGC 4565 is 1°.3.

Figure 57 shows that lines in the spectrum of NGC 3034, better known as M 82, the irregular companion of M 81, are measurably inclined. This is perhaps the best example of an early-type absorption spectrum, infrequently found in the nuclei of spirals. In this case, a spectral class of about A7 is indicated, and it is noteworthy that the same type of spectrum apparently prevails throughout most of the system, with occasional regions where interstellar oxygen emission at 3727 angstroms is conspicuous.

The measured line inclinations were finally used to compute periods of rotation for these thirty-odd systems. To do this the distance of each object must be estimated, and this was done, in most cases, [from the Hubble law] using red shifts in their spectra corrected for the effects of galactic rotation. Eight of these spirals are in the Ursa Major group of galaxies, and one is in the Virgo cluster, so that their distances are

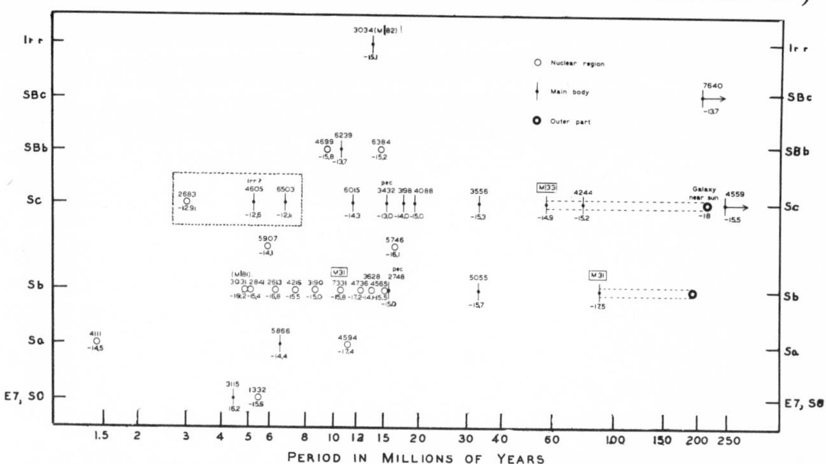

FIG. 58. The rotation periods of the galaxies arranged according to nebular type. Absolute magnitude is shown for each object. The rotation periods of nucleus, main body, and outer regions are designated by different symbols, shown at upper right. The arrows attached to the points for NGC 7640 and NGC 4559 mean that lines in the spectra of these two spirals showed no measurable inclination. (Lick Observatory diagram)

reliable. The computed periods were corrected, where necessary, for the tilt of the principal plane to the line of sight. The absolute magnitude for each spiral was also computed.

The results have been collected in Figure 58, where the periods of rotation are shown according to type of galaxy. The periods are in millions of years, and plotted on a logarithmic scale because of the large dispersion found for the later-type spirals. Periods for the nuclear regions and the main bodies are differentiated, and for comparison those of the outermost parts of M 31, M 33, and the solar neighborhood of the Milky Way Galaxy are included.

The chart does not suggest any very striking generalizations. For example, there is no correlation between absolute magnitude [or luminosity] and period. On the other hand, if the three objects in the dotted rectangle be disregarded because of uncertain distances or a possible reclassification, there is some suggestion of a general increase in period of the main bodies with advancing nebular type [from E7 to Sa, Sb, and Sc].

A sample calculation for NGC 1332 is, briefly: This galaxy has a red shift, corrected for galactic rotation [our motion around the nucleus of the Milky Way], corresponding to +1300 kilometers per second. Since Hubble found that the red shift increases 526 km/sec per million parsecs, the estimated distance of NGC 1332 is 1300/526, or 2.5 million parsecs—about 8 million light years. The period of rotation is then obtained from the simple formula

$$Period/Distance = Constant/Inclination.$$

The constant depends on the telescope's focal length, prismatic dispersion, and other quantities. For the Crossley nebular spectra shown here, and for the H and K lines, the constant equals 12.6. Hence, for NGC 1332, in which the H and K lines are inclined 5°.4, the period of rotation is $(12.6/5.4) \times 2.5 \times 10^6$, or 5.8 million years. Since this galaxy is oriented nearly edge on, no correction for tilt of its fundamental plane of rotation is required.

As already noted, the Hubble constant is now estimated as 100 km/sec per million parsecs, so Mayall's sample calculation now gives the distance of NGC 1332 as 1300/100, or 13 million parsecs—over five times larger than in 1948. It also results in a longer period (30.5 million years), and all the periods in Figure 55 must be multiplied by about 5.26. The significance of these periods of "solid-body rotation" is not yet completely understood. They are certainly related to the average density of matter (stars and interstellar material) in the central regions of galaxies. The faster the rotation (the shorter the period), the greater the density must be. The most recent evidence shows that some galaxies have extremely dense nuclei. In such cases, the spectrum of the nucleus shows emission lines and a broad spread of Doppler-shift velocities. Galaxies with such peculiar nuclei are called Seyfert galaxies, after the late Carl Seyfert, who first studied them at McDonald Observatory in 1940.—TLP

Motions in Galaxy M 77

(*Sky and Telescope*, March 1966)

Occasionally galaxies are found that differ from typical spirals in having especially small, bright nuclei and in showing strong broad emission lines in their spectra. These systems, known as Seyfert galaxies, are often weak radio sources. They have received particular attention in the last two years because of possible similarities to quasars. As D. E. Osterbrock and R. A. R. Parker [University of Wisconsin] comment, "The nuclei

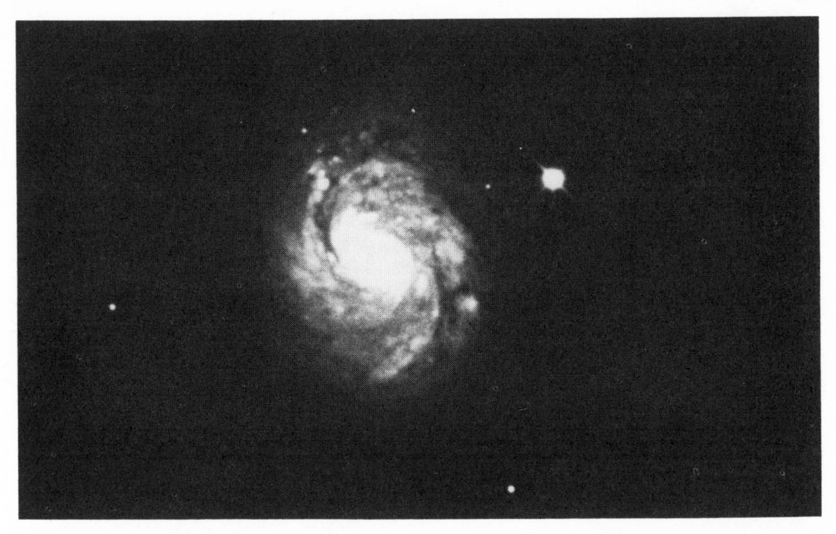

FIG. 59. The galaxy M 77 (NGC 1068), whose overexposed central region hides an exploding nucleus. (Lick Observatory photograph)

of Seyfert galaxies may perhaps be thought of as miniature quasi-stellar radio sources located at the centers of otherwise normal galaxies."

Merle F. Walker of Lick Observatory has made a detailed survey of the central parts of one of the brightest Seyfert galaxies, Messier 77 (NGC 1068) in Cetus. He reported his spectrographic results obtained at the coudé focus of the 120-inch telescope, using a Lallemand electronic camera to reduce exposure times.

Violent motions are apparent in the nuclear region, where discrete clouds up to three or four seconds of arc in extent are being ejected at speeds up to 600 kilometers per second. The linear diameters of these clouds are about 250 to 300 parsecs, and their individual masses may be about 10 million suns, if they are assumed spherical.

"The energy required to accelerate such a cloud to the observed velocity is that of 300,000 supernovae," said Walker. "Since the coupling efficiency must be low, a population of millions of supernovae would seem required to produce the observed result. Thus a new, unknown source of energy in the nucleus of M 77 may be required to account for the observations."

Complicated motions taking place outside the nucleus, within the first half minute of arc, are apparently a combination of outflow and the galaxy's rotation. Farther out, rotational motion predominates. "Thus, the entire central region of the system is disrupted," Walker explained, "perhaps as a result of earlier generations of the type of explosions now observed in the nucleus."

Using the Lallemand electronic camera, he finds that the energy curve of the nucleus is nearly flat from 3500 to 4000 angstroms. In addition, photoelectric observations show that the ultraviolet light of the nucleus is partially polarized, which suggests that some of this light is synchrotron radiation.

Changes in Seyfert Galaxies

GEORGE S. MUMFORD

(*Sky and Telescope,* October 1968)

In 1943, Carl K. Seyfert called attention to the rare variety of spiral galaxies now known by his name. They are characterized by small, bright, very blue nuclei, whose spectra contain broad emission lines. Some of them, such as NGC 1068 and NGC 1275, are moderately strong radio sources, the latter being variable at radio wavelengths. These properties have suggested that the nuclei of Seyfert galaxies are less violent relatives of quasars. Therefore, since many quasars are optically variable [see p. 196], Seyfert galaxies are being tested for light variations.

NGC 4151 in the constellation Canes Venatici is among the brightest and nearest of the Seyfert galaxies, and has a nucleus about one second of arc in diameter. In the spring of 1967, it was measured photoelectrically with the 36-inch Steward Observatory reflector on Kitt Peak by W. S. Fitch, A. G. Pacholczyk, and R. J. Weymann. Their observations showed a magnitude variation of 11.13 to 11.38 in yellow light for the nuclear region, and twice as large a range in ultraviolet light.

At David Dunlap Observatory, T. G. Barnes tested half a dozen Seyfert galaxies for variability between April and December 1967, using a 24-inch reflector and photoelectric photometer. In *Astrophysical Letters* for May 1968, he reports that the variation of NGC 4151 is confirmed. In addition, NGC 1275 appears to have faded, but this is not quite certain, since the galaxy was too faint for accurate measurement with the 24-inch. No variations were found for NGC 1068, 3077, 3516, and 7469.

Changes in the optical spectrum of NGC 4151 are reported by W. W. Morgan, H. J. Smith, and D. Weedman in *Astrophysical Journal Letters* for May [1968]. Spectrograms of the nucleus taken in 1956 with the 82-inch McDonald reflector show, in addition to the bright lines, various absorption features characteristic of spectral type G. But when NGC 4151 was reobserved in February 1968, with the same equipment, the absorptions were gone, except for a faint interstellar K line and a line near 3884 angstroms.

At Mount Wilson and Palomar Observatories, the optical and near-infrared spectrum of NGC 4151 has been studied by J. B. Oke and W. L. W. Sargent. The emission lines originate in at least two physically distinct regions of the nucleus, they have announced in the March [1968] issue of *The Astrophysical Journal.* The nucleus is about 50 parsecs in diameter, and about 1/40 of this volume consists of rapidly moving clouds or filaments of hot gas, from which most of the emission lines arise. These swirls have a mass of about 200,000 suns and a temperature of about 20,000° Kelvin. The remainder of the nucleus consists of low-density gas with a temperature of around 1,000,000°, which produces coronal forbidden lines of iron atoms ionized 9 and 13 times.

These extreme physical conditions in the nucleus of NGC 4151 are evidence of an unstable situation during the release of large amounts of energy—that is, an explosion. This is the "relationship" of Seyfert galaxies to quasars, first noted by E. M. and G. R. Burbidge (San Diego) in 1961. Other evidence of explosions in galaxies will be presented in Chapter 4, and the remarkable quasi-stellar objects are described more fully in Chapter 5. At this point, it is helpful to fit together all these peculiar objects in our broad picture—the "life story of a galaxy" as it was visualized in 1965.—TLP

The Evolution of Galaxies

THORNTON PAGE

(*Sky and Telescope*, January and February 1965)

Except for a supernova here and there, and a few variable stars in the nearer ones, the galaxies haven't changed one whit during the last sixty or seventy years of careful observation. Yet we are virtually certain that they are evolving.

If nothing else, the stars in them are ageing. However, the goal of studying the evolution of galaxies is to account for their many different forms and groupings, as well as their colors, luminosities, motions, masses, and gas content. Over the past decade there has been remarkable progress in research on *stellar* evolution, and a number of astrophysicists are now interested in developing an equally precise and complete theory of galactic evolution. This is a tall order; it brings in cosmology, celestial mechanics, stellar evolution, radio astronomy, the physics and chemistry of interstellar gas and dust—practically all of modern astrophysics—applied to the most distant objects we can observe!

Within range of the 200-inch telescope there are an estimated 100 million galaxies. Each contains many billions of stars, and this provides a certain simplification. We are not particularly concerned with what *one* star will do, but with trends in the group characteristics of *billions* of stars. If different ages, or unlike initial conditions, or various evolutionary processes are significant, there are a great many galaxies of different types and groupings to test the theory.

Observations and Cosmology

Up to forty-five years ago, it was not at all clear how large and distant the galaxies are. In 1920 there was a famous debate in the National Academy of Sciences between Harlow Shapley and Heber D. Curtis as to whether the spiral nebulae could be as large as our Milky Way system, and therefore far outside it. Shapley's studies of globular clusters showed that the Milky Way is a large flat disk, now known to be about 100,000 light years across. By 1936 Edwin Hubble (with the help of Milton Humason and others) had demonstrated that the galaxies are systems comparable to our Milky Way, that they are distributed roughly uniformly in all directions, and that they are generally receding from our Galaxy at velocities proportional to their distances [see Figs. 8 and 61].

This is well described in Hubble's book *The Realm of the Nebulae* (1936). His law of red shifts, $V = HD$ (where V is velocity, H a proportionality factor, and D distance), has become the observational basis for the expanding-universe theory and "Big-Bang" cosmology. However, his distances of the galaxies were all underestimated, so that the constant H came out about five times too large. Later work at Mount Wilson and Palomar Observatories by Walter Baade and by Allan Sandage led to a Hubble constant, H, of 100 kilometers per second for every

FIG. 60. Edwin P. Hubble (1889–1953) devoted more than thirty years of work at Mount Wilson and at Palomar to the problems of galaxies. In 1929 he gave the first convincing demonstration that the red shifts of galaxies are proportional to their distances. (Photograph by Jay Barrie, courtesy of Mount Wilson and Palomar Observatories)

million parsecs, or 19 miles per second for every million light years. That is, a galaxy 100 million light years away appears to recede from us at a rate of 1900 miles per second.

Hubble's counts of galaxies brighter than several selected magnitude limits demonstrated that the number within a distance D is closely proportional to D^3, after correction for red-shift dimming and for the light travel time. Using the current distance scale, this corresponds to one galaxy per 3×10^{19} cubic parsecs (one per 10^{21} cubic light years) on the average. Thus the mean distance between galaxies is about 10 million light years.

The Hubble law can also be used in the form $D = V/H$ to estimate the distance of any galaxy, provided it is more than 10 million light

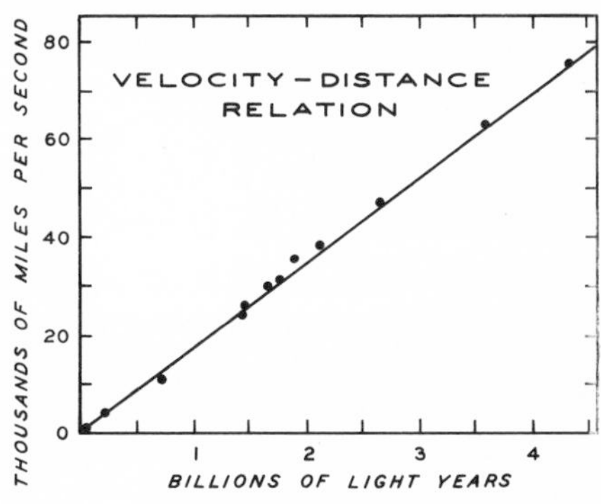

FIG. 61. There is a nearly straight-line relation between observed red shifts and estimated distances of galaxies. So far, red shifts have been measured for almost a thousand galaxies, including members of more than two dozen clusters.

years away (so that the cosmic expansion velocity overshadows any random motion). Measurement of the galaxy's spectrum gives the red shift ($\Delta\lambda$) in a spectrum line normally at wavelength λ. From this, V is found by the Doppler formula, $V = c\,\Delta\lambda/\lambda$, where c is the velocity of light, 186,000 miles per second. (The formula is somewhat more complicated [see p. 185] is V is larger than about one third the velocity of light, as in the largest red shifts measured at Palomar.)

Moreover, when we trace the galaxies back in time, they all seem to have been close together at an epoch t_0 years ago, where $t_0 = D/V = 1/H$. If we make the simplest assumption, that each galaxy has maintained a constant outward velocity, t_0 is the time required to go a million light years at 19 miles per second. This is about 10 billion years.

"Big-Bang" cosmology uses the somewhat more complex reasoning of relativity theory, involving changes in geometry to "explain" Newton's gravitation, as proposed by Einstein half a century ago and confirmed by observation and experiment since. The changes in geometry are often called "curvature of space," the general curvature depending on the average density of matter.

As the galaxies fly apart, the average density gets smaller, the curvature becomes larger, and the total volume of the universe increases. Such an expanding model fits Hubble's law of red shifts well. It is based on the so-called *cosmological principle*—the assumption that the general view of the universe must be the same from every galaxy. That is, we assume that there can be no "edge" to the universe, and that astronomers in another galaxy would find the same law of red shifts that we do, and the same mean density.

This theory leads to an extremely high density in a very small universe some 10 billion years ago, constituting what the Belgian Abbé Lemaître called the "primordial atom" and the American George Gamow called "ylem," a primeval mixture of matter. Both these theoretical physicists studied the explosion that started the expansion of the universe t_0 years ago, Lemaître to explain cosmic rays, Gamow to explain the present relative abundances of the chemical elements. The latter proposed that the elements were formed by nuclear reactions in the very dense ylem during the first few minutes of the history of the universe.

Both of these theories have been supplanted; cosmic rays are charged particles thought to be accelerated to high energy in magnetic fields within our Galaxy, and the chemical elements are believed to have been formed from hydrogen by nuclear reactions within stars. Nevertheless, the Big-Bang theory has a good deal to do with the early history of galaxies, and suggests that they are all of about the same age, having condensed from the expanding gas after it shot out of the primordial explosion and cooled.

However, there are several possible discrepancies between observations and the Big-Bang cosmology. Several years ago a philosophical objection was raised by H. Bondi, T. Gold, and F. Hoyle. These English astrophysicists pointed out that time and space are treated so similarly in Einstein's theory that the cosmological principle should logically be extended. In addition to having no edge and no center,

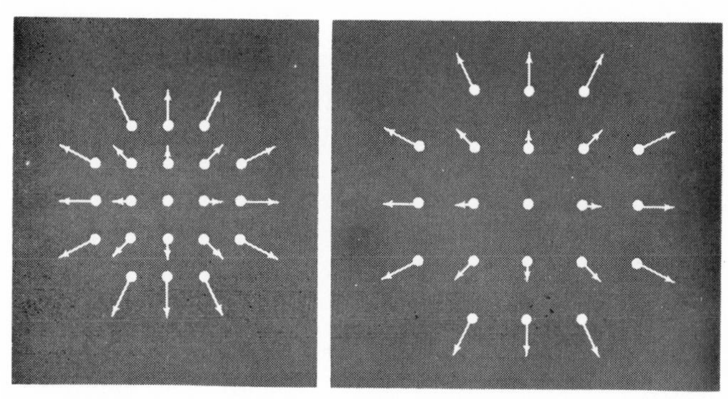

FIG. 62. A. These two sketches describe "Big-Bang" cosmology. Each dot represents a galaxy (our own at the center); each arrow a red-shift velocity, larger at greater distances. Both arrays should be pictured as extending hundreds of times farther than shown here, also above and below the page. The left sketch shows the view 3.5 billion years ago; the right one, today's view.

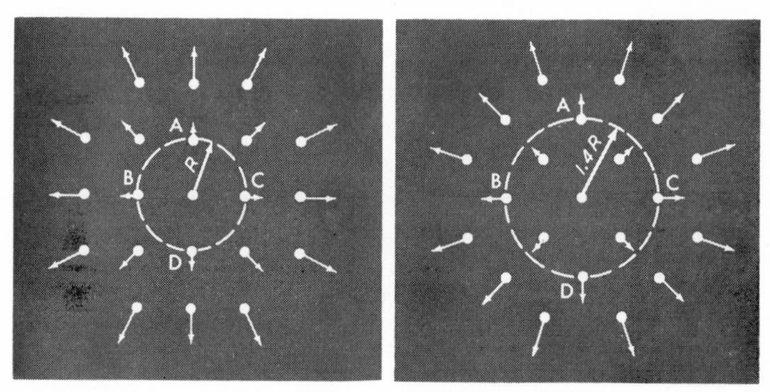

B. In Steady-State cosmology, the view today (right) is the same, generally, as was the view 3.5 billion years ago (left). In the interim, galaxies *A*, *B*, *C*, and *D* (plus one above and one below the page) have increased their distances from us (at the center) by a factor of 1.4 and have speeded up by the same factor. To maintain a constant density out to their distance, fourteen new galaxies must be formed from new matter created as hydrogen atoms.

the universe should have neither beginning nor end. This broader assumption means that the general view of the universe must be *the same at all times*, as well as the same from all places.

That is, the mean density cannot change, even though the galaxies are observed to be flying apart. It was therefore necessary for Bondi, Gold, and Hoyle to assume that matter is being created everywhere at a slow rate (about one hydrogen atom per year per cubic mile) to make up for the outflow of galaxies. This *Steady-State* or *Continuous-Creation* theory also matches present observations well, but it implies that there are galaxies of all different ages within view. Somehow, the newly created hydrogen scattered through space must collect into new systems. The average age of a galaxy, which must be the same for any region of space according to the extended cosmological principle, turns out to be about 3 billion years—much less than the 10-billion-year age of the Big-Bang universe.

Quasi-Stellar Objects (Quasars, or QSOs)

Much of the reasoning above is based on the idea that a cloud of gas condenses to form a galaxy, all the gas eventually forming luminous stars that generate nuclear energy in their hot cores and then "burn out." Another sequence of events was suggested by the Soviet astronomer B. A. Ambartsumian in 1954 and confirmed by a series of remarkable observations at Palomar in 1963.

First, Allan Sandage obtained photographs in red hydrogen-alpha light showing gas clouds being spattered out of the nuclei of several galaxies, particularly Messier 82 [see Fig. 73], as if a huge explosion had occurred there about 1.5 million years ago.

Then Maarten Schmidt, studying spectra of faint blue "stars" at points in the sky from which strong radio emission is coming, found large red shifts, corresponding to recessions of 20 to 50 per cent the speed of light. Next, using the Hubble relation $D = V/H$, he and Jesse Greenstein found that these starlike objects must be very distant, very small galaxies, over 100 times as luminous as the brightest of nearby, normal ones. They appear faint because of their great distances —several billion light years. About a dozen of these queer little galaxies are known, all of them first located by radio telescopes and confirmed by optical spectra. [They are now known as quasars, but were at first called QSOs.]

Several alternative explanations of the QSOs were considered at a special conference held in December 1963 at Dallas, Texas [see p. 266]. Their small sizes (25,000 light years or less in diameter) and emission-line spectra (indicating gas at low density) set an upper limit to the mass, about a million suns or 10^{39} grams—much less than the mass of an average galaxy. The tremendous energy output (at least 10^{45} ergs per second) would use up within a mere 50,000 years or less all the nuclear energy available in such a small mass from the conversion of hydrogen into helium. In fact, these nuclear reactions probably cannot release energy at this rate, and the currently adopted theory ascribes the explosive release of energy to the gravitational collapse of a single gigantic "star."

The theoretical difficulties of this idea are numerous. First, collapse is prevented in normal galaxies by rotation, and in normal stars by gas pressure. Second, how can the gravitational energy released within the star *get out* as light and radio emission? Previous theoretical studies that were based on the general assumption of stability in stars and galaxies must now be revised to apply to short-lived explosions. Several QSOs vary in brightness, according to Harlan Smith and Dorrit Hoffleit's study of old Harvard photographs and Sandage's photoelectric measurements; possibly these objects pulsate.

In addition to posing these problems, the QSOs offer an opportunity to study the characteristics of space much farther from us than was previously possible. If their distances can be determined in other ways, the manner in which the QSOs deviate from Hubble's red-shift–distance proportionality may tell us whether the Big-Bang or the Steady-State cosmology is correct. Eventually, the increase in number of these objects with greater distance or red shift can also be used in this manner. But first many more must be discovered. The brightest QSO known, 3C 273, is of apparent photographic magnitude 12.5 and the rest are 16 to 18. Thus their spectroscopic study is limited to large telescopes, primarily those at Palomar, Lick, and Mount Wilson observatories.

Types and Masses of Galaxies

In addition to establishing the law of red shifts for normal galaxies, Hubble showed deep insight in recognizing the various forms, or morphological types [illustrated in Figs. 24, 25, and 26]. All but the irregular systems have rotational symmetry, but individual specimens are variously oriented to our view, like so many saucers scattered at random. Their structures vary from possibly spherical, smooth-looking E0 galaxies to highly flattened, mottled, "late-type" spirals, Sc or SBc. They look as if they are rotating, and they *are*, as is demonstrated [see Figs. 56 and 57] by inclined spectrum lines. The rotational speeds are 100 kilometers per second or more, but each system is so large that it takes about 100 million years to turn once.

In 1939 Horace Babcock [then at Lick Observatory], first estimated

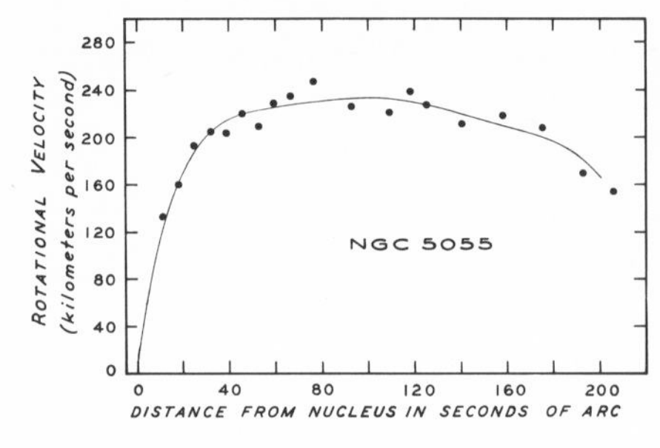

FIG. 63. The rotational velocities of NGC 5055 at various distances from its nucleus, from measurements by G. R. and E. M. Burbidge (San Diego) and K. H. Prendergast (Columbia University). (Adapted from a diagram in *The Astrophysical Journal*)

the mass of a galaxy by measuring the orbital motion of its stars and gas around the center, just as the mass of the sun can be determined from the orbital motion of planets around it. In fact, the curve of orbital velocities at various distances from the nucleus of a galaxy even permits us to determine the distribution of mass within the galaxy, on the assumption that it is stable. The observations are difficult, requiring several long-exposure spectrograms in each case, but such curves [see Figs. 55 and 63] are now available for more than twenty-five galaxies and the Milky Way, from the efforts of several observers. The masses so derived are all about 100 billion suns, or more than 10^{44} grams.

Many of these curves show a straight portion on either side of the nucleus, where the rotational velocity is proportional to distance from the center. In this case the inclined spectrum lines appear straight [see p. 123]. N. U. Mayall, P. O. Lindblad, and I have measured such inclinations in the spectra of ninety galaxies. From these measurements we have deduced the densities in central regions somewhat less than half of each galaxy's full diameter. In fact, for constant slope of the velocity curve, this density is uniform and proportional to the square of the slope. Apparently, a galaxy has central regions of roughly uniform density out to about 0.4 of its radius, beyond which the density decreases.

Another method of finding the masses of galaxies was first applied by Sinclair Smith.[1] He counted the number of members in a cluster of galaxies [such as that shown in Fig. 64] and measured the *differences* of velocity in the cluster. Then he computed the mass of the whole system on the assumption that the largest velocity within it must be less than the velocity of escape from it. This gives an average mass that is ten to a hundred times larger than the values for individual galaxies deduced from their rotations. Of course, the average mass of a galaxy

[1] Also by Fritz Zwicky. Both these astronomers obtained their observations at Mount Wilson about 1931.—TLP

directly affects the mean density of matter in the universe and the inferred space curvature of cosmology.

Why this discrepancy? Do the clusters contain large amounts of unseen matter, or are they breaking up? It may be that the fast-moving galaxies really are not members of the clusters in which they appear on photographs.

Confirmation of the smaller masses comes from studies of *pairs* of galaxies [see Fig. 65], in much the same way that stellar masses are determined from double-star motions. Recently I have shown that elliptical galaxies (types E0 to E7 and S0) are thirty times more massive than spirals, on the average, in fifty-two pairs where the projected orbital motions have been measured. Several other properties are correlated with the type of a galaxy: color, luminosity, ratio of mass to luminosity, ratio of thickness to diameter, central concentration, gas content, and spectral type.

Color measurements show the effect of reddening by interstellar dust, both in our own Galaxy and in edgewise spirals. After correcting for this, E. Holmberg (at Uppsala, Sweden) finds that the average overall color

FIG. 64. The largest objects in the universe are clusters of galaxies. Shown in this half-hour exposure with the Lick 120-inch telescope is the central part of the Corona Borealis cluster of galaxies. The red shift for this cluster is about 21,600 km/sec, indicating a distance of some 700 million light years. (Lick Observatory photograph)

FIG. 65. Valuable information about galaxy masses can be gathered from double systems such as this pair in Virgo, NGC 5432 and NGC 5435. (Lick Observatory photograph)

index ranges from +0.4 magnitude for open spirals (Sc) and irregular galaxies up to +0.85 for ellipticals (E). The luminosities range from 2×10^9 suns for an average Sc system down to 4×10^8 for ellipticals, according to J. Neyman and Elizabeth Scott at Berkeley.

The term *M/L ratio* is widely used by astronomers to indicate a galaxy's total mass (in solar units), divided by the total luminosity (in suns). From my study of pairs, the average value of this quotient is about 2 for spirals, about 90 for elliptical and So systems. Galaxy thickness varies from less than 10 per cent of the diameter in Sc types to more than 30 per cent in E's. As to spectra, emission lines and hot stars are characteristic of the loose Sc spirals, cooler stars of the ellipticals and So systems. Recent radio observations show that cold interstellar hydrogen forms about 5 per cent of the mass in Sc's, only 1 per cent in E's and So's.

It is true that Hubble's original scheme for classification has been improved by other astronomers. The So type was added later, and refinements have been made by G. de Vaucouleurs, W. W. Morgan [see p. 68], and S. van den Bergh. But the significant fact is that many characteristics of galaxies are associated with Hubble's types in an orderly sequence. The barred spirals (SBa, SBb, and SBc), the ring structure stressed by de Vaucouleurs, the odd spikes and tails described by B. A. Vorontsov-Velyaminov, and the bridges between galaxies in pairs emphasized by F. Zwicky may all be minor deviants.

It is evident that the time scale of a galaxy's evolution is very long. Therefore we reason about evolutionary changes from the presently observed forms and internal motions in many different galaxies. We can also reason from the evolution of the stars themselves.

The Milky Way Galaxy and Star Formation in Its Arms

The motions of nearby stars are much easier to observe, both by Doppler shifts and proper motions. Galaxies are so distant that the proper motions (across the sky) of their internal constituents cannot be detected. From our position within the Milky Way system, we can determine its mass from *average* motions of stars in orbits around its center, although the problem is somewhat complicated by the sun's own motion and by uncertainty as to our distance from the center. Over forty years ago Bertil Lindblad in Sweden and Jan Oort in Holland noted evidence for rotation of the Galaxy in statistical analyses of stellar motions.

They found that the average motions of stars at various distances and directions from the sun can be represented by a velocity-distance curve like the one for a distant galaxy in Figure 63. After first increasing very rapidly with distance from the center of the Galaxy, the orbital speeds of stars level off and then decrease at very large distances. Measurements of these orbital speeds show how the mass is spread out around the center, and more details have been added by optical and radio observations of interstellar gas.

The mass of the Milky Way Galaxy is about 1.8×10^{11} suns, and in our general vicinity (approximately 10,000 parsecs from the center) the average density is about 0.075 sun's mass per cubic parsec. Counts of nearby stars indicate a somewhat lower density, and it is now agreed that perhaps one fifth of the mass near the sun is in the form of interstellar clouds of gas and dust. Radio-telescope measurements of 21-cm radio emission confirm this, showing concentrations of hydrogen in three spiral arms near us. All this evidence points to the Milky Way as being an Sb type of galaxy.

The distribution of hot, blue stars in a wider region around the sun also shows the spiral-arm structure. Now the theory of stellar interiors makes it evident that these very luminous blue stars cannot last more than 10 million years or so. Hence it is generally agreed that these stars have formed in the spiral arms from gas clouds by gravitational contraction. Without going into details, it is reasonable to assume that the *rate* of formation of additional stars depends on the gas density, and that stars are forming in the interstellar gas clouds right now.

Maarten Schmidt at California Institute of Technology, George Herbig at Lick Observatory, and many others have studied the star-formation process, using such evidence as the observed numbers and types of nearby stars, the existence of small dust clouds [like that shown in Fig. 66], and the density of interstellar gas. The theory of stellar evolution predicts that a large star, after converting part of its hydrogen to helium, will explode as a supernova, blowing much of its mass back into the interstellar gas clouds and leaving the remainder as a faint white dwarf.

In order to account for the observed types of nearby stars, Schmidt had to assume that over a hundred times as many small stars are formed as large ones. Even so, the interstellar gas is heavily contaminated with "used" gas, containing helium and other chemical elements formed by nuclear reactions inside earlier large stars, which have already exploded as supernovae. Evidently the chemical composition of stars should differ from place to place, depending on how much star formation has occurred, and the observed spectra show this to be the case.

However, the picture is further complicated by motions that are not fully understood. Stars may be moving away from the gas in which they were born. Also, evidence was reported to the International Astronomical Union at Hamburg last summer [1964] that gas is moving outward through our Galaxy in long streamers, some of which suggest an enormous explosion near the galactic center over 100 million years ago.

Such observations raise the question of how a galaxy can maintain its spiral arms over very long time intervals if the gas clouds from which new stars form are exploding outward. One possibility is that magnetic fields due to motions in ionized gas exert a pattern of forces that account for the arms. However, gravitational forces alone may be sufficient to hold the arms together, once they had been formed as denser regions by some initial explosion that was not uniform in all directions.

The point to be made here is that the internal structure of our Galaxy is complex. This has a strong influence on star formation because of density fluctuations in the gas and dust, and possibly through magnetic fields. The detailed structure may reveal a good deal more than we now know about the formation and history of our Milky Way Galaxy. Similarly, the smooth structure of the elliptical galaxies [such as NGC 205; see Fig. 35], probably reflects different conditions during their formation, or a longer life span.

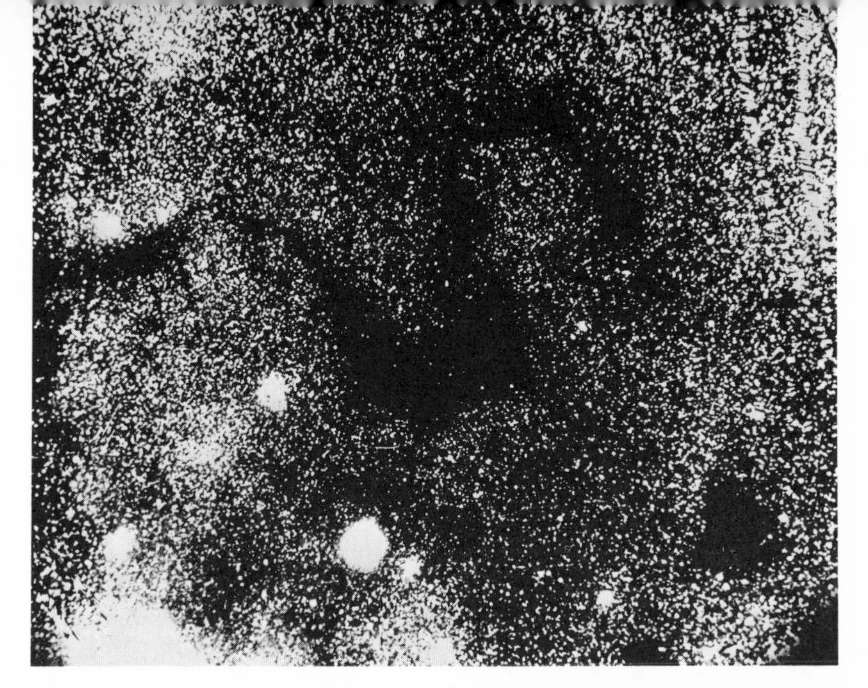

FIG. 66. The Milky Way near 72 Ophiuchi, showing dark globules (lower right) and dark nebulae. A globule is a dust-and-gas cloud from which a star will eventually form by contraction. Photograph by E. E. Barnard, July 6, 1898. The lines at upper right and lower left are cracks in the glass plate. (Courtesy Lick Observatory)

Evolutionary Theory

It has long been recognized that the sequence of galaxy types from E0 through Sc and SBc may reflect evolution. The first idea was that the E types are *old,* Sc and SBc *new.* Then it was discovered that our Milky Way (type Sb) contains old stars in its central parts and young stars in its spiral arms. This showed that evolution down the sequence from Sc to E is probably an oversimplified idea. In fact, if we are correct in believing that E and S0 types are thirty times more massive than the open spirals, and if intergalactic space is clean, it would seem impossible for Sc types to evolve into E0 ones.

However, gas, dust, and young stars are observed in the mottled arms of spirals, whereas only red stars are found in the central parts of spirals, and in E types, where evidently the evolution of stars has proceeded further. On the assumption that all galaxies started as large gas clouds, star formation from the gas must have occurred earlier in the E types, and near the nuclei of spirals, than in the arms of spirals and in irregular systems.

The formation of stars from clouds of gas and dust is now an accepted process to astronomers, but the speed at which it takes place depends on the initial conditions: density, dust content, temperature, turbulence, radiation from outside, magnetic fields, and other factors. As already noted, we can observe in the Milky Way Galaxy two early stages in the process (small dark clouds and newly formed stars in nebulae). Numerous studies of star clusters show later changes in ageing stars. There is every reason to believe that stellar birth and evolution are going on similarly in other galaxies, possibly at different rates.

Schmidt has suggested that the rate of star formation depends primarily on gas density (ρ) raised to some power n, so that $k\rho^n$ is the amount of gas that will collapse to form stars per century per cubic light year, with a consequent decrease in gas density (k is a constant depending on the mass of stars formed.) He expects that the value of n depends upon the average mass of stars formed, and that it equals 1 for stars of solar mass. Higher gas densities would tend to produce more stars of larger mass; that is, high-density interstellar matter is used up more rapidly than low-density matter. Therefore star formation tends to even out large-scale density variations in the gas, though it leaves unchanged a galaxy's overall mass (stars plus gas), as measured by the rotational method [see p. 140].

It is clear that there will be a major difference between a galaxy originally consisting of giant blue stars and one originally of red dwarfs. In fact, the "mix" of newly formed giants, dwarfs, and all sizes in between is undoubtedly of major significance in understanding galaxy evolution. For the region near the sun in our Galaxy, the mix can be inferred by counting the numbers of giant and dwarf stars we can see within a few thousand light years. In other parts of our Galaxy, and in other systems, this mix of newly formed stars may have been very different.

The diagram in Figure 67 shows schematically the evolutionary changes in a condensing gas cloud. It is a time plot of the gas density and the relative numbers of red-dwarf stars, white dwarfs, and blue giants. Explosions of dying giants replenish the interstellar gas for a while, but then the gas is slowly used up and only dwarf stars remain. Since these dwarfs have small luminosities (L) for their masses (M), the mass-luminosity ratio (M/L) for the galaxy as a whole increases with age, and the galaxy becomes redder (its color index grows larger).

All this is consistent with Walter Baade's 1940 concept of young, Population-I stars and old, Population-II ones, and with a simple evolution from Irr and Sc galaxies to So and Eo types. However, the rate of ageing will depend very much on gas density; a galaxy that started with high gas density would have aged more in color and M/L ratio

FIG. 67. The evolution of a gas cloud contracting to form a galaxy. Eventually, most of the gas exists as red and white dwarf stars, while blue giants become fewer.

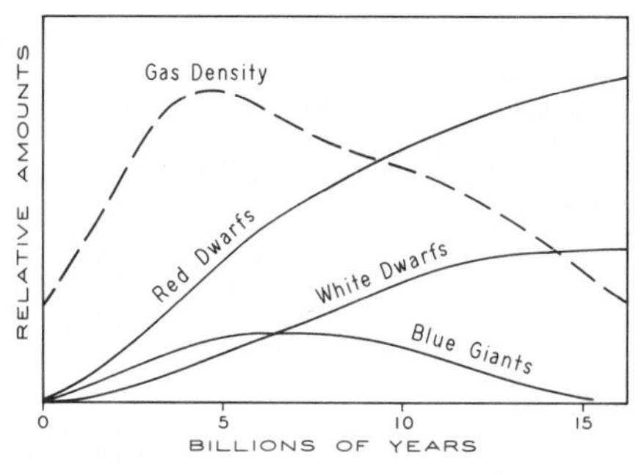

than a low-density system that started at the same time. All the galaxies of large mass must have evolved to So or elliptical types, leaving the lightweights as irregulars and spirals.

Erik Holmberg in Sweden has recently noted a correlation that confirms this expectation. For twenty-five individual galaxies the total masses have been determined from rotation curves. Dividing each mass by the volume gives the mean density of the individual galaxy. As Figure 68 shows, there is a fairly close relation between the color indexes of his galaxies and their densities.

The explanation may be that these galaxies started as gas clouds with different densities. In the denser ones (now So and E systems), star

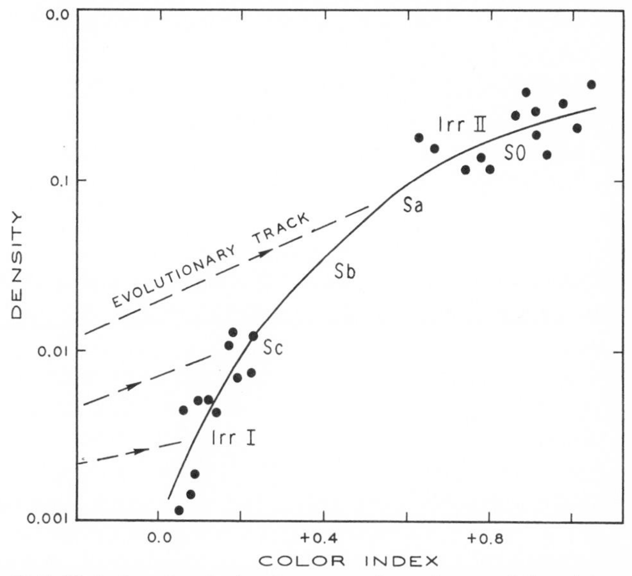

FIG. 68. Erik Holmberg's relation between the colors and densities of twenty-five galaxies. He has divided irregular galaxies into two groups: loose Irr-I types, and compact Irr-II types. The evolutionary tracks are possible changes in contracting gas clouds.

formation and ageing have increased the color indexes more rapidly than in the low-density ones (now Sb and Sc types). Holmberg notes that the lack of very dense galaxies with small color indexes on this plot argues that all galaxies are of the *same age,* contradicting the Steady-State cosmology described above [p. 138]. However, it seems unlikely that the galaxies are still the same size and overall density today as when they first formed. If they have contracted, the density has increased, and the lines representing evolution should slant upward in Figure 68, thereby weakening the evidence for a single age. Moreover, the *rate* of evolution along any one line may vary.

In fact, R. W. Michie at Kitt Peak has calculated how a large cloud of hydrogen gas would collapse. He finds that a cold cloud equal in density to 100 billion suns spread over 10^{18} cubic light years would take more than 10 billion years to collapse—more than the Big-Bang age of the universe. If the initial density is somewhat higher, the contraction starts very slowly, then goes through a number of "bumps" as the gas is heated by compression and resists further contraction. A few billion years after the start, several condensations form, and may lead to a *group* of galaxies rather than one.

This brings us to the idea that small groups and pairs of galaxies may somehow show signs of a common origin. Twin galaxies are presumably the same age. Are they of the same type? Are their axes of rotation lined up? W. Zonn and I, together with several others at Berkeley, have been studying this for the past three years, using measurements of several hundred pairs on the Palomar Sky Survey [see p. 31] prints.

The problem is more complex than it would seem, due to the effects of selection (round-appearing galaxies are missed on the photographs more often than elongated ones of the same brightness), uncertainties in classifying small galaxy images, and systematic errors in measuring the dimensions. Certainly there are many mixed pairs, but preliminary results show a preference for pairs of the same type, as would be expected if both members of the pair condensed from one gas cloud and started with the same gas density. If they were formed by one cloud breaking apart, twin galaxies might also be expected to have their rotational axes lined up, but this is not the case in the few hundred pairs studied.

The rotation of a gas cloud that is condensing into a galaxy must have an effect on the evolutionary process. More precisely, the angular momentum remains constant if no mass is added or lost, and this means that the contracting cloud will spin more rapidly as it shrinks, thus forming a thin disk. It can therefore be reasoned that the spirals were formed from rotating gas clouds, and that elliptical galaxies were formed from clouds with less angular momentum relative to mass. My measures of spectral-line inclinations [see p. 140] tend to confirm this, although the measurements of galaxy sizes are not accurate enough to settle the matter.

The compact galaxies announced by Fritz Zwicky last year and also the QSOs (quasars) are presumably extreme cases of low angular momentum. In fact, the collapse of a large gas cloud to the tremendously high density of a QSO can occur only if the total angular momentum is zero or very close to it. Of course, it is possible that the collapsing cloud spun off some of its mass and got rid of angular momentum. But it seems more likely that the original clouds were of various sizes and spins; the very few with almost zero spin evolved to the few QSOs we see today.

Conclusion

Can we describe the full life history of a galaxy? The answer is still no, although portions have recently become much better understood. Cool gas clouds contract under their gravity; stars form, redden, and die. The shape of the galaxy at any time during this process depends on the initial size and angular momentum of the original cloud. The color and luminosity depend on the density history and the sizes of stars formed. Here is the most serious gap in present understanding: What controls the sizes (or masses) of forming stars? Perhaps rotation, or turbulence, or magnetic fields, or temperature differences have a major effect on where dwarf stars can be born.

Measured masses of galaxies are generally larger than expected from luminosities, and this implies "unseen mass" in the form of very small dwarf stars. On the other hand, the combined mass of all galaxies is much too small to account for space curvature in cosmology, implying vast amounts of unseen matter *between* the galaxies. This dark material might be the remains of dead galaxies that evolved rapidly long ago. Furthermore, it is theoretically possible that a QSO with zero angular momentum could collapse to such high density that it "wraps curved space around itself" and disappears from view.

As this article shows, a wide variety of considerations is necessary in any theory of the evolution of galaxies. The subject is in somewhat the same state of development as the theory of stellar evolution was twenty-five years ago. Ultimately, it will probably have an even greater effect on man's understanding of the sidereal universe.

FURTHER READING

H. Bondi, *The Universe at Large*, Anchor Books, 1960.

J. Greenstein, "Quasi-Stellar Radio Sources," *Scientific American*, December 1963.

E. P. Hubble, *The Realm of the Nebulae*, Yale University Press, 1936 (Dover paperback edition, 1958).

M. L. Humason, N. U. Mayall, and A. R. Sandage, "Red Shifts and Magnitudes of Extragalactic Nebulae," *Astronomical Journal*, **61**, 97, 1956.

W. W. Morgan and N. U. Mayall, "A Spectral Classification of Galaxies," *Publications* of the Astronomical Society of the Pacific, **69**, 291, 1957.

T. Page, "The Evolution of Galaxies," *Science*, **146**, 804, November 6, 1964.

T. Page, ed., *Stars and Galaxies*, Prentice-Hall, 1962.

A. R. Sandage, *The Hubble Atlas of Galaxies*, Carnegie Institution of Washington, 1961.

O. Struve, "Galaxies and Their Interactions," *Sky and Telescope*, **16**, 162, 1957.

O. Struve, "Interacting Galaxies," *Sky and Telescope*, **23**, 16, 1962.

4

Peculiar Galaxies

In a way, every galaxy is "peculiar," since there is no other just like it. It has already been noted (p. 67) that a few per cent of galaxies cannot be fitted into the Hubble sequence of types (Figs. 24, 25, and 26). For instance, a barred spiral should have a straight bar across its nucleus, like the SB galaxies shown in Figure 25; in some spirals, however, the bar is irregular, or has a kink at the center; these galaxies are called SB(pec). Elliptical galaxies should be smooth; a few are mottled —possibly because another faint galaxy is in the foreground—and these are classed E(pec).

These are "mildly peculiar" forms. In this chapter, several more radically peculiar galaxies are described (all in comparatively recent articles, because of the interest in peculiar galaxies engendered by the discoveries of quasars in 1963). Even these odd cases can be typed, and we have tried to follow in this chapter a sequence from peculiar types in close pairs, through exploding galaxies, to dwarf and compact galaxies.

Radio emission, another type of peculiarity, is covered in Chapters 5 and 6. So much work has been done on radio galaxies and quasars that they have now become quite familiar and no longer qualify as "peculiar," although there is still some controversy about their true nature.

It should be noted that other peculiarities (not reflected in the morphology) are observed in the spectra of galaxies—in particular, the spectra of Seyfert galaxies, with broad emission lines (see p. 130). Also, *some galaxies have colors different from those expected for their morphological types, although this may be due to reddening by nearby interstellar dust clouds in the Milky Way. At least two galaxies are strong X-ray sources.*

The distortion of galaxies in close pairs is easily understood. It may be mild if the two galaxies are separated by two or three diameters, or it may be extreme if a near-collision is going on. Actually, collisions are rare; the sizes, relative speeds, and average separations are such that there is an estimated 0.5-per-cent chance of one collision among the 500,000 nearest galaxies during an interval of a million years. A pair like NGC 5432-5435 (shown in Fig. 65) is therefore not likely to collide, nor are the members likely to be just passing one another; they are in orbit, one around the other, as assumed in the calculations of mass (p. 141). There are thousands of pairs like this, presumably formed as pairs, since "captures" are so unlikely. Some are so close as to be in contact, and others show evidence of being blown apart.—TLP

Possible Star Exchanges between Galaxies

GEORGE S. MUMFORD

(*Sky and Telescope*, October 1965)

The gravitational attraction between the two components of a very close binary may distort their atmospheres so much that they come in contact, probably with an exchange of gas taking place through the common point. Now, three University of California astronomers suggest that in a similar manner stars may be streaming between the thirteenth-magnitude elliptical galaxies NGC 4782 and 4783.

E. M. Burbidge, G. R. Burbidge, and D. J. Crampin used in their study the four photographs in Figure 69. At an estimated distance from us of 180 million light years, the galaxy centers seen on the short exposures are about 35,000 light years apart. But longer exposures reveal the outer parts of each system, and in the 45-minute picture (top) there appears to be strong distortion of each galaxy toward the other, possibly with a bridge of luminous matter between them.

Microphotometer tracings of the plates were made at Yerkes Observa-

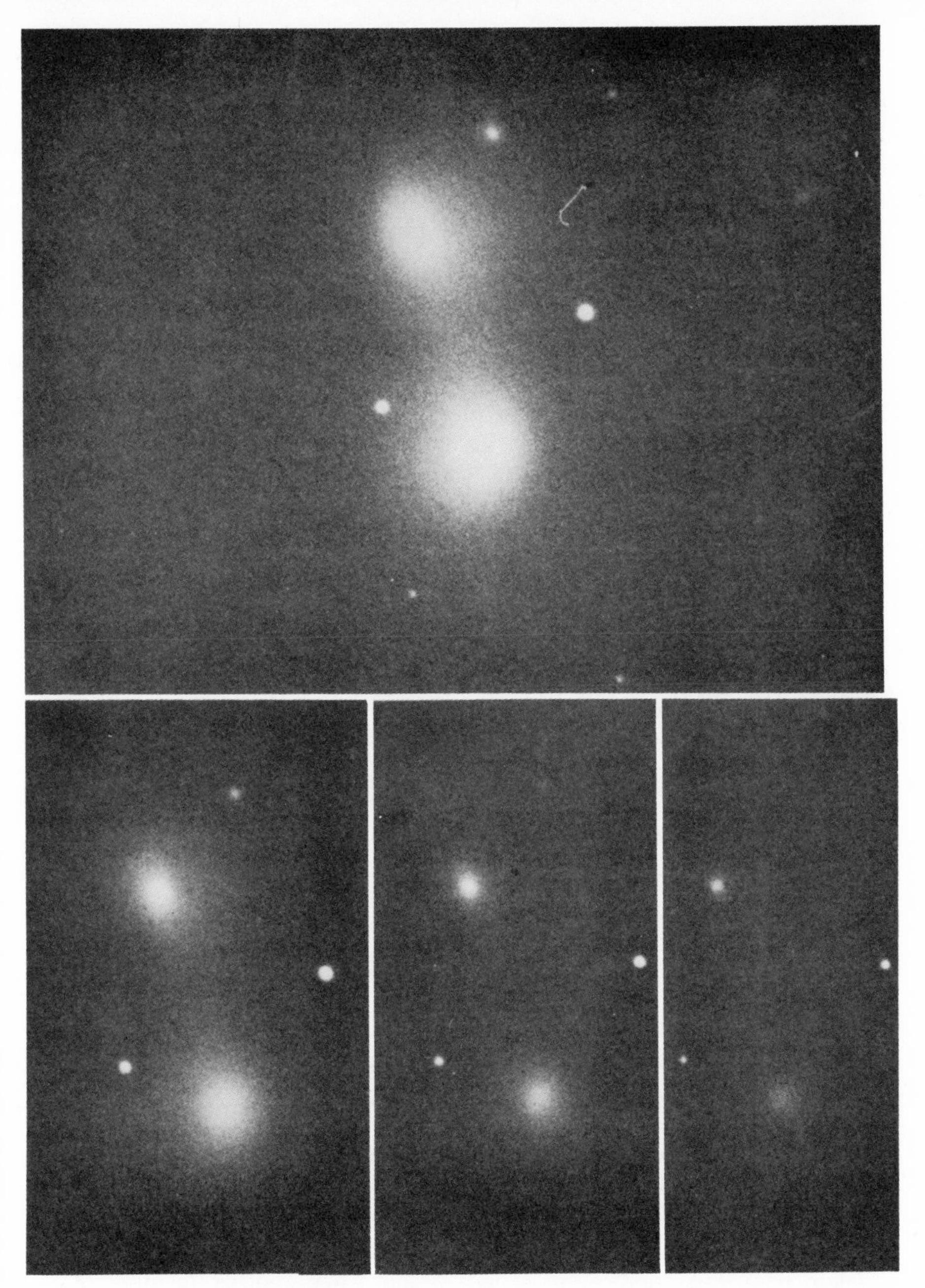

FIG. 69. Prime-focus photographs of NGC 4782 (upper) and NGC 4783, in Corvus. Across the bottom, from the left, the exposures lasted 15, 5, and 2 minutes, respectively, on Eastman 103aD plates through a GG 11 filter (visual-light range). The fourth frame is a 45-minute exposure that brings out the systems' mutual contact. Scale is 1.25 seconds of arc per millimeter. North is at the top, east to the left. (From *The Astrophysical Journal*)

tory. The resulting contours of equal brightness indicate the existence of a gravitational crossover point (through which stars might be exchanging) slightly nearer NGC 4783 than 4782. The former has a much brighter central region, yet three fifths of the total mass (at least 530 billion suns) is probably in NGC 4782.

In *The Astrophysical Journal* [November 1964] these authors attach no special significance to the identification of NGC 4782–83 with the cosmic radio source 3C 278.

A Remarkable Chain of Galaxies

(*Sky and Telescope*, October 1968)

In 1959 the Soviet astronomer B. A. Vorontsov-Velyaminov called attention to an unusual string of five seventeenth-magnitude galaxies near Lambda Draconis. They range in diameter from about 20 to 6 seconds of arc, and form a line 63 seconds long. This system, VV 172, appears in Figure 70 as recorded by the 200-inch Palomar telescope for H. Arp's *Atlas of Peculiar Galaxies.* [In 1968] Wallace L. W. Sargent used the 200-inch telescope to obtain spectrograms of all five components and measured their red shifts.

Designating the galaxies from north to south by the letters **a** to **e**, Sargent finds that four of them have quite similar line-of-sight velocities: **a**, 16,070 kilometers per second; **c**, 15,820; **d**, 15,690; and **e**, 15,480. This group is about 210 million parsecs distant from us, if the Hubble constant is assumed to be 75 kilometers per second for each million parsecs. The projected distance between the centers of galaxies **a** and **e** is then about 60,000 parsecs. Because the red shifts have a systematic trend with position in the chain, it is highly likely that the four form a rotating system, requiring about 600 million years to make a turn.

Strangely enough, galaxy **b** has a very much larger red shift, 36,880 kilometers per second, even though it looks like another member of the same group. Sargent lists three interpretations: First, **b** could be a background galaxy at a distance corresponding to its red shift. However, the probability of one such object being superimposed on the area of the group is very small, perhaps one in 5,000. Second, **b** could lie at the

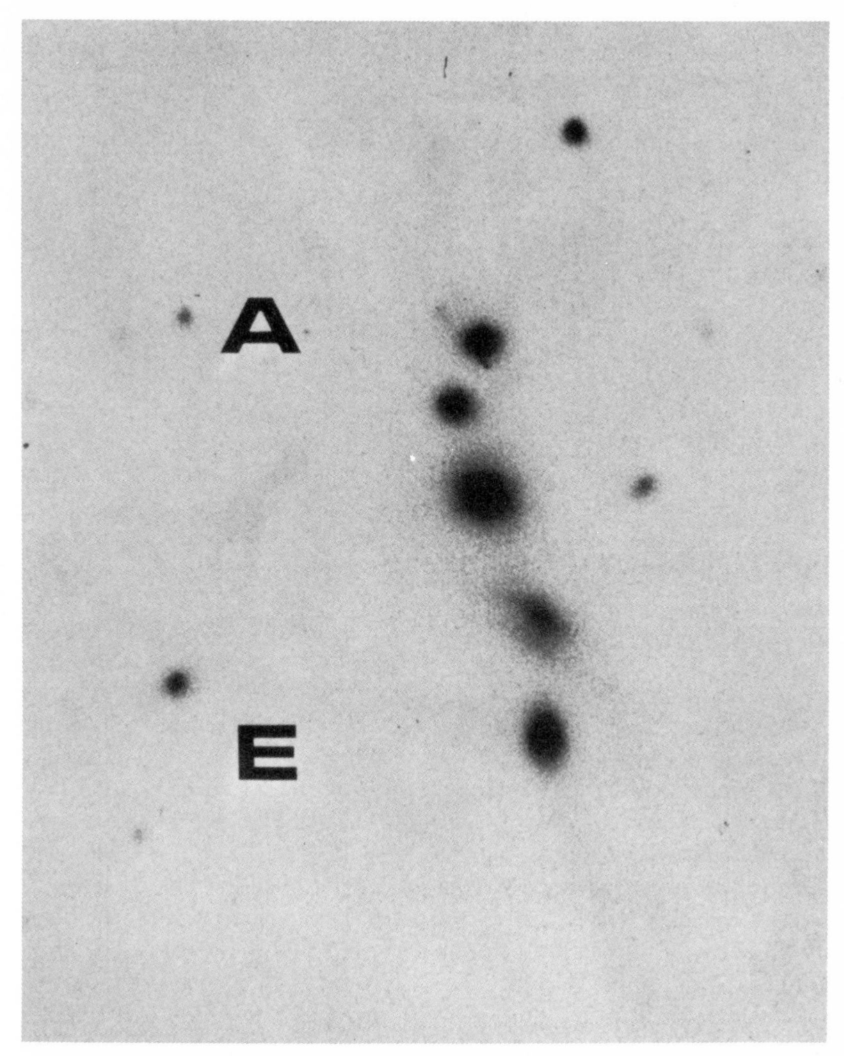

FIG. 70. H. C. Arp's 200-inch photograph of the VV 172 chain, with component **a** at the top and **e** at the bottom. The second galaxy, **b,** has the anomalous radial velocity. North is up, west to the right, and there is a defect on the image of **a.** (Reproduced from *Astrophysical Journal Letters*)

same distance as the others but have a very large gravitational red shift. This alternative is ruled out by the presence of a forbidden emission line of ionized oxygen, which can appear only when the gas density is very low.

A remaining possibility is that the excess red shift is produced by actual motion of galaxy **b** relative to the others. In the August [1968]

Astrophysical Journal Letters, Sargent comments guardedly that VV 172 might illustrate B. A. Ambartsumian's idea that some galaxies may originate from violent explosions in the nuclei of other galaxies.

This puzzling quintet of galaxies needs further study.

NGC 4038–39: An Exploding Galaxy?

GEORGE S. MUMFORD

(*Sky and Telescope*, February 1967)

In western Corvus is a remarkable extragalactic object of about the eleventh magnitude bearing two NGC numbers, 4038 and 4039. The former applies to the brighter northern part, the latter to a fainter southern portion a few minutes of arc distant.

In 1940 Harlow Shapley and J. Paraskevopoulos of Harvard classed NGC 4038–39 as a one-armed spiral galaxy with a ring-tailed structure. It was one of the first extragalactic radio sources to be discovered, and its odd form used to be mentioned as support for the colliding-galaxy hypothesis of cosmic radio sources. And in fact, the rather overexposed images of NGC 4038–39 in the Palomar Sky Survey do give an impression of two spirals in contact, each with a long, curved faint tail.

But on short-exposure plates taken with the 82-inch Struve telescope at McDonald Observatory, the object clearly does *not* consist of two regular spirals. E. M. Burbidge and G. R. Burbidge of the University of California at San Diego find that the regular outline is due to a smooth unresolved background, somewhat reddish, in which the bright ring-tailed structure is embedded (see Fig. 71).

Using the same telescope, these two astronomers have now completed the first detailed spectroscopic study of this system. By placing the spectrograph slit successively in various positions across the prime-focus image of the galaxy, they measured the radial velocities at many points. From these the Burbidges could map the line-of-sight internal motions of NGC 4038–39 in detail.

The pattern of motions is unlike that to be expected from mutually revolving galaxies or from a single rotating galaxy. The smooth gaseous background is approaching us relative to certain of the brighter parts.

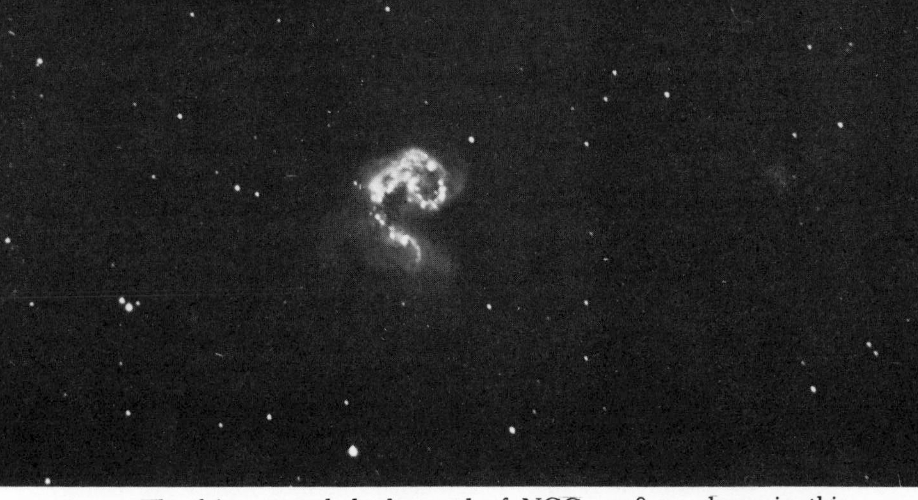

FIG. 71. The faint, smooth background of NGC 4038–39 shows in this 40-minute blue-light exposure with McDonald Observatory's 82-inch reflector. North is up, east is to the left. (From *The Astrophysical Journal*)

of the system. The evidence suggests two centers of expansion, one north of the brightest part of NGC 4039, the other in the bright knot at the northwest edge of 4038. The largest expansion velocities, amounting to several hundred kilometers per second, are found along the eastern edge of the combined system. This fact may mean that the fastest moving material has progressed farthest from a main center of ejection that is hidden in the dusty center.

If NGC 4038–39 is indeed an exploding galaxy, it is a less violent case than some others known. The ionization of its gases is no more than would be produced by the radiation from its *O* and *B* stars, whereas in the exploding galaxy M 82 [see p. 167] the ionization is attributed to ultraviolet synchrotron radiation. "Further, in distinction from the Seyfert galaxies, the emission lines in NGC 4038–39 are not appreciably broad, so that no large turbulent motions appear to be present," the Burbidges note in *The Astrophysical Journal* [September 1966].

The observed velocity of recession of NGC 4038–39 is 1674 kilometers per second, but corrected to 1500 by removal of the effect of our Galaxy's rotation (which carries the sun at 250 kilometers per second toward Cygnus). The Burbidges assume a value of 75 for the Hubble constant and divide this into 1500 to get a distance of 20 megaparsecs (65 million light years) for NGC 4038–39. The full length of the "ring-tail" (measured as an angle of 4° in the sky) is therefore about 67,000 light years across.

FIG. 72. A supernova is marked by the white line in the left photograph of NGC 4725, taken in May 1940. It had faded almost to obscurity by January 1941, when the other picture was taken. (Mount Wilson and Palomar Observatories photographs)

Explosions are sudden releases of energy that "push things apart," or show as sudden increases in luminosity. It is difficult to explain the "pushing" of stars, but interstellar gases could have been forced outward by radiation pressure if there were a sudden brightening of some region in NGC 4038–39.

Suddenly brightening stars (called novae since the days when men thought that they were new stars forming) have been observed many times—see Volume 5 in this series, Starlight. *Novae brighten by a factor of several hundred in an interval of a few days, then die down in a year or so. Much less frequent are the supernovae, stars that brighten by a factor of several billion (see Fig. 72), reaching a luminosity comparable to that of a whole galaxy. Supernovae explosions are the most violent that have been observed in the universe.*—TLP

One Hundred Fifty-two Supernovae

(*Sky and Telescope,* October 1964)

According to a new check list compiled by Fritz Zwicky, Mount Wilson and Palomar Observatories, a total of 152 supernovae have been discovered and clearly identified as such since 1885, when S Andromedae exploded near the nucleus of Messier 31.

As a sample of the information contained in Zwicky's list, take the last object, designated 1964f because it was the sixth supernova found in 1964. This appeared in the galaxy NGC 4303, a barred spiral in the Virgo cluster having an apparent velocity of recession of 1671 kilometers per second. Maximum light was reached in June 1964 at somewhat fainter than magnitude 14.0, and the light curve was of Type I [see p. 160]. The discoverer was L. Rosino of Asiago Observatory in Italy.

The brightest supernova in the Zwicky catalogue is S Andromedae itself, for which he gives a photographic magnitude of 7.2 at maximum. The next two brightest are Z Centauri (8.0), which appeared in NGC 5253 during 1895, and the supernova of 1937 in IC 4182, which attained magnitude 8.2. On the other hand, some very recent discoveries never became as bright as magnitude 19.

Up until 1930, fewer than twenty supernovae had been discovered, all of them by chance. The first systematic search was begun by Zwicky in 1933. After a wartime interruption, the search was resumed in 1954 with the 18-inch Schmidt telescope on Palomar Mountain, and in 1959 the 48-inch Schmidt became available for this program. By 1961 discovery patrols were organized on an international basis, and now observatories in eleven nations are participating.

The total number of definitely established supernovae would be increased to 155 if we added those that appeared in our Milky Way Galaxy in the years 1054, 1572, and 1604. There is also fragmentary evidence of other supernovae in A.D. 185, 369, 827 and 1006.

Supernova in Messier 83

A. D. THACKERAY

(*Sky and Telescope*, November 1968)

On the evening of July 16, 1968, a South African amateur, J. C. Bennett of Pretoria, reported to Radcliffe Observatory that the nucleus of the spiral galaxy M 83 in Hydra (Fig. 73) appeared brighter and more starlike than he had seen it before.

Within four hours, two low-dispersion spectra of an 11- or 11½-magnitude object five seconds of arc west of the galaxy's nucleus had been obtained with the 74-inch Radcliffe reflector. The spectrum was

FIG. 73. M 83 (NGC 5236) in Hydra, one of the brightest spiral galaxies in the southern sky, is the site of the supernova discovered in July 1968 by J. C. Bennett. The brightest star at the top of Figure 74 appears in this view 28 millimeters from the left edge and 24 from the top. This Harvard Observatory photograph was taken with the 60-inch reflector at Bloemfontein, South Africa, exposure 90 minutes. South is at the top, and the scale of the reproduction is 12″.4 per millimeter. (Courtesy of Harvard College Observatory)

nearly featureless, but possibly that of a Type-I supernova contaminated by light from the bright nucleus. (Type-I supernovae differ from Type II in having a weak ultraviolet continuum, higher luminosity, and tending to occur in elliptical galaxies or the central parts of spirals.)

A half hour later, the announcement of Bennett's discovery of the brightest supernova since 1962 was sent to the International Astronomical Union's telegram bureau at the Smithsonian Astrophysical Observatory in Cambridge, Massachusetts. Bennett has supplied this account of his discovery:

"While sweeping for comets on July 16th with a 21-power Moonwatch apogee scope, I came across an object at 17:40 Universal time that appeared to be a star of about 9th magnitude in the center of a faintly luminous patch. Convinced that I had never observed such an object before, I sighted along the telescope tube and found to my surprise that it was in the position of M 83. I recalled that several weeks before I had observed M 83 without noticing the striking feature of a central star.

"During the past four years I have been listing and describing southern-hemisphere objects that appear cometlike in small telescopes at low power, so these would not be misleading in my searches for comets. My index card for M 83 had no drawing, such as I had made for other objects. Yet, if the starlike core were a permanent feature, why had it not been mentioned?

"I next consulted the *New General Catalogue*, where M 83 is listed as NGC 5236, and described as 'extremely suddenly brighter in the middle to a nucleus.' To test whether the bright object was really the nucleus, I used my only other telescope, a 3-inch refractor.

"At magnifications of 35 and 100, the object (bluish in color) appeared starlike; at 150X it appeared to have a slightly ragged edge. After consulting some more literature, I informed Dr. Thackeray."

All previous Radcliffe photographs of M 83 have been 15-minute exposures or longer, showing a heavily overexposed nucleus. Dense prints of the most lightly exposed of these plates seem to show a brightening of the nucleus in the position of the supernova. We do not know how much brightening has occurred, and lightly exposed plates from other observatories are needed.

The middle and right frames of Figure 74 are short exposures by P. J. Andrews on July 24 [1968], and reveal the fantastic complexity of the nucleus. Some 150 years ago, John Herschel described the center of M 83 as "equal to a 9th-magnitude star, 8″ in diameter, and of a resolvable character like a globular cluster."

It is remarkable that Bennett's supernova is the fourth and brightest one discovered in M 83, and also the closest to the nucleus. How many other supernovae in the nuclear region have gone undetected in the last 50 years?

By August 6, 11 Radcliffe spectra had been obtained by M. W. Feast, A. Menzies, R. Wood, and the writer. Spectra taken on August 1 and 4 extend into the red, and confirm that the supernova was probably of Type I. They show the broad band at 4650 angstroms more prominently than revealed by early spectrograms that were probably taken near maximum light.

The first estimates of the supernova's magnitude require correction because of the added light of the nucleus. Andrews' photographs on July 24, supplemented by his photoelectric magnitudes of comparison stars, suggest a photographic (blue) magnitude of 12.2 at that time.

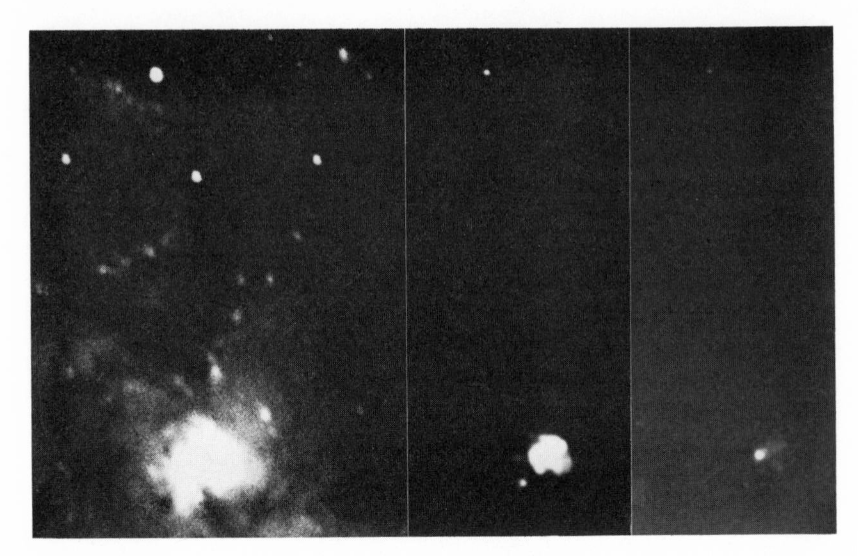

FIG. 74. The nuclear region of M 83, photographed with the 74-inch reflector at Radcliffe Observatory, Pretoria, South Africa. At left is a 15-minute blue-light exposure by D. S. Evans on April 17, 1950. The other two are by P. J. Andrews on July 24, 1968, with the telescope aperture reduced to 44 inches. In the middle picture, a 16-minute blue plate, the supernova blends into the bright image of the nucleus (the star below and left is merely a field object). Only in the last view, a 30-second yellow plate, does the 1968 supernova stand out sharply, with almost no nuclear structure discernible. (Courtesy of Harvard College Observatory)

This would imply an absolute magnitude of −15.2, which is equivalent to the luminosity of 100 million stars like the sun.

The prompt cooperation between amateur and professional astronomers is a happy feature of this important discovery by J. C. Bennett.

Nests of Supernovae

GEORGE S. MUMFORD

(*Sky and Telescope*, May 1965)

There are seven external galaxies in which more than one supernova have been observed: NGC 3938 and 4157 have had two each, while NGC 3184, 4303, 4321, 5236, and 6946 have had three each.

Moscow astronomer B. V. Kukarkin points out that these systems are strikingly similar. Six of them are spirals of type Sc, while the seventh (NGC 4303) is an SBc barred spiral. All are giant galaxies ranging in absolute magnitude from —19 to —20, and having loosely wound arms.

The average frequency of supernova appearances in them is one every fifteen years. This is many times the usually adopted rate for galaxies in general—about one in every three hundred years. The greater tendency for Sc systems to be the sites of supernova explosions may have important implications.

What Causes a Supernova Outburst?

GEORGE S. MUMFORD

(*Sky and Telescope*, October 1967)

In the cataclysmic phenomenon of a supernova, a star blows up, for a short time becoming comparable in luminosity to the combined light of all the other stars in its galaxy. Many astrophysicists believe that this event begins with a violent collapse (implosion) of the star, which reverses into an explosion. It has been suggested that instability leading to such an implosion would occur when the temperature of a star's core rose to 5 or 6 billion degrees Kelvin, causing iron nuclei to dissociate into alpha particles and neutrons. But the reversal of the implosion is difficult to explain.

A different suggestion has been advanced by Z. Barkat, G. Rakavy, and N. Sack, who are theoretical physicists at the Hebrew University, Jerusalem. Their calculations show that stars more than twenty to thirty times as massive as the sun become dynamically unstable at somewhat lower central temperatures. The instability sets in just after the formation of oxygen by combining helium nuclei (alpha particles) under the extremes of temperature and pressure in the core of an ageing star. Sudden decrease in pressure results from pair formation (the conversion of a photon of light to an electron and positron when the photon traverses a strong electric field).

After the instability sets in, the star contracts faster and faster, and the rate of thermonuclear reactions increases by many orders of magnitude. There is an energy release corresponding to the consumption of several solar masses of oxygen, resulting in an explosion of the star.

Detailed computations were carried out for a star model composed of forty solar masses of oxygen. The implosion heats the center to about 3.2 billion degrees, and a tenth of the oxygen is consumed before the explosion occurs. In all, some six solar masses of oxygen are burned, releasing about 6×10^{51} ergs of energy—more than enough to disrupt a star.

The Israeli scientists write in *Physical Review Letters:* "The end result of the dynamical calculations is that the star is completely dispersed into space, the gases moving with velocities up to several thousand kilometers per second."

Identification of Cassiopeia A?

GEORGE S. MUMFORD

(*Sky and Telescope*, October 1967)

The strong radio source Cassiopeia A is believed to be the result of a supernova explosion about three centuries ago, this age being indicated by the diverging motions of faint shreds of nebulosity. But until very recently no record of such a supernova was known.

P. Brosche, a German astronomer at the Astronomisches Rechen-Institut in Heidelberg, points out that two separate Korean records state that a "guest star" was seen to the west of the constellation Wang-Liang from December 4, 1592, until the following March. As the star group Wang-Liang consists of Beta, Kappa, Gamma, Eta, and Alpha Cassiopeiae, the position and the age more or less fit the conjectural supernova.

There is a time discrepancy of a century, but in the case of the Crab nebula the date of its supernova (A.D. 1054) is also earlier than its expansion age indicates.

Tycho's brilliant supernova of 1572 appeared in the same part of the sky. Could the Korean statement possibly be incorrectly dated records of Tycho's star?

Supernova remnants, such as the Crab nebula, are described in Volume 5 of this series, Starlight. *In addition to a gas cloud several light years across, blown out by the supernova and moving away from it like an expanding bubble, there is a strong radio emission resulting from high-speed ions (mostly electrons and protons) that follow helical paths in an interstellar magnetic field. This "synchrotron radio emission" lasts for centuries after the supernova outburst—evidence of the vast energy release. Many of the nearby galactic radio sources (in the Milky Way) are found to have the radio spectrum characteristic of synchrotron emission. Other radio emission (with a different spectrum) comes from high-temperature regions in the Milky Way, and the sum of all these would be detected by radio telescopes (if there are any) in other galaxies as a faint radio emission from our Galaxy.*

Some other galaxies have strong radio emission and show other evidence of enormous explosions, much larger than a supernova outburst, emitting light and radio energy in much the same way.—TLP

Explosion in M 82?

(*Sky and Telescope*, July 1963)

The irregular galaxy M 82 in Ursa Major (Fig. 75) has long been known as a radio source. Now, new photographic observations by C. R. Lynds (Kitt Peak Observatory) and A. R. Sandage (Mount Wilson and Palomar Observatories) reveal that this galaxy contains a peculiar system of filaments.

These are located along the minor axis of M 82, and extend 3000 parsecs above and below its fundamental plane. Their spectra show emission lines characteristic of gaseous nebulae. The filaments appear to be expanding along the minor axis with velocities ranging up to 1000 kilometers per second. Apparently they are material expelled from the central regions of M 82 about 1.5 million years ago in a gigantic explosion.

From the brightness of the filaments in hydrogen-alpha light and from their dimensions, it is calculated that the visible mass of the expanding material could be as much as 5.6 million suns. The filaments form fragments of loops, and seem to indicate lines of magnetic force, which suggests that optical synchrotron radiation is being emitted. This proposal is supported by Lowell Observatory findings that light from the filaments is strongly polarized.

FIG. 75. This picture of M 82 was taken many years ago with the 60-inch telescope of Mount Wilson Observatory. The dark dust clouds that dapple the system may account for its generally yellowish light, comparable to the sun's and at variance with its overall spectral type, A5. (Mount Wilson and Palomar Observatories photograph)

By combining radio and optical data, the two astronomers estimate that the present rate of synchrotron radiation is 2×10^{42} ergs per second. If this rate is assumed to have been constant over the 1.5 million years since the explosion, then 9×10^{55} ergs of energy have been released in this catastrophic outburst.

Cataclysm in Messier 82?

(*Sky and Telescope*, November 1963)

In the last few years, evidence has been collecting that the irregular galaxy Messier 82 in Ursa Major is the seat of a gigantic explosion now in progress.

This ninth-magnitude system is about 10 million light years away and is a strong radio source. In 1961 C. R. Lynds found that its radio spectrum is similar to that of the Crab nebula in Taurus, known to be the debris of a supernova explosion.

Next, at Lowell Observatory, Aina Elvius and John S. Hall found that the light from M 82 was strongly polarized in a pattern that suggests the existence of a magnetic field of galactic dimensions.

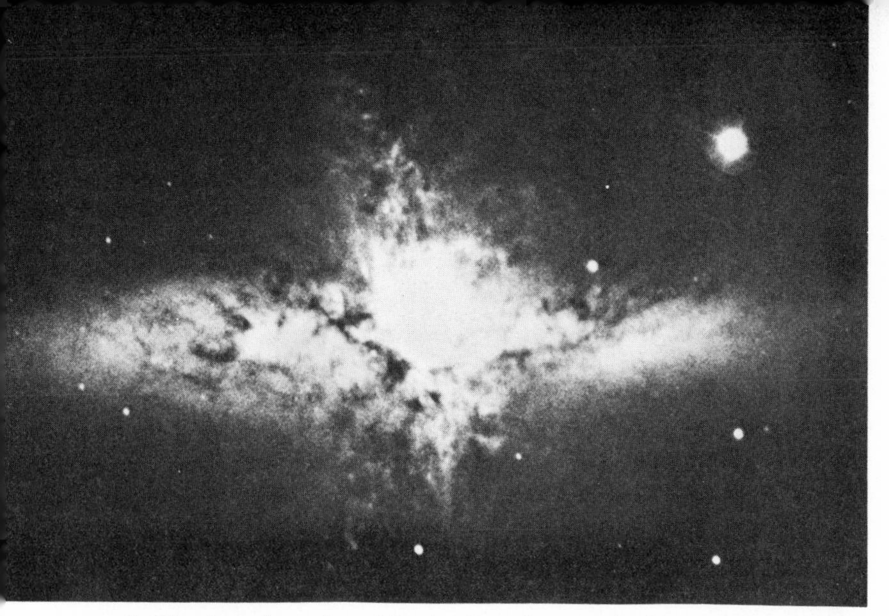

FIG. 76. The irregular galaxy Messier 82, photographed in red light of hydrogen by Allan R. Sandage with the 200-inch Palomar reflector. The enormous puff of luminous hydrogen gas escaping from the nucleus of this system stretches over 10,000 light years above and below the disk of the galaxy, which is about 20,000 light years across. (Mount Wilson and Palomar Observatories photograph)

Direct evidence of the explosion came in a photograph (Fig. 76) taken last spring [1963] by Allan R. Sandage with the 200-inch Palomar telescope. He used a red filter and red-sensitive emulsion to record M 82 in hydrogen-alpha light. Enormous jets of material are streaming from the galaxy's nucleus, above and below the flattened main body.

Spectroscopic observations of these jets were made by Lynds at Lick Observatory. His spectrograms showed that the outward velocities in the jets were as high as 1000 kilometers per second at points most distant from the system's center. From these velocities it was calculated that the explosion started 1.5 million years ago.

Sandage and Lynds have estimated that the material being expelled amounts in mass to about 5 million suns. An explosion of this magnitude may well be a major source of high-energy cosmic rays.

The new observations of M 82 add to recent evidence of explosions on a galactic scale. Radio and optical astronomers have identified five intrinsically luminous galaxies that appear to be in the early stages of great explosions. Until all this evidence became available, most astronomers thought that explosions could not occur on such a tremendous scale.

More about the Explosion in Messier 82

(*Sky and Telescope*, June 1964)

Allan R. Sandage and William C. Miller of Mount Wilson and Palomar Observatories now find an extensive new system of very faint filaments in the outer regions of M 82. Parts of the filamentary structure had been discovered last year by Hugh Johnson with the Crossley reflector at Lick Observatory.

These tangled wisps, which extend 4000 parsecs above and below M 82's central plane, differ from the previously known jets in being photographable in blue and yellow light, as well as red. The light from filaments is nearly 100 per cent polarized, as demonstrated by photographs taken with the Palomar 200-inch telescope through a Polaroid filter in different orientations.

Describing their work in *Science* for April 24, 1964, the California astronomers state, "The most likely interpretation of these results is that a magnetic field is present and is aligned predominantly along the minor axis of M 82. . . . It would seem that M 82 may be the first galaxy in which large-scale magnetic-field structure has actually been observed."

Sandage and Miller suggest that the filaments shine by optical synchrotron emission, caused by very fast electrons spiraling in the magnetic field. To produce radiation in visible wavelengths, the electrons would need energies of 10^{12} to 10^{13} electron volts, if the magnetic field is between 10^{-5} and 10^{-6} gauss.

The origin of this magnetic field is unknown, but the Mount Wilson astronomers suggest, "It may have originally been present in the main body of the galaxy and have been drawn out by the plasma thrown poleward in the initial explosion along the minor axis."

Producing photographic prints showing both the extremely faint filaments and the details inside brighter regions was a difficult darkroom problem. To make the faintest features plainer, composite prints were made. . . .

Unusual Feature near Spiral Galaxy M 81

GEORGE S. MUMFORD

(*Sky and Telescope*, June and September 1965)

Attempts to photograph extremely faint outlying features in galaxies have recently been made at Mount Wilson and Palomar Observatories by Halton C. Arp. During last year's sunspot minimum, the terrestrial night sky was often exceptionally dark, and exposures as long as 50 minutes were feasible with the Palomar 48-inch Schmidt telescope. To avoid the night-sky emission line at 5577 angstroms wavelength, Arp restricted his observations to the spectral region between 4700 and 5400 angstroms.

Various galaxies were explored in the hope of finding connections or distortions in the very faint outermost regions of members in groups. In some cases, several identical negatives were viewed stacked together in front of a strong light. While thus inspecting the bright Sb galaxy M 81 in Ursa Major, Arp found an unexpected and puzzling feature—a very faint ring around one end of the spiral (see Fig. 77).

On ordinary photographs the diameter of M 81 is about 15,000 parsecs, but the ring is about 20,000 parsecs from the nucleus of M 81; it is strongest in the direction of nearby M 82, a peculiar exploding galaxy [see p. 167].

Arp speculates that high-energy electrons emitted from the latter system are impinging upon the extensive magnetic field of M 81. He writes in *The Astrophysical Journal* [January 1965], "these electrons will be bent from linear flight paths and radiate their energy according to the strength of the magnetic field they find themselves in." Only a few microgauss would be needed to account for the observed glow; in fact, too strong a field would cause the electrons to lose their energy too quickly to produce this effect.

Assuming that the electrons in the ring were ejected from M 82 at about the speed of light, Arp calculates that the great explosion in that system occurred some 400,000 years earlier. The age of the ring is about 300,000 years. Further observations of this peculiar feature in M 81 may help explain the intriguing and powerful explosion in M 82.

These neighbor systems are about 10 million light years from the sun. . . .

Figure 77 is a composite from three plates, plus an insert. The very bright annulus is the completely burnt-out image of M 81 on an extremely long exposure. The oval insert is a familiar view of the galaxy

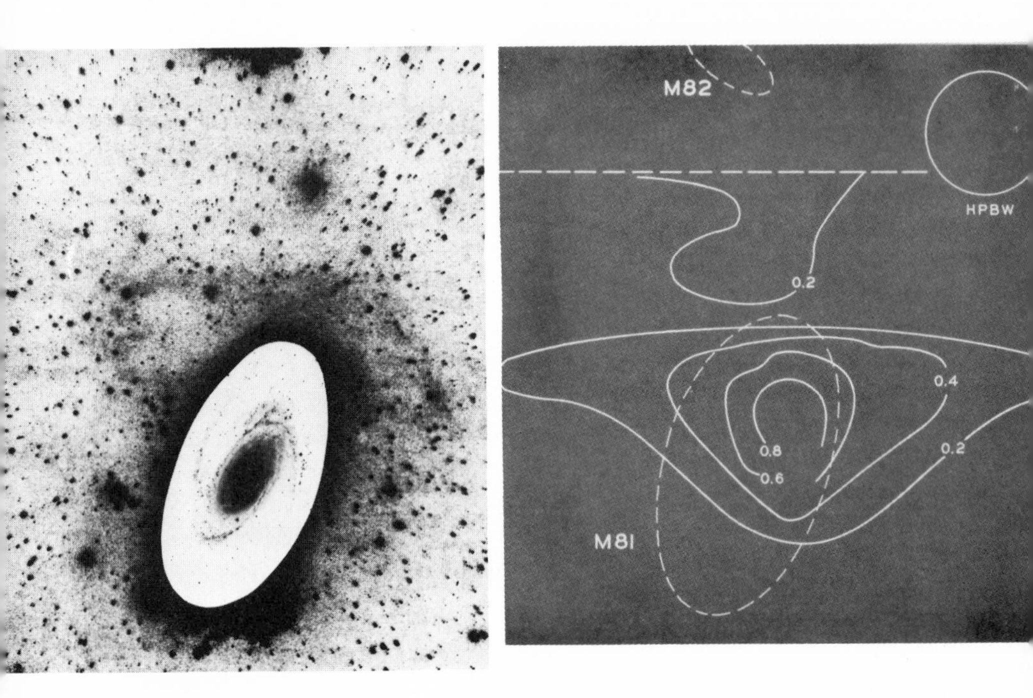

FIG. 77. A. Three long-exposure photographs taken with the 48-inch Schmidt telescope on Palomar Mountain have been superimposed to strengthen the faintly luminous ring around the northern end of M 81. Inserted on this negative is a normal view of the central region. (Mount Wilson and Palomar Observatories photograph)
B. Contours of 21-cm radio flux near M 82 measured by M. L. De Jong with the 300-ft telescope of the National Radio Astronomy Observatory.

on the same scale. The so-called ring is the long, nearly horizontal, irregular strip of light just above the top of the insert. . . .

The bright strip at top center of the picture is an outlying portion of the neighboring galaxy M 82, recently recognized as an exploding object. It may be the source of the energy causing the ring to shine.

Nature of the Ring around M 81

(*Sky and Telescope*, January 1968)

Concerning the nature of the large, faint, elliptically shaped ring around the spiral galaxy M 81, its discoverer points out that radio emission in this region of the sky conforms very well to the outline of the ring. Halton C. Arp obtained the 21-cm measures of hydrogen radio emission, contoured in Figure 77B from the National Radio Astronomy Observatory. He writes:

"Not only do the radio measures indicate nonthermal emission from a plasma in the region of the ring, but there is a tongue of radio emission observed to be coming down (southward) from M 82, which is the source of the synchrotron radiation in my original interpretation of the ring. It is not often that differences in interpretation can be settled so rapidly and completely, in this case as a result of the critical radio observations made by Marvin L. De Jong at Green Bank, West Virginia.

Intergalactic Dark Cloud?

GEORGE S. MUMFORD

(*Sky and Telescope*, March 1965)

The galaxy NGC 4189 is a 12.5-magnitude member of the Virgo swarm, at a distance from us of about 38 million light years, according to S. van den Bergh, David Dunlap Observatory, Toronto. It seems a normal spiral, except that just within its western edge there is a strangely open circular area with a bright central knot.

In *The Astronomical Journal* [November 1964], Charles T. Kowal of Mount Wilson and Palomar Observatories offers this interpretation: "Apparently NGC 4189 is being partially obscured by a dark cloud moving across it. This cloud is dense enough to almost completely obscure the parts of NGC 4189 behind it, yet it has little or no light of its own, except for the small nucleus in its center."

Most of the dark clouds in our own Galaxy lie very near the plane of the Milky Way. Because NGC 4189 is in galactic latitude +74°, not very far from the north galactic pole, it is probable that the dark marking is not part of our system, but in intergalactic space.

The hole could, of course, be a feature of NGC 4189 itself, possibly due to an explosion. But the regular shape of the hole and the apparently undisturbed spiral arm seen faintly through it suggest no catastrophic explosion there.

A third possible interpretation is that the dark marking is caused by the interposition of a dwarf galaxy. However, other dwarf galaxies, such as the Sculptor and Fornax systems, are virtually transparent and do not have nuclei. . . .

The Sculptor and Fornax Dwarf Galaxies

PAUL W. HODGE

(*Sky and Telescope*, December 1964)

In 1937 Harlow Shapley, then director of Harvard Observatory, and his assistant, Mrs. Eric Lindsay, made a surprising discovery. While examining a photographic plate for galaxies, Mrs. Lindsay found a peculiar object, unlike anything she had seen before. The plate had been taken in 1935 with the 24-inch Bruce refractor at the Boyden station [later Boyden Observatory] in South Africa.

The particular emulsion used happened to be unusually sensitive, recording stars about 1.5 magnitudes fainter than normally reached by that telescope. Therefore when Shapley attempted to find the peculiar object on other plates, its faintness made it almost invisible, and difficult to confirm. Yet, after a search through the immense plate file of the observatory, another photograph was located that showed the object well.

Surprisingly, this second plate had been taken thirty years earlier with a camera of only 3-inch aperture, at the Harvard southern station in Peru. The total exposure time was 72 hours, obtained during a week whenever the sky was clear. On the discovery plate the object looked

FIG. 78. In this large-scale photograph taken by W. Baade and E. P. Hubble with the 100-inch Mount Wilson telescope, the Sculptor dwarf galaxy appears as an open swarm of faint stars. It and the similar system in the neighboring southern constellation Fornax are both members of the Local Group of galaxies. (Mount Wilson and Palomar Observatories photograph)

like a loose grouping of unusually faint stars arranged nearly evenly within a circular region. The confirmation plate had a much smaller scale, and there was just an unresolved smudge of dim light. Lying in the constellation of Sculptor, the object was called the Sculptor system (see Fig. 78).

When announcing this discovery, Shapley mentioned several possible interpretations. The system could be a nearby cluster of low-luminosity stars, a distant cluster or unexpectedly small galaxy of normal stars, or an exceedingly distant cluster of small galaxies. Favoring the second possibility, he described the Sculptor object [in *Telescope* for July–August 1938] as probably a new type of stellar system: "The final test of the nature of the images came when photographs were made with the 60-

inch telescope on Harvard Kopje. They showed definitely that the images are stellar, and that the system cannot be made up of galaxies.

"A careful count reveals that the group covers nearly two square degrees of the sky, and that within this area there are photographed about ten thousand stars belonging to it. How many more than that there may be, too faint to have left their mark on the photographic plate, we cannot say."

Soon thereafter, at Mount Wilson Observatory, E. P. Hubble and Walter Baade set about to study this problem, even though Sculptor is far south and difficult to observe from the Northern Hemisphere. On nights when the southern sky was exceptionally clear, they turned the 100-inch telescope almost to the horizon to obtain a series of plates that confirmed Shapley's discovery.

FIG. 79. The dwarf galaxy in Fornax, as recorded in blue light with the ADH Baker-Schmidt telescope at Boyden Observatory. North is at the top. The field is about 1.4° across, and Lambda² Fornacis is the bright star at the right. The system has five globular clusters, discernible on this photograph. (Courtesy Boyden Observatory)

In the meantime, at Harvard, assistants were searching through other plates for undiscovered objects of the Sculptor type. Not far away, in the constellation Fornax, a second system was found, but with even fainter stars (see Fig. 79). With the 100-inch, Baade and Hubble verified that this was a more distant counterpart of the Sculptor system.

When many plates of the Sculptor system had been taken, the Mount Wilson astronomers found about forty variable stars near magnitude 19.5. Since these are cluster-type pulsating stars of known intrinsic luminosity, their distances could be deduced from their apparent brightnesses. It turned out that the Sculptor system was 84,000 parsecs away (approximately 270,000 light years), placing it well outside of our own Galaxy.

Baade and Hubble, when publishing these results in 1939, stated that the new object was definitely a very small galaxy near ours, and that it resembled an enlarged globular cluster. Their statement that it was really not a new type of system at all was regretted later by Baade, who admitted in 1958 that the Sculptor system had been a potential clue to the riddle of two stellar populations, a concept that did not develop until years after this object was found.

The Sculptor system is, in fact, an aggregation made up entirely of what were later to become known as Baade's Population-II stars, but in the early days it was not recognized as such.

So distant is the Fornax object that its cluster-type stars, if it should contain any, would be too faint to be detected with the 100-inch telescope. However, there are a few globular clusters associated with the Fornax system, and Baade and Hubble based a distance determination on the brightness of the stars in these globulars. The distance was found to be a little more than twice the distance of the Sculptor system, or 188,000 parsecs.

Since both the Sculptor and Fornax objects are very faint and extended, it is not an easy matter to measure their total brightnesses. Shapley estimated very roughly from visual inspection of the Harvard plates that each has a total apparent magnitude of about 9. As of this writing, no more acccurate value of their apparent magnitudes has been published; if we accept Shapley's rough guess and use the Baade-Hubble distances, we find these galaxies are both dwarfs. The absolute photographic magnitude for Sculptor turns out to be only —10.6, and that for Fornax —12.4. That is, the Sculptor system has only 1/10,000 the intrinsic luminosity of our Milky Way Galaxy. More recent data on both the apparent magnitudes and the distances of these systems confirm these conclusions.

As indicated above, Shapley made extensive star counts of the Sculptor system to study its structure. A few years ago I counted stars on six

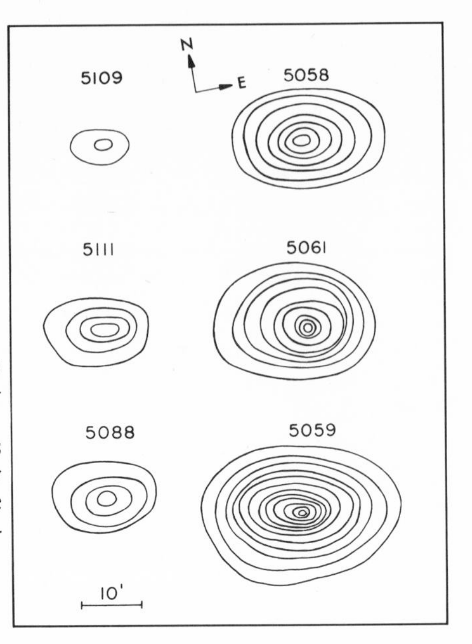

FIG. 80. Counts of stars on each of six Boyden Observatory photographs (numbered 5058 to 5111) of the Sculptor galaxy, plotted as contour lines of equal star density (number of stars per unit surface area). (*Astronomical Journal* diagram)

new plates taken with the ADH Baker-Schmidt telescope at Boyden Observatory, confirming Shapley's conclusion that the Sculptor system has a surprisingly large angular extent for so faint an object. The best of these plates reaches photographic magnitude 19.1, and in the richest portion of the galaxy the star density rises to eight stars per square minute of arc.

The counts show (see Fig. 80) that Sculptor is elliptical in shape, extending 1°.5 by 1°. The large apparent size, however, does not mean an intrinsically large galaxy. The linear diameter is only 2000 parsecs, based on the Baade-Hubble distance; thus Sculptor was the smallest galaxy known at the time of its discovery and measurement.

Our own Milky Way, as well as most of those galaxies familiar to us, is at least ten or fifteen times larger. A similar study of the Fornax system showed that it, too, is large in apparent size, its apparent length of almost 2° leading to a linear diameter of 3000 parsecs. In shape, Fornax is almost identical to the Sculptor system; both have a 0.65 ratio of minor to major axis.

The stellar distribution [density of stars in the various parts] in these two systems is strikingly different from that in normal elliptical galaxies. The reason for this difference appears to be their proximity to our very massive Milky Way Galaxy, which produces tidal effects in the dwarf systems that strongly influence their shapes. In both cases, the stellar distribution along the major axis falls off abruptly at just the point where stars would be pulled away from the dwarf system by tidal effects. (In 1957 S. von Hoerner at Bonn, Germany, showed how to calculate such limiting radii.)

Baade and Hubble suggested in their 1939 paper that the Sculptor and Fornax systems were rather similar to globular clusters. A recent investigation of the color-magnitude diagram for the Sculptor galaxy (Fig. 81) fully confirms this. Colors and brightnesses can be measured for only its brightest stars, but for these the diagram looks identical to that of a rich globular cluster [such as M 3; see Fig. 50].

From this diagram it has been possible to show that the chemical composition of most of the Sculptor stars is probably similar to those in Messier 3, which are moderately deficient in heavy elements. There are a few blue main-sequence stars that may be members of the Sculptor system, but they do not seem to be any different from the few unexplained blue stars found in globular clusters.

In fact, only in its exceedingly low star density does the Sculptor galaxy seem to differ noticeably from a normal globular cluster. Its total mass is estimated to be about fifteen times that of M 3. The total number of RR Lyrae (cluster-type) variables has been determined by A. D. Thackeray in South Africa to be about seven hundred, some five times the number in M 3.

The distance of the Fornax system is so great that resolution into stars is difficult except on plates taken with the largest southern telescopes. Only its brightest giant stars have been photographed and studied, the RR Lyrae variables being too faint. However, one unusual feature of the Fornax galaxy is obvious.

It contains five globular clusters, all very normal objects of their class. Why should Fornax have globulars while Sculptor does not? Probably because Fornax' mass is about ten times that of Sculptor; the mass estimates are 20 million suns for Fornax, and 2 million for Sculptor.

The brightest star in the Fornax galaxy proper is about half a magnitude brighter than the brightest members of its own globular clusters,

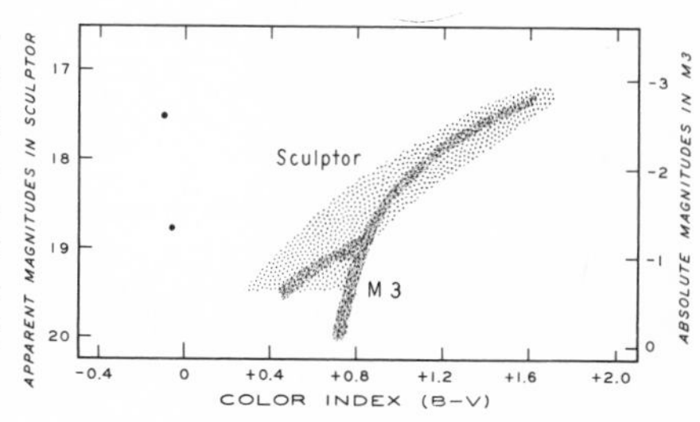

FIG. 81. The color-magnitude diagram for stars in the Sculptor galaxy superimposed upon a similar plot for stars in the globular cluster M 3. Note their close correspondence. The distance modulus $(m-M)$ for Sculptor is 20.1 magnitudes, corresponding to a distance of about 350,000 light years (see p. 60).

suggesting that perhaps there is a difference in composition between the dwarf galaxy and globulars. Or is there a spread in age within that galaxy, or still other causes? We can learn about this only from a color-magnitude diagram. To obtain the necessary data for it we need a really large telescope in the Southern Hemisphere.

The Fornax galaxy, which bridges the gap between dwarf galaxies near enough to be resolved and normal elliptical systems too far away to be studied in detail, may provide clues to the mystery of the composition, age, and evolution of elliptical galaxies.

New Dwarf Galaxy?

(*Sky and Telescope*, June 1964)

Recently, at Mount Stromlo Observatory in Australia, D. Sher found an unusual object in the Sky Survey atlas of the 48-inch Palomar Schmidt telescope. Appearing as a smudgy patch under two minutes of arc in diameter, this feature is located at right ascension 5^h10^m, declination $-33^\circ.0$, in the constellation Columba. A photograph taken with the Uppsala Schmidt at Mount Stromlo is reproduced in *The Observatory* for December 1963.

The Sher object has the very low surface brightness and lack of central condensation that characterize dwarf galaxies. Thus it is evidently an intrinsically faint, relatively nearby stellar system, but its distance is unknown. Although the discoverer states he will not study it further, the new dwarf should be within reach of several large reflectors in the southwestern United States.

Dwarf elliptical galaxies, like the Fornax and Sculptor systems, may be very numerous; they have such low surface brightness that nearby ones can easily be missed, and their total luminosity is so small that they cannot be detected at large distances.

There is another reason for missing galaxies in surveys of the sky; if they are starlike, they may be quite bright but so condensed that they give almost a point image in a telescope. Such "quasi-stellar" objects recently came into prominence because of their radio emission, to be described in Chapter 5.—TLP

Compact Galaxies

(*Sky and Telescope*, September 1964)

Using the large telescopes on Palomar Mountain, Fritz Zwicky has confirmed observationally his prediction that strange galaxies exist which are characterized by extremely high density and small size.

The Cal Tech astrophysicist based his anticipation of galaxies more compact than hitherto known upon considerations of statistical mechanics. Over an immensely long time interval, stars will tend to aggregate in ever denser clusters. He calculates that a compact swarm about ten light years across and containing a million million luminous stars could be formed in this way, before the radiation between the stars became so intense and the interstellar gas so agitated that a cosmic catastrophe would occur. . . . Such compact galaxies would be observationally almost indistinguishable from stars.

Zwicky searched plates taken with the 48-inch Palomar Schmidt and found such objects by the hundreds. On the average, there is one compact galaxy at least as bright as magnitude 17 in each five square degrees of the sky.

About thirty of these objects have already been studied with the 200-inch Hale telescope. They generally appear on direct photographs as tiny, sharp-edged disks of high surface brightness. Commonly they have exceedingly faint spiral arms, jets, or halos, and many of them occur in pairs or multiple systems.

Much variety is shown by the spectra of compact galaxies, some with remarkable properties. The spectral classes range from K to O, the last being most uncommon among normal galaxies. In some cases, the absorption and emission lines are greatly broadened, indicating internal motions of thousands of kilometers per second—an explosive expansion.

Such internal velocities would indicate masses of the order of 10^{10} to 10^{12} suns for these systems, whose luminosities are estimated at 10^{8} to 10^{10} times the sun's. Thus the mass-to-luminosity ratio is about a hundred times greater, in the more extreme cases, than that of normal spiral galaxies, indicating that some of the compact systems may contain large amounts of dark matter. In fact, Zwicky points out that an aggregation of dark material can ultimately become far more condensed than one of luminous matter (stars) before disruption occurs.

The existence of compact galaxies is important to cosmologists. For example, there is the well-known problem of the masses of clusters of galaxies. The mass derived from the sum of a cluster's members is typically less than the value deduced from the velocities inside the cluster. This difference might be explained, at least in part, by the nonluminous matter in compact galaxies.

Another suggestion by Zwicky concerns a possible connection between compact galaxies and quasars (quasi-stellar radio sources, such as 3C 273). Conceivably, a luminous compact galaxy may become unstable and change rather rapidly to a still more compact dark system, with the release of an enormous flood of energy.

One of the predictions of Einstein's general theory of relativity, formulated half a century ago, is that light will suffer a strong bending in the presence of an intense gravitational field. Because of their exceedingly high density, this "lens" action by compact galaxies (and some quasars) should be very strong—as much as one minute of arc deflection for light passing along the surface of a compact galaxy.

Therefore Zwicky is checking the neighborhood of each compact galaxy and of certain stellar radio sources for possible double or doughnut-shaped images produced by this relativistic effect on the light of even more remote galaxies lying practically in the same line of sight. . . .

Tiny Galaxy

GEORGE S. MUMFORD

(*Sky and Telescope*, August 1966)

While photographing unusual galaxies with the 200-inch telescope, H. C. Arp of Mount Wilson and Palomar Observatories chanced upon an object (see Fig. 82) that appeared, except for its fuzzy edges, to be a faint double star of magnitudes 17.8 and 17.9. Spectroscopic observations showed emission lines, displaced by amounts indicating a recession of 1330 kilometers per second.

Apparently this body is far outside our Galaxy, for its red shift corresponds to a distance of 13 million parsecs. That far away, the 1.1-second-of-arc diameter of each component indicates a linear diameter of 70 parsecs, and the absolute magnitudes are —12.7 and —12.8.

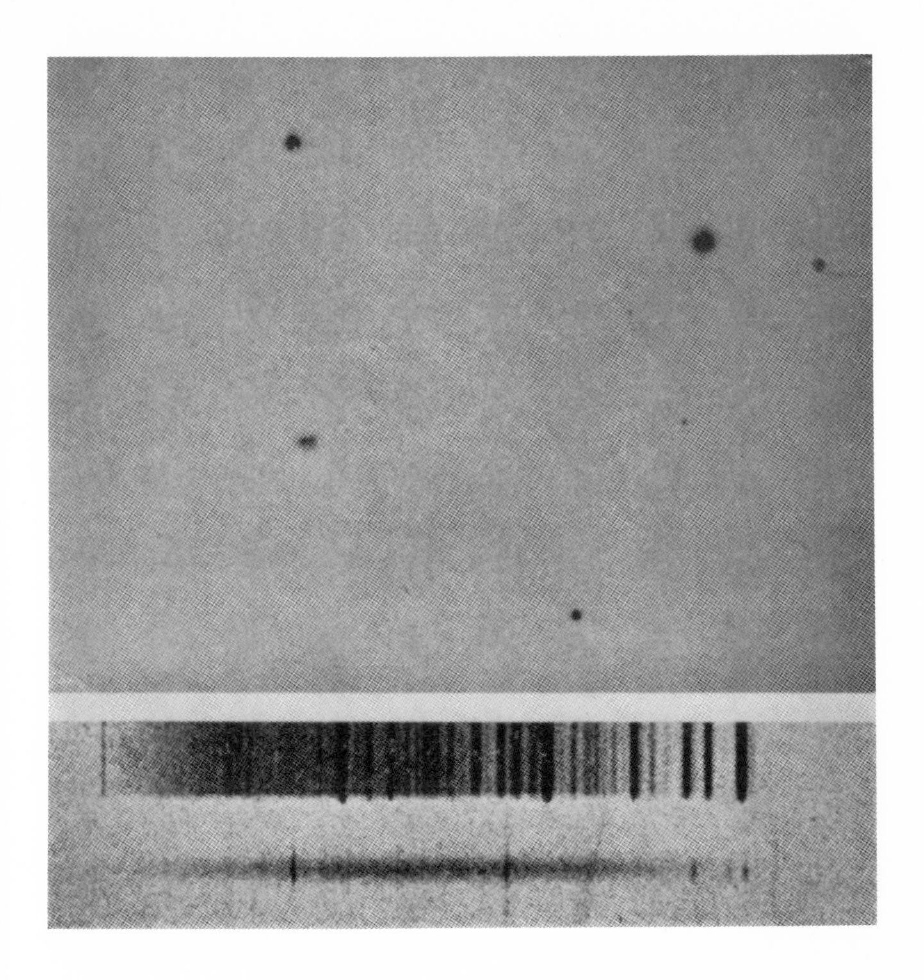

FIG. 82. In this direct photograph, a 3-minute exposure with the 200-inch Hale Telescope, Halton Arp's condensed galaxy is the double image at left center. Its spectrum appears beneath a comparison spectrum and shows emission lines of hydrogen and ionized oxygen. (From *The Astrophysical Journal*)

Hence this strange system is about ten times more luminous than a globular cluster, though of about the same size. On the other hand, it is about ten times fainter intrinsically than what is ordinarily called a dwarf galaxy, and is almost a hundred times smaller.

Arp believes that this puzzling object is best described as a very small, condensed galaxy. He suggests that galaxies of this kind are common but unrecognized because of their starlike appearance.

Fragments of Galaxies

GEORGE S. MUMFORD

(*Sky and Telescope*, June 1967)

In general, elliptical galaxies are many times more massive than spirals, and in some parts of the sky we see a single very large elliptical in association with one or more spiral or irregular systems. The members of such groups seem to be moving away from each other, as if expanding from an explosion. This possibility is discussed by José Luis Sersic of Córdoba Observatory, Argentina, in the March [1967] *Leaflet* (453) of the Astronomical Society of the Pacific.

He and H. A. Abt of Kitt Peak National Observatory have proposed that many galaxies may be fragments of larger masses that exploded some time ago. Statistical data indicate that it is the giant ellipticals which produce the fragments seen now as smaller elliptical, spiral, and irregular systems. Sersic writes:

"It therefore seems that in the early stages of the universe the only galaxies in existence were massive ellipticals. These undoubtedly in some way resulted from earlier cosmological evolution. Instabilities then developed in these high-density systems, causing explosions that ejected fragments of various sizes. . . . For a brief period during each explosion there was intense radio emission, which we now observe in the radio galaxies and quasi-stellar objects."

5

Quasars

New techniques of observing are expected to reveal new aspects of the universe, but radio astronomy outdid all other twentieth-century developments with the dramatic discovery of the quasi-stellar objects. Everything about them, from the quaint name "quasar" to their extremes of distance, density, and relativistic effects, has attracted the attention of the public, of mathematicians, and of physicists, as well as of astronomers. As recently as 1960 most of the older astronomers could not bring themselves to believe that radio telescopes were "seeing" parts of the universe previously invisible or unrecognized; as recently as 1960 the present concept of quasars would have been ridiculed as science fiction.

The story, an epic of the sudden expansion of our knowledge of the universe, starts with the discovery of extraterrestrial radio sources by U.S. radio engineers in the 1940s, and the recognition of small-size radio sources by Australian and British radio astronomers ten years later. Optical astronomers were slow to accept the extragalactic nature of small radio sources. Not until 1960 were these surveyed systematically with large optical telescopes.

One difficulty was caused by the poor angular resolution of radio telescopes. The radio astronomers could not specify accurate positions of the radio sources in the sky, and did not know, in fact, how small ("quasi-stellar") some radio sources were until radio measurements were made in 1962 as the moon occulted one. Larger radio "dishes" and long-base-line interferometers have since provided accurate radio directions, now matching those provided by optical photographs.

Another reason for delay in recognizing the astronomical significance of radio sources was the difficulty in obtaining data on the radio spectrum of any one source. Radio telescopes were constructed for surveys of the sky at one frequency. As other instruments were built for other frequencies, each could measure relative "brightnesses" of various sources with good accuracy, but the absolute calibration of each radio telescope was difficult and time-consuming. Until this calibration was done, the radio measurements at different frequencies could not be expressed in absolute units of "flux density," watts per square meter per cycle per second (w/m^2 cps, or $w\ m^{-2}Hz^{-1}$, where Hz stands for Hertz, the unit of frequency, one cycle per second). Of course, similar difficulties had existed in the 1920s, when accurate optical measurements were made of star colors and the distribution of energy in stellar spectra. In both cases, "standards"—stars or radio sources with accurately known spectra, to which other sources (radio or optical) could be compared—had to be established. Although "radio magnitudes" were used for a while, radio astronomers have wisely avoided the clumsy scale of optical magnitudes (defined by $m = 2.5 \log b_0/b$, where b is the optical brightness and b_0 that of a zero-magnitude star), and express their measurements of flux density (s) which corresponds to optical brightness. The flux density is measured in "flux units" (1 f.u. $= 10^{-26} w\ m^{-2}Hz^{-1}$) at frequency f.

In summary, the historical developments leading up to the discovery of quasars involved the work on galaxies discussed in previous chapters of this book, plus the following steps in radio astronomy, partly covered in Volumes 4, 5, 6, and 7 of this series:

1940s: Bell Telephone Labs' detection of extraterrestrial radio waves (Karl Jansky); identification of Milky Way as a source (Grote Reber); detection of solar radio emission (J. S. Hey and others in England).

1950s: Australian interferometer measures of small radio sources (B. Y. Mills); discovery of cold hydrogen emission at 21 cm (H. van de Hulst); British construction of large radio telescope (Bernard Lovell); Cambridge survey of small radio sources (Martin Ryle); Dutch survey of interstellar H-I (Jan Oort); synchrotron theory of radio spectra (I. Shklovsky); Cal Tech optical survey of small radio sources (Thomas Matthews).

1960s: Occultations of small radio sources (C. Hazard, in Australia); discovery of large red shifts in optical spectra of quasars (Maarten Schmidt); discovery of optical variability of quasars (Harlan Smith); discovery of radio-quiet blue galaxies (Allan Sandage); construction of theoretical models of quasars (J. L. Greenstein, G. R. Burbidge, W. A. Fowler, F. Hoyle).

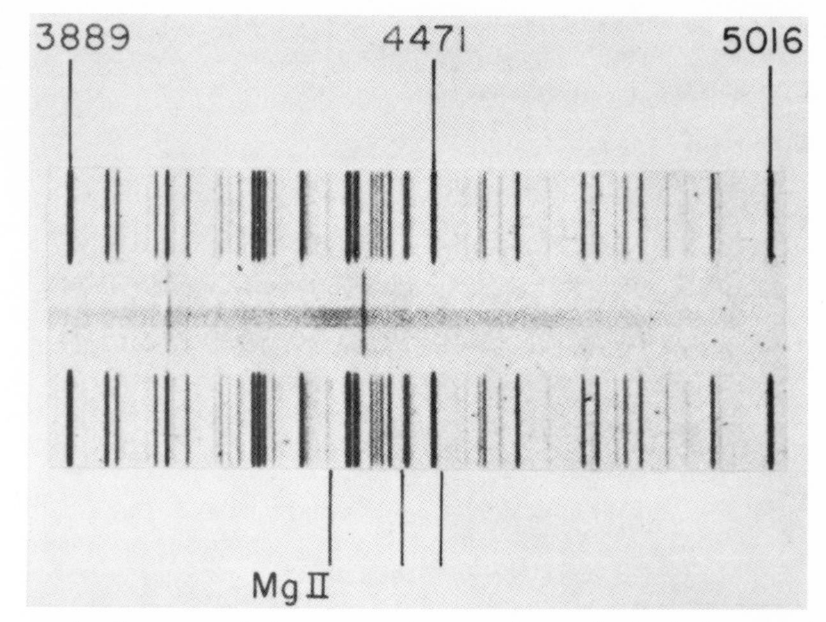

FIG. 83. Spectrum of the quasar 3C 279, taken with the Lick 120-inch reflector and prime-focus spectrograph. The fuzzy gray emission line in the center of this negative is due to ionized magnesium (Mg II), normally at 2798 A, red shifted to 4308 A ($z = 0.536$). The strong narrow line just to the right is atmospheric, and there are two other faint quasar lines to the right of it (marked at bottom). Wavelengths of lines in the comparison spectrum are marked at the top. (From *The Astrophysical Journal*)

The climax, in late 1963, was Maarten Schmidt's discovery of large red shifts in the spectra of 3C 273[1] *and other quasars. He measured several fuzzy emission lines (like those in Fig. 83) and identified them with spectrum lines far to the ultraviolet, lines that previously had never been observed because ozone in the earth's atmosphere is opaque to ultraviolet light. Red shifts ($z = \Delta\lambda/\lambda$) as large as 0.2 and 0.4, according to the Hubble law, imply very large distances, which further imply enormous luminosities of these small, dim quasars. When z is larger than about 0.1, the simple Doppler formula ($v = cz$) is no longer accurate, and the relativistic equation must be used:*

$$v = cz \frac{2 + z}{2 + 2z + z^2}$$

—TLP

[1] 3C 273 indicates radio source No. 273 in the *Third Cambridge Catalogue* of small radio sources.—TLP

Most Distant Object?

(*Sky and Telescope*, May 1964)

A small but intense radio source in the constellation Auriga is farther away than anything hitherto measured in the universe, according to a March 30 [1964] announcement by Mount Wilson and Palomar Observatories. This object is the quasi-stellar source 3C 147.

Maarten Schmidt and Thomas A. Matthews, who had identified 3C 147 with an eighteenth-magnitude optical feature, have studied its spectrum with the 200-inch telescope. Two definite and three probable spectral lines, due to ionized oxygen and neon, showed the largest red shift yet established for any celestial object. If the shift is interpreted as due to Doppler effect, the apparent velocity of recession is 76,000 miles per second, calculated from the relativity formula.

The previously most distant known source, 3C 295, has an apparent recession of 67,000 miles per second from the same formula. Since the numerical relation between red shift and distance is uncertain for extremely remote objects, Mount Wilson and Palomar astronomers have not announced specific distances. They state only that 3C 147 and 3C 295 are several billion light years away, and that the former is 10 to 20 per cent more distant than the latter.

Schmidt and Matthews also report spectrum observations of another very distant quasi-stellar radio source, 3C 47. Seven lines of hydrogen, ionized oxygen, neon, and magnesium indicate a velocity of 63,000 miles per second. For two other, similar objects, 3C 48 and 3C 273, the respective velocities are 56,000 and 27,000 miles per second.

Long exposures with the 200-inch telescope are required to record the spectra of these remotest sources. At best, the spectrograms show no features except traces of a few emission lines, whose identification is made even more difficult by the huge shifts in wavelength.

The two quasi-stellar sources, 3C 147 and 3C 295, are intrinsically the most intense emitters of radio energy in the sky. Thus far, more than a dozen quasi-stellar radio sources are known, and Matthews estimates that perhaps a quarter of the thousands of radio sources in the universe are of this kind. The origin of the enormous amounts of energy they radiate remains a challenging mystery.

Quasi-Stellar Galaxies

(*Sky and Telescope*, August 1965)

Many objects hitherto regarded as very faint blue stars inside our own Milky Way are actually a strange variety of extremely distant galaxies, according to Allan Sandage of Mount Wilson and Palomar Observatories. These objects (which he calls QSG, for *quasi-stellar galaxies*) differ from the famous quasars (quasi-stellar radio sources) in *not* being strong sources of radio emission.

In our Milky Way, the vast majority of faint stars are yellow and red dwarfs. But back in 1947 M. L. Humason and F. Zwicky (both of Mount Wilson and Palomar) pointed out that a sprinkling of abnormally blue stars as faint as magnitude 15 occurs near the north galactic pole. Very extensive searches made by W. J. Luyten (University of Minnesota) revealed that these dim blue objects occur by the thousands in many parts of the sky, to as faint as magnitude 19. At Tonantzintla Observatory in Mexico, G. Haro, B. Iriarte, and E. Chavira discovered many more of these oddities.

Up until this year, astronomers generally believed that the faint blue objects were stars, some of them being relatively nearby white dwarfs, the others members of an extensive halo surrounding the Milky Way Galaxy. Nevertheless, there were a few hints pointing toward an extragalactic nature. Already in 1947, Humason and Zwicky noted that No. 46 in their list was evidently a fifteenth-magnitude galaxy, and in 1958 Iriarte pointed out that another example, Tonantzintla No. 730, was recognizable on photographs as a galaxy rather than a star. Last year, Zwicky reported on the widespread existence of very compact galaxies, some of them blue, and even anticipated that they may be related to quasars [see p. 179].

Sandage's discovery is that there are two kinds of faint blue "stars." Those brighter than magnitude 14.5 are for the most part truly stellar, whereas the majority of fainter ones are enormously distant, superluminous galaxies. In *The Astrophysical Journal* for May 1965, he explains how he came to this conclusion.

During the last year or two he has been systematically photographing the positions of quasi-stellar radio sources, and identifying many of them by their excess ultraviolet brightness (see Fig. 84). But, curiously, he

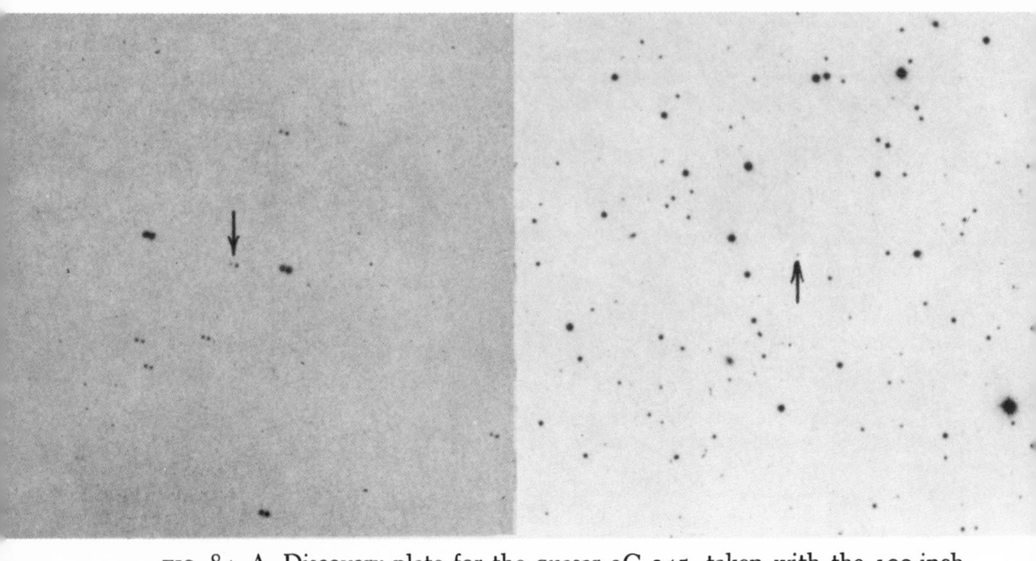

FIG. 84. A. Discovery plate for the quasar 3C 245, taken with the 100-inch reflector at Mount Wilson. This negative shows two images of each star, the left one in blue light (3900 A to 5000 A) and the right one in ultraviolet (3200 A to 3900 A). The arrow shows 3C 245, which has "UV excess" because the right (UV) image is much the stronger.
B. Quasar 3C 9, on a plate (right) taken with the Palomar 48-inch Schmidt. (From *The Astrophysical Journal*)

found many objects that imitate the ultraviolet excess of the true quasars, but are not near any radio-source position. A study of plates taken with the 48-inch Schmidt telescope showed that such "interlopers," down to a limiting magnitude of 18.5, averaged three per square degree. This suggested a connection with the Luyten-Haro objects, which occur with a frequency of four per square degree in high galactic latitudes to a limiting photographic magnitude of 19.

Sandage's second piece of evidence came from color measurements. For each object the magnitude was measured in ultraviolet light (**U**), in blue light (**B**) and in visual (yellow) light (**V**), and two different color indexes could be determined: **U** — **B** (ultraviolet minus blue) and **B** — **V** (blue minus yellow). It is possible to plot these in a two-color diagram, as shown in Figure 85. Sandage found that there was a startling difference when he made separate plots for those objects brighter than visual (**V**) magnitude 14.5 and those fainter.

In the former case, the plotted points fell very close to the curving line characteristic of normal stars; in the latter, most of the points were well above that curve, and instead lay close to the straight line charac-

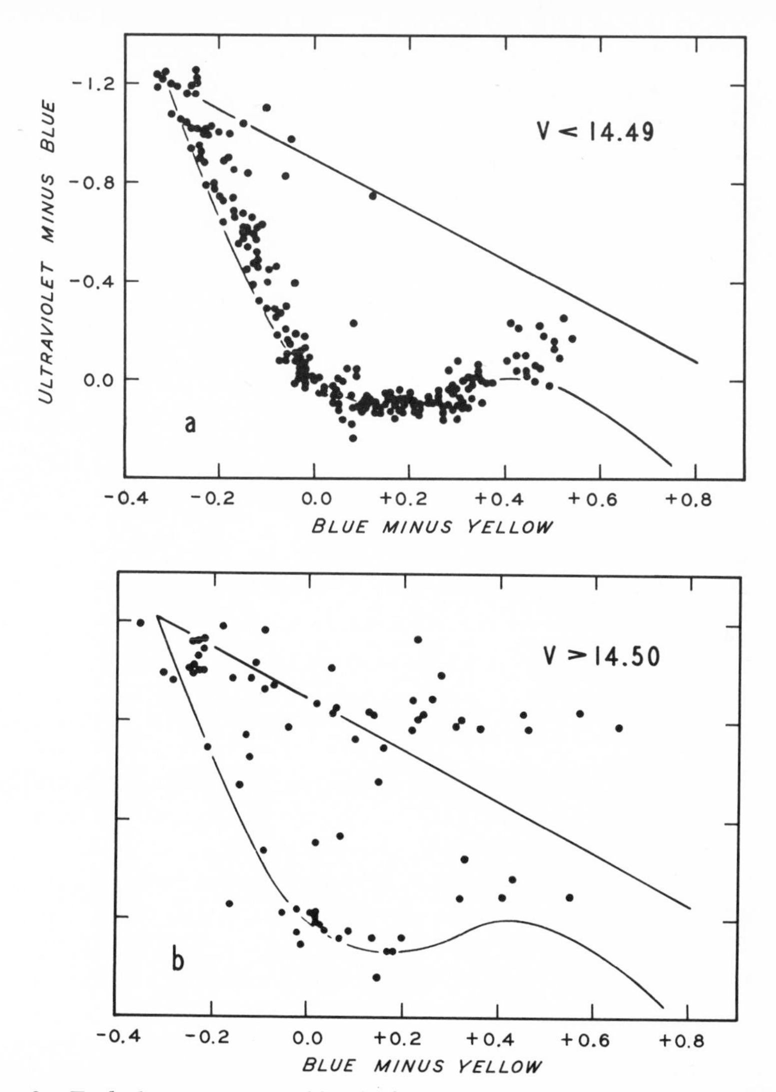

FIG. 85. Each dot represents a blue high-latitude object in these two charts, where **U—B** color index is plotted against **B—V** index. In the right-hand chart, for objects fainter than visual magnitude 14.5, many of the high points are identified by Allan Sandage as quasi-stellar galaxies (QSG). (Mount Wilson and Palomar Observatories charts)

teristic of quasars. The implication was clear: the brighter high-latitude blue objects are stars, the fainter ones quasarlike.

Supporting evidence came from counts to different limiting magnitudes for high-latitude blue objects, from the catalogue of 8746 of them by Haro and Luyten. Sandage's diagram (Fig. 86) shows a slow increase in their numbers between magnitudes 10 and 15, then a much more

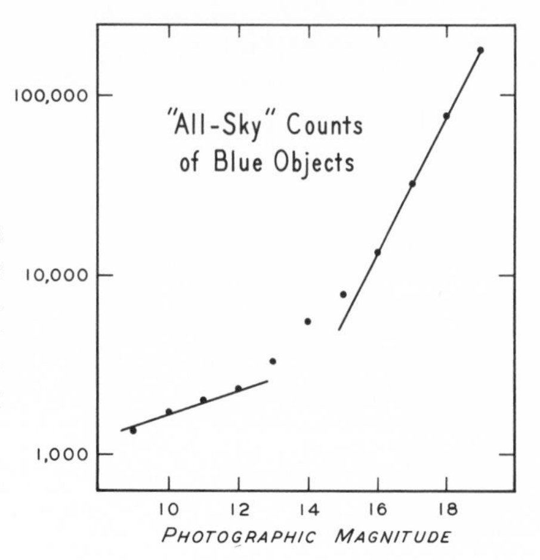

FIG. 86. The Haro-Luyten catalogue lists 8746 blue objects in one twentieth of the sky. Here are Allan Sandage's counts of the total numbers brighter than various magnitude limits, scaled up to represent the entire sky.

rapid rise between 15 and 19. The change in trend again indicates two different types of object. Moreover, the slope of the steep part of the curve matches rather well the slope predicted for analogous counts of quasars.

The statistics indicate that down to photographic magnitude 12 halo stars outnumber quasi-stellar galaxies by about twenty-five to one; by magnitude 15 their numbers are roughly equal; and at magnitude 19 the QSGs dominate by about twenty-five to one.

As a final check, Sandage and Maarten Schmidt measured the red shifts of six suspect QSGs with the Palomar 200-inch reflector during late April and early May [1965]. Of the six, one turned out to be a star, and two showed no lines. The remaining three, however, showed very large red shifts.

The first of these three, a seventeenth-magnitude blue "interloper," tentatively labeled BSO 1, showed two very broad emission lines, at 3473 and 4279 angstroms. Sandage identifies these as the carbon lines 1550 and 1909, shifted from the far ultraviolet to the observable region of the spectrum. The shift amounts to $z = 1.241$. In the spectrum of the second object, Tonantzintla No. 730, three oxygen lines, two of neon, and four of hydrogen could be recognized. These indicate a shift of $z = 0.0877$, with no ambiguity. Clearly this object is extragalactic, and if its distance is estimated from the ordinary velocity-distance relationship, its photographic absolute magnitude comes out about —22.2, exceptionally luminous for an optical galaxy.

The third object was sixteenth-magnitude Tonantzintla No. 256. Its spectrograms show three lines each of oxygen and neon, and four of hydrogen, with $z = 0.1307$, a very large shift for a normal sixteenth-magnitude galaxy.

These three objects resemble quasars in every respect, except that they are radio quiet.

Sandage gives some preliminary estimates of the properties of QSGs. There should be about eighty-three of them brighter than photographic magnitude 12 over the entire sky. From the red-shift data, the distance of a twelfth-magnitude QSG is about 230 million parsecs (based on a Hubble constant of 75 kilometers per second for each million parsecs). Inside a sphere of this radius, there are about 20,000 times as many normal galaxies as QSGs. Thus the latter are cosmically rare phenomena.

Nevertheless, they are much commoner than the quasars, by a factor of about five hundred. Sandage suggests that the quasars are in fact quasi-stellar galaxies going through a temporary stage of intense radio emission.

Sandage's article stirred some controversy about who had actually "discovered" blue stellar objects. In particular, Fritz Zwicky (also at Cal Tech–Mount Wilson) wrote to Sky and Telescope in December 1965:

". . . In the past few years I have reported on the widespread presence of very compact galaxies, some of them blue. It has been repeatedly stressed that many of the blue starlike objects found in large-scale surveys by G. Haro, W. J. Luyten, and myself are not galactic stars, but compact blue extragalactic systems, many of them radio quiet. Some time ago my former collaborator J. Berger of Paris Observatory showed that some quasars are identical with PHL (Palomar-Haro-Luyten) objects; notably, 3C 9, for which Maarten Schmidt has measured the largest red shift yet, is PHL 2871.

"The five spectra of blue quasi-stellar galaxies investigated by Sandage resemble in every respect those of some of the many blue compact galaxies on which I have reported. Thus his paper does not represent a discovery.

"As yet I know of no definite answer to the question of what percentage of all starlike faint objects in high galactic latitudes are compact extragalactic systems. The extensive statistical investigations of their apparent brightnesses, colors, and sky distribution that Luyten, Berger, and I have made in the past seven years indicate that most of the blue PHL objects are closer than 500 million parsecs. This contradicts Sandage's conclusion, from a much more limited analysis, that the majority of PHL objects are at greater distances than this."

There is still some disagreement about the distances of the blue stellar objects and the quasars (see p. 245), and the statistical investigations by Zwicky and others do not give as clear-cut conclusions as he implies. The PHL objects, selected for their blue colors on Palomar 48-inch Schmidt photographs, were catalogued by G. Haro (Director of the Tonantzintla Observatory near Mexico City) and W. Luyten (Professor at the University of Minnesota). These blue stars, numbering over 8000, are all at high galactic latitudes. Sandage later agreed that his estimate of the number of radio-quiet blue stellar galaxies was too high. Other spectra of blue objects, by T. D. Kinman at Lick Observatory, showed a smaller fraction to be galaxies with large red shifts. Thus, Sandage's estimates on page 191 should be reduced by a factor of one half or one third.

Nevertheless, there is a class of radio-quiet small blue galaxies, and their peculiar color provides a way to recognize very distant objects. The search, and the measurement of red shifts, continued to provide ever more striking results.—TLP

The Most Remote Objects Ever Identified

(*Sky and Telescope*, July 1965)

Five quasi-stellar radio sources that recently were studied optically with the 200-inch Palomar telescope have turned out to be the remotest known objects in the universe—both in time and distance. Each of the five, reports Maarten Schmidt, is considerably farther than 3C 147, which previously held the record.

The most extreme of the five is 3C 9, which on photographs appears as an eighteenth-magnitude speck of light in the constellation Pisces. From its spectrum Schmidt finds that it appears to be receding from the earth at 80 per cent of the velocity of light, or 149,000 miles per second!

All these five quasars were originally detected in surveys with large radio telescopes, four of them (3C 9, 3C 245, 3C 254, and 3C 287) at Cambridge University, England, the other (CTA 102) at California Institute of Technology.

There are only a few dozen known quasars, but the number is mounting steadily as the result of an intensive cooperation between radio and optical astronomers. In the case history of a typical well-observed example, a radio source first becomes a suspect because of its small angular diameter. The right ascension and declination are then measured as accurately as possible with a large radio telescope, and this position is inspected on long-exposure photographs taken with large telescopes.

Instead of the fuzzy image of a galaxy in this spot, there may be found a faint, practically starlike image, recognizable as a probable quasar from its ultraviolet brightness [see Fig. 84]. The color is checked by comparing photographs or by measurement with a photoelectric photometer. The final piece of positive evidence is a spectrogram showing a few emission lines, all strongly displaced toward longer wavelengths. Of the more than forty quasars known, nine now have measured red shifts.

Schmidt's five objects are all very faint, and it was difficult to obtain spectrograms of them even with the prime-focus spectrograph of the 200-inch telescope. Exposures of four to five hours were needed. All five spectra showed two or more emission lines, generally about thirty or forty angstroms broad. Since these lines were enormously shifted in wavelength, *from the far ultraviolet into the visible region,* their identification needed care.

Fortunately, the spectrum of 3C 254 contained as many as five observable lines, at 3314, 4845, 5948, 6466, and 6705 angstroms. Comparing these wavelengths with a list of thirty-seven lines that could conceivably be encountered, Schmidt could identify them with lines of carbon at 1910 angstroms, magnesium at 2798, neon at 3426, oxygen at 3727, and neon at 3869. These identifications are confirmed by the consistent results for red shifts, all five lines giving $z = 0.734$.

Similarly, Schmidt's spectrograms of 3C 287 showed three emission lines, at 3192, 3916, and 5753 angstroms, which could be attributed with assurance to 1550 (carbon), and 2798 (magnesium), red-shifted by $z = 1.055$.

Only two lines each were shown by 3C 245 and CTA 102. They could be assigned to features also present in the preceding objects, at laboratory wavelengths of 1909 and 2798 angstroms. The numerical red shifts for these quasars are $z = 1.029$ and 1.037, respectively.

The largest displacement by far is that of the two spectrum lines of 3C 9, measured at 3666 and 4668 angstroms. The former is actually the Lyman-alpha line of hydrogen, normally at 1216 A; the latter 1550 A (carbon). This is the first time that Lyman-alpha in an astronomical spectrum has ever been observed from the earth's surface. For 3C 9, the numerical red shift is fully 2.012.

When we are dealing with speeds that are only a small fraction that of light, we can compute the radial velocity of an object by the simple formula V = cz, where c is the speed of light and z the numerical red shift. This is appropriate for such relatively slow-moving objects as stars and the nearer galaxies. But with speeds that are a sizable fraction of c, this formula breaks down. Instead, the radial velocities of the quasars must be computed by a complicated formula based on Einstein's special theory of relativity [p. 185]. As mentioned earlier, the relativistically computed velocity of recession for 3C 9 is 149,000 miles per second. The corresponding figures found by Schmidt for the other four quasars are: 3C 254, 93,000; 3C 245, 113,000; CTA 102, 114,000; and 3C 287, 115,000 miles per second.

The larger the red shift, the greater the distance, but it is a difficult matter to specify the actual distances of the quasars. Only for much nearer objects (the majority of observable galaxies) is there a simple linear formula connecting distance with red shift: $D = Hcz$, where H is Hubble's constant and c the velocity of light. At very large distances, the relation is no longer linear, and a more complex equation is needed.

Tantalizingly, this equation is ambiguous, because it contains the so-called deceleration parameter, q_0, whose numerical value depends on our choice of a cosmological model. For example, if the universe matches the Steady-State model, q_0 is -1; for the Big-Bang model it is $+1$.

To the cosmologist, quasi-stellar sources offer the fascinating possibility of eventually enabling him to pick out the proper model. For this purpose, it is very desirable to discover as many quasars as possible and to probe their properties more deeply.

Both cosmologists and theoretical astrophysicists are currently much occupied with quasars. So remote are these objects that even their modest apparent brightnesses imply enormous intrinsic luminosities. A quasar may radiate 10 trillion times as much energy as the sun.

Some of the latest thinking on the origin of such stupendous floods of energy was presented in April [1965] to the National Academy of Sciences by W. A. Fowler, of California Institute of Technology. He has discarded his previous theory that the energy source of a quasar is gravitational collapse on a large scale. Instead, he now believes that thermonuclear reactions can suffice. He has calculated that the optical energy released by a quasar is well within the nuclear resources of a "star" 100 million times more massive than the sun. In such a hypothetical superstar, conversion of 7 per cent of its hydrogen to helium would provide the energy observed optically—about 10^{59} ergs.

Are Quasars Truly Remote?

GEORGE S. MUMFORD

(*Sky and Telescope*, June 1966)

Many astronomers believe that the quasi-stellar radio sources (quasars) are the most distant objects observed as yet, because of the enormous red shifts of lines in their optical spectra. However, several do not accept this view, among them the Mount Wilson and Palomar astronomers Halton C. Arp and Fritz Zwicky.

In the March 11 [1966] issue of *Science*, Arp reports a preliminary study of objects in his forthcoming *Atlas of Peculiar Galaxies*, in which he calls attention to giant elliptical systems, such as NGC 2397, that are surrounded by strongly disturbed matter. Arp finds that these galaxies show a tendency to be situated between pairs of radio sources 2° to 6° apart on the sky. Among the radio sources in nineteen probable and ten possible configurations of this kind, there are at least eight quasars! This implies that the quasars and the peculiar galaxies are physically related, and that their distances from us are comparable.

One of these eight quasars is 3C 273. On the opposite side of its associated elliptical galaxy is the very intense radio galaxy Virgo A (Messier 87). Arp suggests that both 3C 273 and Messier 87 were ejected from the central object many millions of years ago.

How, then, are the enormous red shifts of the quasars to be explained if these objects are at moderate distances? Interpreted literally as velocities of recession, these red shifts indicate speeds of up to 80 per cent that of light. Arp notes that the large red shifts might be the Einstein gravitational red shift, or Doppler shifts due to rapidly collapsing material, or they may be a new phenomenon.

The first alternative is a likelier choice, according to Zwicky. In *Comptes Rendus* of the French Academy of Sciences, he suggests that both quasars and certain radio-quiet, compact galaxies are so dense they exhibit strong Einstein gravitational red shifts. This, in turn, implies that quasars may be relatively close objects, cosmologically speaking.

The Einstein gravitational red shift is caused by a strong gravitational field. In order to produce $z = 0.5$ *or* 2.0*, the gravitational field must be enormous—as could possibly exist at the edge of a superdense star.*

The trouble is that the spectra of quasars show emission lines, which are known to be produced only in tenuous gas clouds. In the strong gravitational field near the surface of a superdense star there can be no tenuous gas cloud; a nebula would be compressed to a high-density skin showing no emission lines.

There is more to come on the problem of distances to the quasars. Oddly enough, the distance debate involves the variability of their brightnesses. Variability seems to be a characteristic of quasars and of radio-quiet quasi-stellar galaxies.—TLP

Quasar 3C 446 Erupts

GEORGE S. MUMFORD

(*Sky and Telescope*, September 1966)

In the most spectacular instance yet of optical variability in a quasi-stellar radio source, 3C 446 has recently brightened by 3.2 magnitudes, according to Allan R. Sandage, Mount Wilson and Palomar Observatories.

This object is located in Aquarius, at right ascension $22^h23^m11^s$, declination $-5°12'.0$ (1950 coordinates). Late in 1964 at Mount Wilson, J. D. Wyndham first located its blue, starlike image on photographs. With the 200-inch reflector, Sandage measured its photoelectric **V** magnitude on October 3 and 5, 1964, as 18.42 and 18.36, respectively. But in 1966 his measures gave 15.14 on June 24 and 15.27 on July 12.

The outburst must have occurred after September 1965, when Maarten Schmidt, in the course of spectroscopic observations, estimated 3C 446 as not brighter than eighteenth magnitude.

Brightness Variations of Quasars

GEORGE S. MUMFORD

(*Sky and Telescope*, June 1967)

A number of astronomers are currently studying the light variations of the optical objects associated with quasi-stellar radio sources. One quasar of special interest is 3C 446 in Aquarius, first recognized on photographs in 1964 as a blue object resembling an eighteenth-magnitude star. It attracted wide attention last summer when Allan Sandage found that it had brightened to magnitude 15.2.

In England, at the Royal Greenwich Observatory, R. D. Cannon and M. V. Penston have been photographing 3C 446 with the Herstmonceux 26-inch refractor, as part of a program of monitoring the brighter quasars. Since last October A. J. Wesselink and J. Hunter, Jr., have used Yale University's 40-inch reflector at Bethany, Connecticut, for photography of this same object. And at Palomar Mountain in California, J. B. Oke has made photoelectric measurements of both 3C 446 and 3C 279 with the prime-focus spectrum scanner on the 200-inch reflector.

The nineteen Herstmonceux plates of 3C 446 between August 1966 and January 1967 show that it varies between magnitudes 15.9 and 17.0. Cannon and Penston noted three maxima, all of equal brightness, whereas the minima were unequal. From this they conjecture that the drops in brightness may be caused by absorbing clouds intermittently masking a compact light source.

The British astronomers have measured the position of this quasar on four of their photographs, relative to seven nearby stars, to detect any possible shift in the quasar's location between the times it is bright and faint. [Such a shift might occur if one small off-center portion of the quasar brightened while the remainder remained constant. The center of light of the small image would then shift as the off-center portion brightened.] No such shift as large as 0.1 second of arc could be found.

The Yale observers report in *Science* [for April 7, 1967] that significant variations in the light of 3C 446 can occur over intervals of a day. This is also the experience of T. D. Kinman at Lick Observatory, who has been studying this object with the 120-inch reflector.

On July 19, 1966, Oke found that 3C 446 was half a magnitude fainter than a week earlier. It continued to fade by 0.1 magnitude per day while becoming redder. Oke also reports that 3C 279 has fluctuated by nearly two magnitudes during a little more than a year, on one occasion brightening by 0.25 magnitude in twenty-four hours. This object, like 3C 446, becomes moderately redder when it is fainter.

One implication of these studies is that the time scale for substantial optical changes in the continuous spectrum of quasars is one or two days. From consideration of the travel time of light within the quasar, Oke suggests that a large part of the continuous optical radiation comes from a region about 2000 astronomical units across.

Probably the most significant of Oke's findings is that emission-line intensities remain constant while the continuous spectrum of a quasar changes. Perhaps the small central region providing the continuum is surrounded by a large nebula producing the emission lines. This view is consistent with that held by the British astronomers.

Sorting Out Faint Blue Objects

(*Sky and Telescope*, November 1967)

Photographic surveys have shown the existence of thousands of faint blue starlike objects in high galactic latitudes. These are a tantalizing mixture of quasars, other peculiar extragalactic objects, and stars in our own Galaxy. There has been a good deal of discussion about the exact proportions of nearby stars and distant galaxies [see p. 191].

In Prague [at the XIII General Assembly of the International Astronomical Union in late August 1967] East German astronomer Nikolaus B. Richter proposed a simple method for preliminary sorting, based on the fact that the blue starlike objects tend to vary in brightness.

If a series of photometric observations shows a light curve typical of some well-known type of variable star, the object is a star belonging to our Galaxy. If there are very rapid oscillations (0.5 magnitude or more in a night), the object is presumably of stellar dimensions, and hence a member of our system. [If it fluctuates within one hour, an object can be no larger than 1 light hour, or 670 million miles.] If the light variations are slow, the candidate can be either galactic or extragalactic. Finally, if a well-observed object shows *no* variation, it is likely to be extragalactic, and should be tested for red shift.

Using photographs taken with the 52-inch Schmidt telescope of Karl Schwarzschild Observatory, Richter applied this test to a sky field in Andromeda containing thirty-seven known blue objects. He had ultraviolet, blue, and yellow plates covering a span of about four years, which were measured with an iris photometer. Three of the objects are clearly galaxies and one is a close double star. Of the remaining thirty-three, nineteen are variable, ten suspected to be variable, and only four appear constant. Of the nineteen variable objects, nine changed rapidly, and hence are stars.

Since optical studies of quasars and radio galaxies require the use of giant reflecting telescopes, Richter's method for screening the thousands of possible cases should be of much value.

Variable Radio Emission from Quasars

(*Sky and Telescope*, November 1965)

Optically, the brightest of the quasi-stellar sources is 3C 273 in Virgo, which appears about magnitude 12.5 on photographs. Hence there are images of it on thousands of photographs taken since the late 1880s at Harvard, Heidelberg, Moscow, and other observatories. When the optical variability of 3C 273 was discovered in 1961, Harlan Smith and Dorrit Hoffleit at Yale were able to trace the light variations on Harvard photographs taken during the preceding seven decades. There are slow oscillations of about half a magnitude amplitude which require a decade or so per cycle, overlaid by more rapid lesser fluctuations of irregular character.

The question of whether analogous radio variations take place had to await new observations, and their discovery has now been announced by W. A. Dent of the University of Michigan. Beginning in July 1962, the Michigan 85-foot radio telescope has been used for repeated measurements of the flux densities of some thirty-five cosmic nonthermal sources, at a wavelength of 3.75 centimeters (frequency 8000 megacycles per second).

These observations show that the radio emission of 3C 273 increased more or less steadily by about 40 per cent in almost three years. In each of the forty-seven separate sets of measurements, comparison was made with the intense source Virgo A, only about 10° north of the quasar. As a check on the constancy of Virgo A, it was frequently compared with Cygnus A, without finding significant change.

3C 273 is a double radio source, with components about 20 seconds of arc apart: 3C 273A is an optically faint jet, and 3C 273B is the quasi-stellar object. Their radio spectra differ in such a way that at a wavelength of 3.75 centimeters most of the flux comes from the B component, which is evidently the seat of the variation.

For ten months in 1963 Allan Sandage at Palomar Mountain measured 3C 273B at three wavelengths with a photoelectric photometer on the 200-inch telescope. While the radio flux was increasing, the optical brightness faded 0.2 magnitude.

The Michigan survey indicates that two other quasars (3C 279 and 3C 345) are also radio variables. During the past year, each has decreased relative to Virgo A by nearly 20 per cent. At high frequencies, both resemble 3C 273B in having a flat-topped radio spectrum (spectral index near zero), and Dent conjectures that this may actually be a general property of quasi-stellar objects that are radio variable.

Summarizing his findings, in *Science*, June 11, 1965, Dent comments: "The time scale of the observed radio variation of 3C 273B is short and is probably about the same as the thirteen-year period of the optical variations. Since large fractional variations in the emission must occur over a time scale greater than the 'light-travel' time through the source, the upper limit to the linear size of the radio component of 3C 273B is less than about thirteen light years, or four parsecs. Thus knowledge of the angular diameter of the varying component will give an upper limit to the distance of the source."

Before Dent announced his results, no variation in radio emission had been established for any extragalactic source, with the highly doubtful exception of CTA 102, a quasar in Pegasus. Early in 1965 the Soviet radio astronomer G. B. Sholomitskii claimed that CTA 102 underwent large intensity changes with a 100-day period, from his observations at 940 megacycles per second.

His finding has been criticized in *The Astrophysical Journal* [July 1965] by P. Maltby and A. T. Moffet of Cal Tech, who point out that it appears to rest on only nine observations, of limited accuracy. Moreover, measurements at nearly the same frequency were made between 1959 and 1961 at Cal Tech's Owens Valley Radio Observatory; these show that the flux density of CTA 102 was constant to within ±4 per cent during those particular years.

Cooperation in Observing Variable Radio Sources

GEORGE S. MUMFORD

(*Sky and Telescope*, May 1968)

Many extragalactic radio sources (particularly the quasars) are variable over an extremely broad span of wavelengths: optical, infrared, millimeter, and radio. Years of observational data may well be needed to unravel the complexities of these variations.

Clearly, it would be advantageous for observers to concentrate their coordinated efforts on a limited number of carefully selected sources. A working list was compiled last November [1967] at a conference in Ann Arbor, Michigan, where nineteen astronomers of American and Canadian observatories met at the invitation of F. T. Haddock (University of Michigan) and T. Kinman (Lick Observatory).

Ten sources were selected for continuing intensive study, most of them already having shown marked variability (Table 4). To obtain

TABLE 4. SELECTED VARIABLE RADIO SOURCES

Designation	*R.A.* h	m	*Dec.* °	′	*Type*	*Mag.*	*Chart*	*P*
3C 84*	3	16.5	+41	20.0	S	12.8	———	4
CTA 026	3	37.0	−01	56.3	Q	19.0	**147**, 848	9
3C 120	4	30.5	+05	15.0	S	14.7	———	3
3C 273	12	26.6	+02	19.7	Q	13.0	**140**, 1	1
3C 279	12	35.0	−05	31.2	Q	15-18	**141**, 328	7
PKS 1510-08	15	10.1	−08	54.8	Q	16.6	**145**, 951	10
3C 345	16	41.3	+39	54.2	Q	15-17	**141**, 328	2
3C 371	18	07.3	+69	49.0	N	14-15	**144**, 459	8
3C 454.3	22	51.5	+15	52.9	Q	15-17	**144**, 1,234	5
3C 446	22	23.2	−05	12.3	Q	15-18	**141**, 328	6

The positions (*R.A.* and *Dec.*) are in 1950 coordinates. *Type:* S, Seyfert galaxy; Q, quasi-stellar source (quasar); N, N-type galaxy. *Magnitude* is approximate photographic. *Chart* references are to *The Astrophysical Journal*. *P* = priority on observing program.

* 3C 84 is also known as NGC 1275 and as Perseus A.

sufficient time resolution, it was recommended that radio observations at wavelengths 10 to 20 cm should not be more than a month apart, while infrared and optical observations should be made daily. The light variations of the brighter of these sources can be monitored photographically with relatively small telescopes. . . . Radio instruments specially designed for patrolling these variable sources would be advantageous. The University of Michigan group proposed a fixed-elevation-antenna telescope (FEAT) of 50-meters (about 165 feet) aperture. Designed to have a very stable gain, such a telescope could measure flux density and polarization of one hundred sources each day, in the wavelength range 9 millimeters to 15 centimeters.

These many reports, some of them in conflict, show that quasars received a great deal of attention from 1965 to 1968, and that many quasars vary both in optical brightness and in radio flux. The rapidity of these variations ("time scale" of a few days) implies that a small region is varying—a region with dimensions equal to a few "light days," or 0.01 light year, or 1000 a.u., or about 100 billion miles. A galaxy is usually 10,000 to 100,000 light years in diameter, and the first idea, as described above, was that only part of it varied. This implies some structure—nucleus, disk, halo, and jets, perhaps. Of course, as the term "quasi-stellar" implies, the small angular size of quasars makes any study of structure difficult, especially so because radio telescopes have poor resolving power.

The "radio scintillation" described below is analogous to the twinkling of stars and does not refer to real brightness changes or variability. Scintillation indicates small angular diameter. In order to measure the angular diameter, an even better scheme is to watch the moon occult a quasar.—TLP

Twinkle, Twinkle, Little Quasar

GEORGE S. MUMFORD

(*Sky and Telescope*, January 1965)

Recent studies at Cambridge, England, show that several quasi-stellar radio sources fluctuate randomly in intensity from second to second, while other sources, of larger angular diameter, do not. This investigation is described in *Nature* by A. Hewish, P. F. Scott, and D. Wills, who report that normal effects in the earth's atmosphere cannot explain their observations.

Possibly this scintillation arises from irregularities in the interplanetary medium, like those known to exist and to scatter radio waves in the outer solar corona. Such scintillation would only be important for sources having angular diameters of less than half a second of arc, when observed at wavelengths greater than one meter. If this interpretation proves correct, the Cambridge University astronomers may have found a new technique for recognizing radio sources of small angular size, as well as for studying variations in the interplanetary medium.

Structure of 3C 273

GEORGE S. MUMFORD

(*Sky and Telescope*, August 1965)

When the moon occults a radio source, the moving lunar limb serves as a high-resolution scanning device. Hence the occultation of a quasi-stellar source can provide valuable information about its structure. However, one complication arises for sources less than about ten seconds of arc in size: the recorded changes in intensity are dominated by diffraction at the moon's edge.

P. A. G. Scheuer, Cambridge University, has obtained a practical solution to the mathematical problem of reconstructing the original distribution of brightness across a small-diameter source. In a recent issue of the *Monthly Notices* of the Royal Astronomical Society [**129**, 199], he applies his methods to occultations of the important source 3C 273 observed by C. Hazard on August 5 and October 26, 1962.

It was previously recognized that 3C 273 is double. Scheuer's analysis of recordings at 1410 megacycles shows that 3C 273A is about five seconds broad, with a core 0.9 second in diameter. The B component has a diameter of 0.5 second. There are uncertain indications of other nearby, weaker components of small size.

Structure of Radio Sources

(*Sky and Telescope*, March 1967)

At the Owens Valley Radio Observatory operated by California Institute of Technology, mapping of the detailed structure of cosmic radio sources is a major activity. For these observations, twin 90-foot steerable antennas (radio telescopes) are used together as an interferometer with variable spacing. At the December 1966 meeting of the American Astronomical Society, Edward B. Fomalont reported the results of a study of 532 sources, of which 514 are extragalactic, at a frequency of 1425 megaHertz, which is a little less than the frequency of the neutral hydrogen 21-cm line.

The observations were made at nine spacings of the two antennas along an east-west line, ranging from 144 to 2626 wavelengths (100 feet to a third of a mile). From a combination of the nine measurements, the east-west brightness profile of each source could be deduced. It was possible to interpret without ambiguity source structures as complicated as a triple.

According to Fomalont, three fifths of the sources were unresolved (under 40 seconds in diameter), and the rest were about equally divided between partially and fully resolved. Of the completely resolved sources, five were single, thirty-two double, thirty triple, four complex, and twenty-seven consisted of a halo and core.

In a typical double source, the two components have nearly equal intensities and sizes, and are placed nearly symmetrically on either side of the optically identified galaxy. A halo-core source consists of a small radio core, probably complex, that coincides with a galaxy, and a much larger, usually simple, region of emission. This halo may not be symmetrical around the core, and in some cases is quite displaced from it.

Halo-core sources are more commonly associated with normal elliptical galaxies than with peculiar ellipticals or quasi-stellar objects. Many of the triple and more complex sources are genetically related to double and halo-core varieties.

In discussing the theoretical significance of these observations, Fomalont described a model for the evolution of a double source, in which the components' separation grows about five times as fast as their diameters. He also mentioned reasons for believing that a halo is of the same age as its core, instead of being the result of a previous outburst, as is commonly supposed. One finding of importance is the relative scarcity of simple sources characterized by one region of emission.

Radio Interferometers with Very Long Base Lines

(*Sky and Telescope*, September 1967)

Measuring the radio diameters of quasars is one of the most important and difficult tasks facing radio astronomers today. The vanishingly small angular diameters of these cosmic radio sources call for extremely high resolving power. Yet the resolving power of a telescope of specified

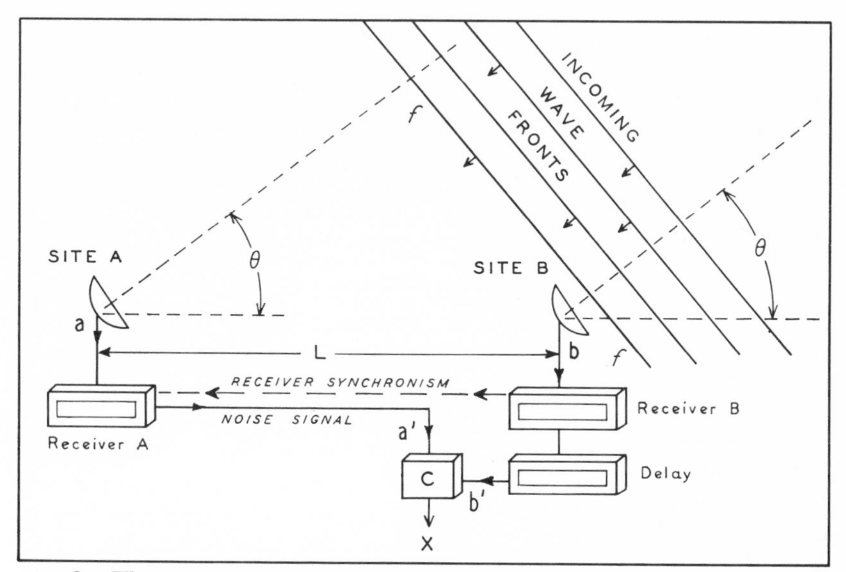

FIG. 87. The manner in which two dish-antennas, at *A* and *B*, can be used together as a long-base-line radio interferometer, when the two stations are linked. Now, at distances too great for direct linkage, atomic clocks are used to synchronize the receivers *A* and *B*.

aperture is, in principle, inversely proportional to wavelength, and radio astronomers work with wavelengths often a million times those of visible light. Hence radio telescopes of prohibitive size would be called for, and radio interferometers are used instead.

In place of a single huge dish many miles across, the interferometer consists of two smaller dishes separated by a corresponding distance, with provision for comparing the phases of the signals received at the two sites. [Figure 87 shows that waves from exactly the same direction (a point source very far away) will arrive at Site A later than at Site B. The combined signal at C can be sharply cancelled by interference when the phase delay is adjusted. But if there are two sets of waves from slightly different directions (the angular diameter of the source), the interference will be destroyed. The angular diameter can be calculated from the delay adjustment.]

In early experiments, two radio telescopes were connected by coaxial cable or by microwave radio link to permit this phase comparison. In 1965, antennas at Jodrell Bank and at Defford, eighty miles away, were joined by microwave links, and measurements at 6-cm wavelength established that several quasars are smaller than 0.025 second of arc.

But with conventional interferometers using cable or microwave links, it is difficult to preserve the phases of the signals for distances greater than a hundred miles, and it is troublesome to compensate for the relatively large and variable time delay between signal receptions at the two telescopes.

An exciting breakthrough, permitting much longer base lines and superior resolution, was described at the Yerkes meeting of the American Astronomical Society [in June 1967]. J. L. Locke of the National Research Council of Canada reported on work done with eight Canadian colleagues at the Dominion Radio Astrophysical Observatory (DRAO), Queen's University, and the University of Toronto. B. F. Burke of Massachusetts Institute of Technology spoke for an eight-man team that included scientists at Lincoln Laboratory and the National Radio Astronomy Observatory (NRAO).

In Canada, an interferometer was formed by radio telescopes 1910 miles apart: the 150-foot dish of Algonquin Radio Observatory at Lake Traverse, Ontario, and the DRAO 84-footer near Penticton, British Columbia.

These telescopes were not linked. Instead, each had its own atomic clock (a very accurate rubidium frequency standard), and incoming radio flux, as well as the atomic clock "ticks," recorded on a wide-band magnetic tape recorder. In this way the phase relationship between the two sets of signals was preserved. Later the two recordings were played back together (with a suitable very small time offset) to obtain a tracing of the interference fringes illustrated in Figure 88.

In this manner, the Canadians observed quasars 3C 273B and 3C 245 at a frequency of 448 megahertz (wavelength 67 centimeters). Although the base line was 3074.359 kilometers (about 4.6 million wavelengths), the fringes were strong, indicating that the sources both remained unresolved (as if they were point sources). Their diameters at this wavelength are therefore less than 0.02 second of arc. A more stringent test would have been possible at shorter wavelength, but a long wavelength was chosen because some quasar theories predict that the diameter should decrease with frequency.

In addition, nine quasars have been measured at 448 megaHertz with antennas 114 miles apart—the 84-foot Algonquin telescope and the 60-foot dish of the Defence Research Telecommunications Establishment, near Ottawa. These sources turn out to be at most 0.3 second in size.

The American method was similar, but applied to an interstellar hydroxyl cloud (OH molecules that emit 18-cm radio waves) in our Galaxy, rather than to a quasar. The two telescopes used as an interferometer were the 120-foot Haystack Hill antenna in Massachusetts and the NRAO 140-foot in West Virginia, 525 miles apart. Each

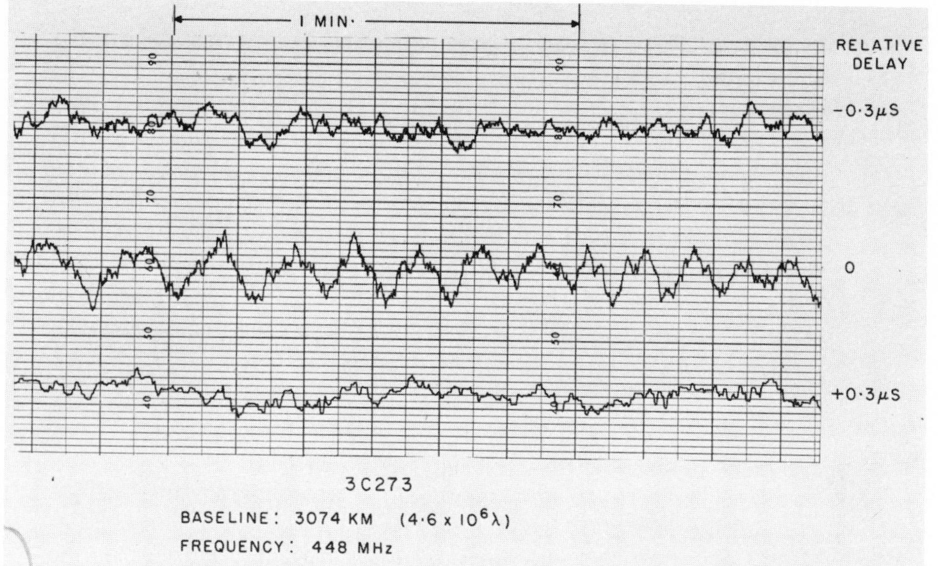

FIG. 88. The middle record shows radio-interference fringes of the quasar 3C 273B, formed by combining the outputs of two independent antennas, 1910 miles apart. The records above and below it (one signal delayed 0.3 microseconds more, the other 0.3 microseconds less, than the other) show that even a very small phase change blurs the fringes.

station had its own atomic frequency standard, synchronization being achieved through the Loran-C navigation system.

The particular object observed, a radio source named W3, is a knot of OH with a radial velocity of —45.1 kilometers per second. At the working frequency of 1665 megaHertz, the interferometer base line was 4.7 million wavelengths, yet the persistence of fringes indicated non-resolution. "This implies that the apparent angular size of the object is less than 0.01 second of arc," Burke announced.

In *Science* for July 14, 1967, C. Bare of NRAO and four collaborators report a third successful test of radio interferometry with independent local frequency standards. They used telescopes at Green Bank, West Virginia, and at Maryland Point, Maryland, 137 miles apart, and found that four quasars were unresolved (angular diameters less than 0.″05).

The future possibilities of the new technique are virtually unlimited, the Canadian scientists point out. "There is nothing to prevent simultaneous observations being made soon from points at opposite sides of the globe . . . some 8000 miles apart. A decade from now, they may be made simultaneously from two observatories, one on the earth, and one 240,000 miles away on the moon."

With this use of widely separated radio telescopes, radio resolution has been improved to match or exceed the best possible on optical photographs. The results show that several quasi-stellar radio sources are less than 0.001 second of arc in diameter, corresponding to 20 light years if they are at distances of about 4 billion light years, and it may be possible to push this down to two light years (but not to the "few light days" indicated by the light variations, p. 197).

Spectrum analysis can also show evidence of quasar structure—both in optical and radio wavelengths.—TLP

Spectra of Quasars

(*Sky and Telescope*, June 1966)

Because of the faintness of almost all quasi-stellar objects (quasars), their spectra have been difficult to photograph. But now many excellent spectra have been obtained by image-tube techniques with the 72-inch reflector at Lowell Observatory, and with the 84-inch reflector at Kitt Peak National Observatory. This work was reported by W. Kent Ford, Jr., and Vera C. Rubin, of the Department of Terrestrial Magnetism, Carnegie Institution of Washington.

They used a cascaded image tube [electronic light-amplifier] developed by the Radio Corporation of America for the Carnegie Image Tube Committee. The tube is incorporated into a simple Cassegrain spectrograph. In thirty minutes, the unwidened spectrum of an object as faint as visual magnitude 18.0 can be recorded at a dispersion of 425 angstroms per millimeter. The spectrograms extend from 4500 angstroms to 8500, and hence include the red region where normal photographic plates are relatively insensitive.

Figure 89 shows image-tube spectrograms of three relatively bright objects. NGC 1068 is not a quasar but a tenth-magnitude Seyfert galaxy—that is, a peculiar galaxy with an emission spectrum akin to a quasar's. Because it is a nearby object (as intergalactic distances go), its hydrogen-alpha line is very little red-shifted from its rest wavelength of 6563 angstroms. In 3C 273, the quasi-stellar object with the smallest known red shift ($z = 0.158$), the hydrogen-alpha line is shifted to about 7600 angstroms.

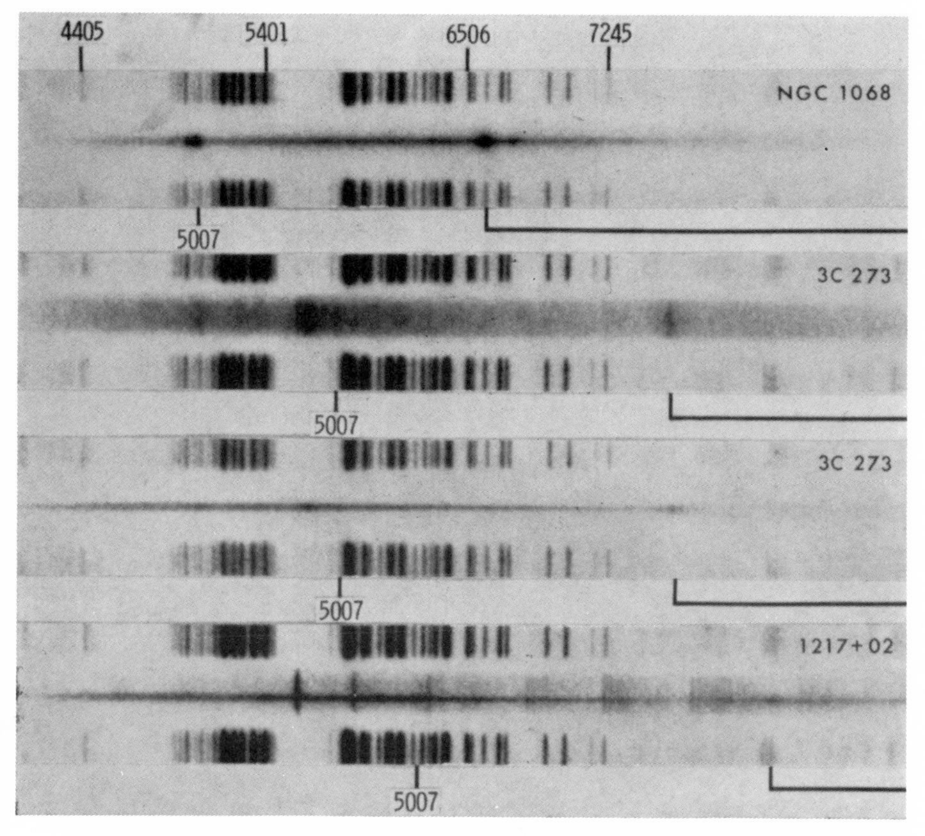

FIG. 89. Image-tube spectra of a peculiar galaxy (NGC 1068) and two quasars (3C 273 and 1217 +02), each with helium-neon comparison spectrum above and below it. This is a negative reproduction, emission lines appearing dark. The upper spectrum of 3C 273 has been artificially broadened. Pen marks (right) indicate the position of the red-shifted hydrogen-alpha line. In the 1217 +02 spectrum, the three strong, long lines at 5577 A, 5892 A, and 6300 A and the bands to the right are terrestrial. (Carnegie Institution photograph)

The last of the four spectrograms is of the source 1217 +02. This quasi-stellar object (discovered by J. G. Bolton in Australia) is within 2° of 3C 273 on the sky, but at magnitude 16.5 is about four magnitudes fainter. In its spectrum, the hydrogen-alpha line is displaced to 8146 angstroms, well into the infrared. Ford and Rubin have determined the red shift of 1217 +02 as 0.240, from measurements of nine lines. For six other quasi-stellar objects, having no previously published red shifts, five or six spectrograms each were obtained.

Radio-Source Spectra at Centimeter Wavelengths

(*Sky and Telescope*, April 1965)

Recently the radiation of thirty-five nonthermal cosmic sources has been measured at a wavelength of 3.75 centimeters (8000 megacycles per second frequency) with the University of Michigan's 85-foot radio telescope. Comparison with previous measurements at longer wavelengths showed, in most cases, the expected steady drop in intensity with increasing frequency.

Two notable exceptions were the radio sources 3C 84 (NGC 1275) and 3C 279, which W. A. Dent and F. T. Haddock believe may have a new type of radio spectrum. Their fluxes at 3.75 cm and 1.83 cm (8000 and 16,400 megacycles) were nearly ten times greater than would be expected from extrapolation of measurements at longer wavelengths, as shown in Figure 90. The first of these sources is the thirteenth-magnitude galaxy NGC 1275. Very recently A. Sandage and J. D. Wyndham have identified 3C 279 on Palomar photographs as a blue star-like image of magnitude 16.8, with a faint jet visible.

The Michigan astronomers have adopted a distance of 175 million light years for NGC 1275, from which they find 100,000 light years for its overall diameter and 8500 light-year's diameter for the associated radio source. One of the models they propose involves an extended uniform cloud of electrons, 170 per cubic centimeter, whose motions correspond to a temperature of 4500°K. The amount of ionized hydrogen required would be about 10^{10} solar masses (a tenth the mass of our Galaxy). The mechanism by which such an enormous amount of matter would be ionized is not known, but it could be related to the physical processes responsible for quasi-stellar radio sources.

NGC 1275 belongs to the Seyfert class of galaxies, distinguished by a bright, small-diameter nucleus and having an optical spectrum consisting of a strong continuum and very broad emission lines. This type of spectrum is also characteristic of quasi-stellar objects.

Figures 89 and 90 show the wide differences between optical and radio spectra, most of them due to the techniques involved. Optical spectra show more detail, and little attention so far has been given to measures of the amounts of energy at different wavelengths. (Such measures of

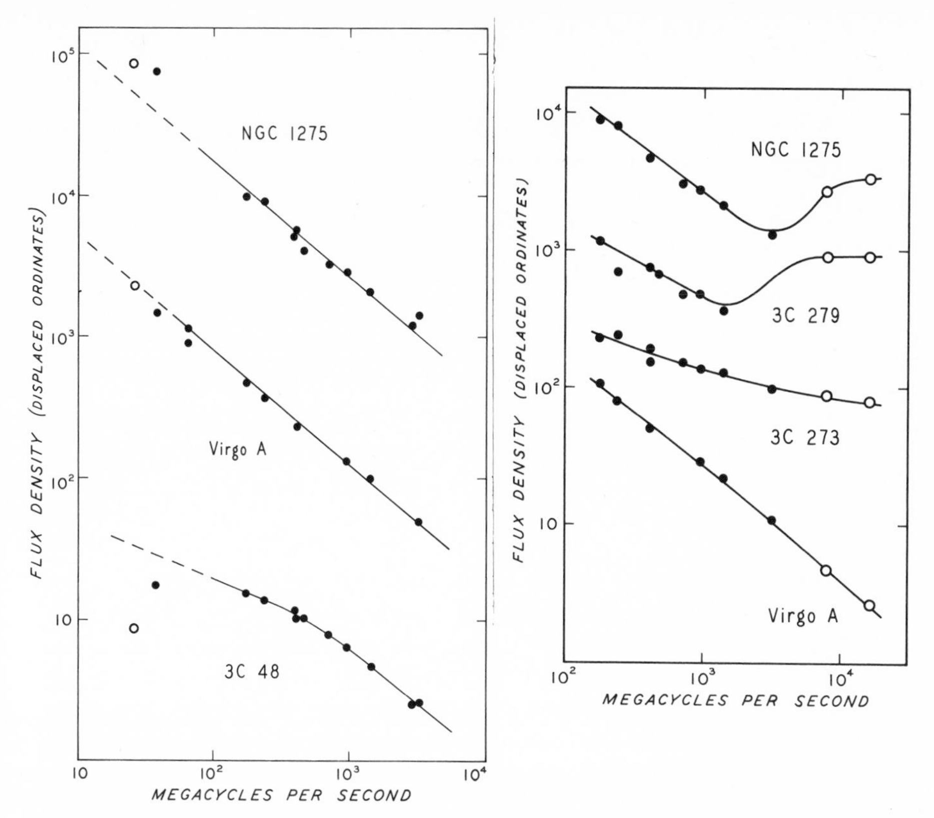

FIG. 90. Radio spectra of five sources. Those at the left were measured at the University of Maryland, and those at the right at University of Michigan. Circles represent new observations at very short and very long wavelengths. The curves have been arbitrarily displaced along the vertical (logarithmic) axis to keep them separate. NGC 1275 is also known as 3C 84.

*"color" are made by photometers using filters for ultraviolet (**U**), blue (**B**), and yellow (**V**) light, rather than from spectra like those in Figure 89.)*

The radio data in Figure 90 extend over a much wider range in wavelength (λ)—almost 300 meters on the left to 1.83 cm on the right—and are generally plotted on a frequency scale ($f = c/\lambda$, where c is the velocity of light, 3×10^{10} cm/sec). Logarithmic scales are used on such a plot, and the radio spectral index, n, is the slope of log s versus log f, where s is the flux density (see p. 184).

The spectral index expected from synchrotron radio emission is about —0.7, and the general shape of the log s — log f curve depends on the temperature, magnetic field, extent, and density of the charged particles whose motion gives rise to the synchrotron emission. Thus the structure and size of a quasar are to some extent revealed by its radio spectrum.

Of course, the spectrum of a quasar (or galaxy, or star) extends from long radio waves through infrared, visible light, and ultraviolet, to X rays. Very recently the techniques of X-ray astronomy have been developed to a point where the X-ray emission of galaxies can be measured. The first X-ray sources discovered were, however, all in our Milky Way.—TLP

Galactic X-Ray Astronomy

C. STUART BOWYER

(*Sky and Telescope*, November 1965)

Questions concerning the birth and death of stars seem to have a special fascination for scientists and laymen alike. For example, what is the ultimate fate of stars that have exhausted their nuclear fuel? It is possible that important clues to this question will be provided by the recent detection of X rays from the night sky.

Fifteen years ago X rays from the sun were observed by some of the first rockets that carried instruments above the absorbing layers of the earth's atmosphere. While it was realized that other stars should emit X rays, simple calculations showed that the intensities to be expected would be far too small to be detected. Indeed, experiments carried out by the Naval Research Laboratory in 1960 verified that nonsolar X-ray emission was below the sensitivity of the instruments then available.

This was the background when a quite unexpected discovery was made in June 1962 by R. Giacconi, H. Gursky, F. R. Paolini, and B. B. Rossi, of American Science and Engineering, Inc. They used a rocketborne X-ray detector, sensitive to radiation of 1 to 8 angstroms wavelength, in an attempt to measure lunar X rays, expected to be produced by fluorescence as solar X rays impinged on the moon. While such lunar radiation was not detected, an intense source of X rays was discovered in the night sky.

Because the detector used in the experiment was not collimated, it was difficult to determine the size or location of the source, but it appeared to be outside the solar system. . . . In April 1963 a Naval Research Lab team (E. T. Byram, T. A. Chubb, H. Friedman, and the writer) at the Hulburt Center for Space Research flew a sensitive X-ray detector with the field of view restricted to 10°. As the rocket spun on its axis, the detector scanned most of the night sky above the horizon. Two sources of X rays were clearly recognized in this experiment.

The more intense of the two, in the constellation Scorpius, was probably the same one detected by the American Science and Engineering group. Its angular diameter was found to be less than 0°.2. Surprisingly, it did not match the position of any radio source or unusual optical object. The second X-ray source coincided with the Crab nebula in Taurus, the well-known remnant of a supernova explosion.

One pressing question about the nature of the X-ray sources was their exact sizes. Because of the extreme difficulty in focusing X rays, this information is not directly available. But in the case of the Crab nebula, nature has provided us with an occulting disk—the moon. On July 7, 1964, our group conducted a rocket experiment to monitor the X-ray emission as the moon passed directly over this source.

The visible nebula is so large that it took over twelve minutes to disappear. Since this was longer than the observing time possible from the rocket, the firing was arranged to permit measurements while the moon's edge was crossing the central part of the nebula. A pointing mechanism kept the detector aimed at the Crab during the course of the flight. From the decrease in X-ray intensity as the moon progressed, it was found that this radiation was coming from a diffuse central region almost one light year across.

In June and November 1964 our group continued its search for additional cosmic X-ray sources by launching rockets carrying large-field X-ray detectors. Two new sources were found in Cygnus, and a great deal of radiation was observed to be coming from a large area in the general direction of the galactic center (Sagittarius). While the collimation of the detectors was not adequate to resolve this area with certainty, the radiation probably emanated from five neighboring sources of small diameter. . . .

All of the ten sites (Table 5) of cosmic X-ray emission discovered to date lie close to the central line of the Milky Way (Fig. 91). Although the number of sources is small, this concentration indicates that they are situated inside the Milky Way Galaxy and are not extragalactic.

There has been lively discussion among astrophysicists as to the physical nature of cosmic X-ray sources, and a number of theories have been advanced. Recent considerations have centered on three possible explanations: neutron stars, thermal Bremsstrahlung, and synchrotron emission.

Neutron stars are postulated by some astrophysicists to be an end point in the evolutionary development of massive stars. In a star of large mass, nuclear burning proceeds most rapidly in the core. Energy is obtained from the conversion of hydrogen to helium, until most of the hydrogen in the core is exhausted. When this happens, the core con-

TABLE 5. OBSERVED X-RAY SOURCES

Source	*R.A.(1950)* h	m	*Dec. (1950)* °	*Counts*[1]
[2]Tau XR-1	05	31.5	+22.0	2.7
Sco XR-1	16	15	−15.2	18.7
Sco XR-2	17	08	−36.4	1.4
Sco XR-3	17	23	−44.3	1.1
[3]Oph XR-1	17	32	−20.7	1.3
[4]Sgr XR-1	17	55	−29.2	1.6
[5]Sgr XR-2	18	10	−17.1	1.5
Ser XR-1	18	45	+05.3	0.7
Cyg XR-1	19	53	+34.6	3.6
Cyg XR-2	21	43	+38.8	0.8

[1] Counts per square centimeter per second.
[2] Within 1′ of Crab nebula's optical center.
[3] 1°.1 from supernova 1604 (Kepler's star).
[4] 2°.7 from galactic center.
[5] 2°.0 from M 17, the Omega nebula.

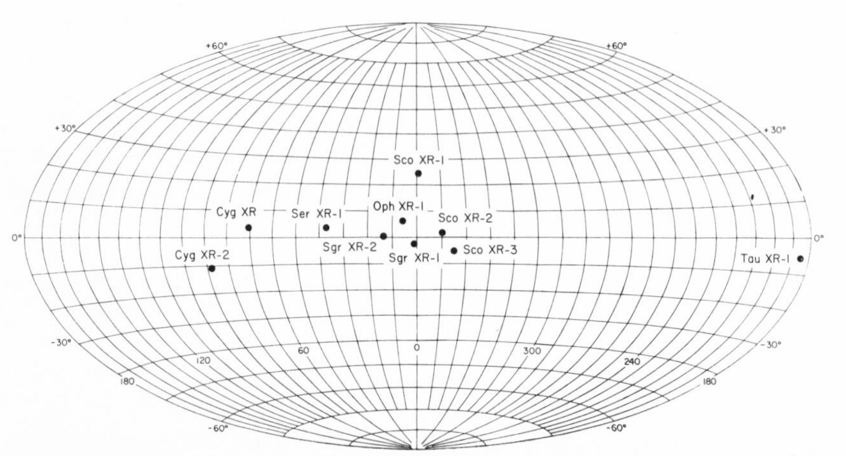

FIG. 91. Known X-ray sources are plotted in this all-sky map, whose horizontal diameter, labeled 0°-0°, represents the galactic equator. Note that the ten sources are concentrated along it rather than being spread over the entire heavens.

tracts, and the temperature rises until helium burning begins. While detailed stellar models have not been calculated much beyond that point, it is generally believed that this cycle—core contraction to a higher temperature, followed by further nucleosynthesis—is continued through various stages involving heavier and heavier atoms—or, rather, completely ionized atomic nuclei. The cycle stops only when iron is formed in the core, which is surrounded by intermediate-weight atomic nuclei for which nucleosynthesis is not yet complete.

Though details of what happens next are unclear, there is general agreement that eventually the star's core collapses, removing support for the star's outer layers, which fall inward to release gravitational potential energy. The sudden increase in the star's central temperature speeds up nuclear reactions in the unburned material until a catastrophic explosion takes place. Becoming a supernova, the star blows off its envelope.

If the mass of the original star was not too great, the stellar remnant of the supernova explosion consists mostly of neutrons, with a small admixture of protons and electrons. And if the density is sufficiently great, hyperons—high-energy particles that are extremely unstable under normal conditions—may also be present. Part of the gravitational energy released in the collapse heats up the remnant to a temperature of the order of 10^{10} degrees Kelvin.

An atmosphere compatible with this core would have a surface temperature of about 10 million degrees, and its radiation would be mainly at X-ray wavelengths. This fact, coupled with the small size of the object, virtually rules out the possibility of optical detection. (Indeed, such a neutron star would have to be nearly inside the solar system to be observable even with the 200-inch Palomar telescope.) Also unobservable as a radio source, it would still be a copious emitter of X rays.

Nevertheless, neutron stars are not likely mechanisms for all X-ray sources. The Crab nebula's diffuse emitting region would seem to require some other process.

Thermal Bremsstrahlung is a mode of X-ray production that astronomers have invoked to explain part of the solar corona's radiation at angstrom wavelengths. In this process, fast electrons experiencing close encounters with heavier ions emit X rays when they are accelerated by the electric fields of the ions. If this mechanism is responsible for the cosmic X rays, there must be a continual supply of energetic electrons to replace those that have lost energy by radiation. Possibly the continual supply could come from radioactive elements formed in a supernova explosion. If a few per cent of the exploding mass became heavy radioactive elements, their decay would produce enough energy to maintain the observed X-ray flux of the Crab nebula. However, one would expect that the entire nebula would be emitting X rays, not just the central part.

Synchrotron radiation is a third possible explanation. Electrons moving through a magnetic field undergo constant acceleration as they spiral around the magnetic lines of force. If the electrons have extremely high energies and travel at nearly the speed of light, detectable radiation is emitted by them. This synchrotron radiation, so named because it was first observed at optical wavelengths from a synchrotron in the laboratory, is one of the chief mechanisms for the emission of cosmic radio waves. In a few cases, such as the Crab nebula and the luminous jet in the galaxy M 87, the synchrotron mechanism is responsible for optical as well as radio emission.

Synchrotron X rays can also be produced if the electrons have sufficient energy. Once more, a supernova explosion can be supposed to supply the initial energy, but again there is a problem of maintenance. If the synchrotron mechanism is to operate for more than a few years, a continual supply of high-speed electrons is needed, either new ones injected into the emitting region or reaccelerated ones.

All the rocket flights so far have detected a "background flux" of X rays in addition to the intense discrete sources. This background is observed over the entire sky, . . . and its origin is uncertain. . . .

X Rays from outside the Galaxy

(*Sky and Telescope*, March, 1968)

Three University of California scientists have just reported on the strange appearance of the sky as viewed by instruments that detect soft X rays, of 44 to 70 angstroms wavelength. Hitherto, X-ray astronomers had recorded a number of discrete sources and the hard X-ray background at wavelengths of 1 to 20 A. Now C. S. Bowyer and his co-workers G. B. Field and J. E. Mack have succeeded in mapping the soft cosmic X-ray background.

Their observations were made with a gas-filled proportional counter that was flown aboard an Aerobee 150 rocket from Natal, Brazil, on December 13, 1966. The detector faced the sky through a metal honeycomb that limited its field to a diameter of eight degrees (edge of the field being defined as half-power point).

Following customary practice, the rocket's spin was controlled to make the detector scan the sky at the proper rate. But the star photometers intended to monitor the changing aspect of the rocket failed to operate, and the moment-to-moment orientation had to be reconstructed laboriously from on-board magnetometer data. Fortunately, both the Crab nebula and Scorpius XR-1 were unambiguously observed on the same scan, making this analysis possible.

The observations extend over a large fraction of the sky north of the galactic equator. The soft X-ray counts were generally least along the central line of the Milky Way and greatest toward its north pole. Thus, to an X-ray detector the Milky Way appears "dark" and the rest of the sky "bright."

This is a situation closely analogous to the so-called zone of avoidance of the extragalactic nebulae, which appear most numerous in high galactic latitudes but are very scarce in low latitudes, where they are severely dimmed by interstellar dust. But in the case of soft X rays reaching us from outside our Galaxy, the zone of avoidance is due to absorption by interstellar hydrogen (and helium) gas.

Therefore, Bowyer and his associates interpret these observations as evidence for an extragalactic origin of the soft X rays. They suggest that this X radiation is either emission by distant and overlapping discrete sources or due to inverse-Compton scattering in intergalactic space of quanta of the three-degree blackbody cosmic background radiation. [That is, low-energy quanta of radio radiation might absorb energy from high-speed electrons while they are near intergalactic protons.]

In the 44- to 70-angstrom interval essentially no counts are detected from the galactic equator, but a significant flux comes from the direction of the north galactic pole. Just as astronomers can determine the extinction coefficient of the interstellar dust layer by comparing numbers of galaxies in different latitudes, so the California scientists could determine the X-ray extinction in the gas layer. The result is puzzling, for the X-ray absorption is much less than would be expected from radio astronomers' measurements of the 21-cm line of interstellar hydrogen. If this discrepancy is real, it can have far-reaching implications.

In addition to the extragalactic background radiation, there is possibly an X-ray–emission component originating inside the Milky Way system. The observations show increased counts over an extended area around galactic longitude 330°. Bowyer and his associates suggest that this radiation may come from a "galactic corona"—an envelope of hot gas surrounding the core of the Milky Way Galaxy. On this interpretation, the Aerobee observations indicate a corona radius of five kiloparsecs (16,000 light years) and a mass of hydrogen roughly equal to 100 million suns. Details of this study appeared in *Nature* [**217**, 32 (January 6, 1968)].

It is likely that X-ray astronomy will develop as radio astronomy has, possibly revealing whole new classes of objects in the universe. The X-ray emission from sources in the Milky Way (see Fig. 91) certainly make our Galaxy an X-ray source, and at least one other galaxy, M 87, has recently been found to emit X rays. This, and several other X-ray sources, are also known to be radio sources. Of course, as the level of X-ray detection is improved, galaxies and quasars may turn out to be X-ray emitters—requiring extreme conditions in the theoretical models now under study to "explain" the quasars.—TLP

6

Masses, Luminosities, and Models

The variety of work on galaxies and quasars during the past four or five years is difficult to present in an organized way. Observations as diverse as X-ray measurements from rockets, optical spectra with image tubes, and radio measurements with 2000-mile base lines are all involved, together with detailed measurements within the Milky Way, dynamical studies of gravitational collapse, and theoretical calculations of nuclear energy generation. Astronomers themselves, as the following article shows, find it difficult to organize discussion of these matters at meetings.—TLP

Galaxies and Quasars: Research Reported at the IAU Congress in Prague, August 1967 [1]

THORNTON PAGE

(*Sky and Telescope*, December 1967, January and February 1968)

Galaxies got more attention than any other topic at the XIII Congress of the International Astronomical Union in Prague, August 21–31, 1967.

[1] Parts of this article did not appear in *Sky and Telescope* because some of the studies reported had been presented there previously.—TLP

There were many other interesting astronomical problems discussed, but throngs of astronomers and other participants (about three thousand in all) crowded the meeting halls for discussions of galaxies, galactic structure, and radio astronomy, and for the excellent invited discussion by Sir Martin Ryle and Allan Sandage. I must admit to some bias in this report—partly because of my own interest, partly because the discussion of galaxies and cosmology includes so much: interstellar clouds, the "mix" of stellar types, the "spread" of stars, gas, and dust, their motions, masses, and luminosities, possible material between the galaxies, explosions of galaxies, the "background light," and the question of cosmogony (how the stuff we see got there in the first place).

From another point of view, the study of galaxies involves the new techniques of X-ray astronomy, radio astronomy, and filter photography, as well as the older methods of spectroscopy, photometry, and photography. The distribution of galaxies in the sky—pairs, groups, clusters, and clusters of clusters—involves statistics; and the study of their forms, "morphology," has generated all sorts of assumptions and complex calculations.

This year, the key word was "rings." There were rings of stars in the Milky Way, rings of magnetic field in quasars, rings of hydrogen gas around the nuclei of other galaxies, and similar rings in which the ratio of mass to luminosity differed systematically. By contrast, "distance" roused less interest. A highly touted debate on the distances of quasars was won, I thought, by the big-distance men (Britain's Ryle, Mount Wilson's Sandage, and Columbia's Woltjer), and even the opposition speaker (San Diego's G. R. Burbidge) admitted that he didn't believe that quasars are nearby; he only wanted to emphasize that their distances are not yet well established.

After this indication of the cosmically turbulent atmosphere at Prague, I will try to present a systematic review of reported research results under seven headings:

1. Structure and internal motions of galaxies
2. The Milky Way system
3. Colors and spectra of galaxies
4. Masses and luminosities
5. Clusters of galaxies
6. X-ray emission
7. Radio galaxies and quasars

Unfortunately, these are not grouped rationally in the IAU commissions—the organized groups of astronomers interested in a common problem. Although most of these topics belong in Commission 28 (Galaxies), tradition has kept the lion's share in Commission 40 (Radio Astronomy); discussion of the Milky Way Galaxy was split between Commission 33 (Structure and Dynamics) and Commission 34 (Interstellar Matter), and X-ray results were confined to a special session.

In addition to the verbal presentations at the meeting, a report has been published by each commission summarizing research work over the three years since the last IAU Congress, in Hamburg, Germany. The Commission 28 Report, prepared by Rudolph Minkowski, Allan Sandage, E. Margaret Burbidge, George McVittie, and B. A. Vorontsov-Velyaminov, gives more detail on some of the topics covered here.

Structure and Internal Motions of Galaxies

In the past, spiral structure has been explained as a result of the slower rotation at larger and larger distances from the nucleus of a galaxy; the outer ends of an arm would necessarily lag behind the inner parts as the whole galaxy rotates. But this presents several difficulties: how to explain the straight bars in barred spirals (SB), and why there are no arms in ellipticals (E), and how the normal spirals (S) got their two arms to start with.

A great deal of work has recently been done on the morphology of galaxies—the fascinating similarities and differences in their shapes. Edwin Hubble's classification scheme [see Figs. 24, 25, and 26], dating from about 1930, has been modified by several astronomers, most notably by W. W. Morgan (Yerkes Observatory), who take account of the spectrum, or types of stars in a galaxy, as well as its shape. Catalogues by B. A. Vorontsov-Velyaminov (Moscow) and by H. C. Arp (Mount Wilson) list peculiar or "deformed" galaxies that obviously deviate from Hubble's E, S, SB, and irregular (Irr) types. Paul Hodge (University of Washington) finds that Irr types are generally flattened (not completely irregular), and has studied the dwarf galaxies as a separate group.

In spite of these refinements, Hubble's types serve well to classify the vast majority of the galaxies that we can photograph. It is well established that E-type galaxies (and the intermediate So type) contain mostly old, Population-II stars, little or no gas or dust, and few or no young, blue, Population-I stars, whereas SB, S, and Irr galaxies contain many blue stars and much interstellar gas and dust. In Prague, the senior Soviet astronomer, B. A. Ambartsumian, reported on work at Burakan Observatory, Armenia, which revealed "superassociations" (groups of brilliant blue stars) in about 20 per cent of Sc galaxies. These superassociations can be no more than 10 million years old; hence Ambartsumian concludes that there are "spurts" of star formation

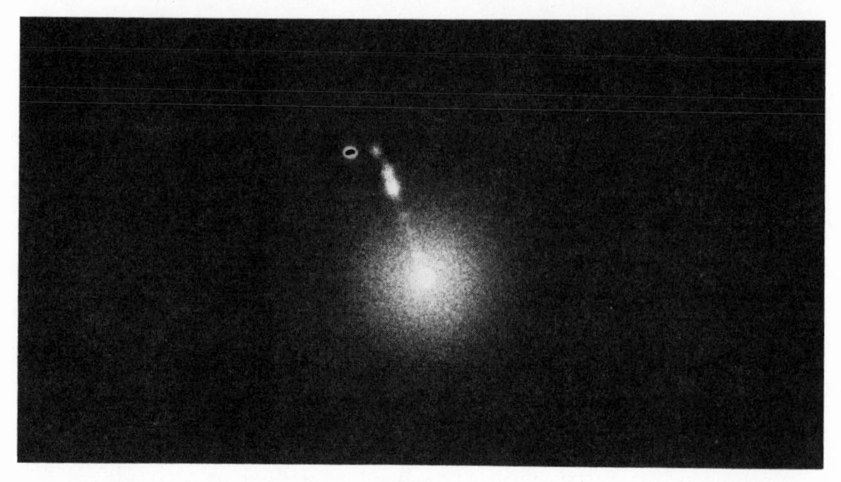

FIG. 92. The E-type galaxy M 87, with a jet (upper right) that is interpreted as a gas cloud ejected from an explosion in the nucleus about 100 million years ago. H. C. Arp has discovered a counterjet on the opposite side (lower left), not visible on this photograph. (Official U.S. Navy photograph)

within limited regions of spiral galaxies. The Burakan astronomers also find wide variety in the nuclei of over six hundred spirals—some are small and bright; others are much less concentrated. By contrast, one of Ambartsumian's co-workers, A. T. Kalloghlyan, reports that the bars of SB galaxies are all of the same surface brightness; this type of galaxy is much more uniform than the spirals.

The high-density nucleus is probably the source of explosions that caused the streamers of gas coming out of M 87 (see Fig. 92) and the jets and filaments that California astronomers have seen being ejected from other galaxies [see Figs. 71 and 76]. It is possible that spiral arms were first formed by this sort of ejection and that in quasars we see the explosions taking place. For a reason not fully understood, such explosions eject two gas clouds in opposite directions. H. C. Arp reported his discovery of a "counterjet" from M 87, a faint, thin streamer, more than 40″ long, opposite the bright jet shown in Figure 92. There are three other E galaxies along the line of these jets, and Arp thinks they are fragments of previous explosions. He has found dozens of cases where peculiar galaxies are lined up like this, but other astronomers are not yet convinced of the connection.

For many years, H-II regions have been used to study motions in the outer arms of spirals. These clouds of ionized hydrogen gas near blue stars have emission-line spectra, strong in the red hydrogen-alpha (Hα) line. In the 1930s Horace Babcock, N. U. Mayall, and others at Lick Observatory measured Doppler shifts showing the rotation of spirals. More recently G. Courtès at Haute Provence Observatory (France) has

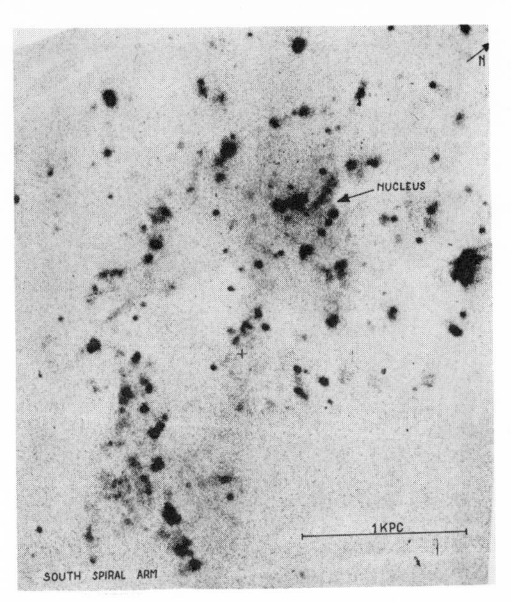

FIG. 93. The Sc-type galaxy M 33, photographed in Hα light (through Ilford 608 red filter on Eastman 103aE plate) with an *f*/1 reducing lens at the Newtonian focus of the 77-inch reflector at Haute Provence Observatory in France. South is at the bottom, east at the left; scale 5′ per inch. The distance of 1 kiloparsec in M 33 is marked. (Courtesy G. Courtès)

photographed through Hα (red) filters the H-II regions in six large nearby spirals. In Prague he reported his use of a Fabry-Perot interferometer on the Hα emission in M 33, as illustrated in Figures 93 and 94. The interferometer consists of two parallel plates of half-silvered glass (an etalon) between lenses designed to give a parallel light beam in front of the plateholder.

Figure 93 is an Hα photograph of M 33 taken with a reducing lens on the 77-inch reflector at Haute Provence Observatory. Figure 94 is the view through the interferometer centered 4.5 minutes of arc south of the nucleus. The circular rings are produced by Hα light and would be exactly circular if all this hydrogen emission were of exactly the same wavelength (6563 angstroms). Small deviations from circularity show Doppler shifts due to different radial velocities of the gas, and this can be measured to about 2-km/sec accuracy. A total of 536 points were so measured on this plate, and 512 other points on three other plates on different centers, for a total of 1048 measures. Because the Hα emission extends all over M 33 (two or three times fainter between the strong H-II regions), Courtès and his coworkers have been able to map with high accuracy the motions of gas in M 33. They find that M 33, unlike several other galaxies and the Milky Way system, has

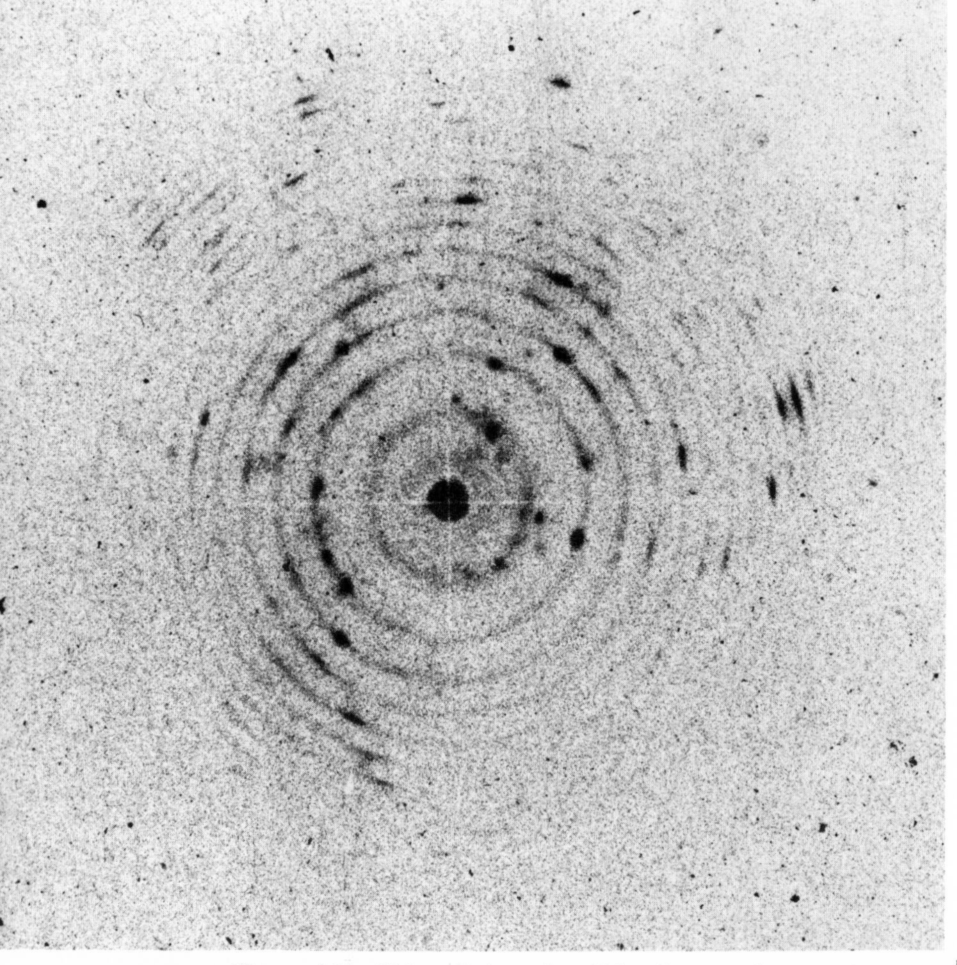

FIG. 94. Photo of $H\alpha$ "fringes" through a Fabry-Perot etalon interferometer in front of an $f/1$ reducing lens. Each ring represents $H\alpha$ emission in M 33. The cross hairs (and central black circle) were centered 4.3 minutes of arc south of the nucleus. Accurate measures of the rings give Doppler shifts at the corresponding points all over the image of M 33. (Courtesy G. Courtès)

little or no expansion velocity away from the nucleus, and that the rotational velocity of the arms exceeds that of the "disk gas" between the arms by 15 to 20 km/sec. What's more, this disk gas is at higher temperature than the H-II regions in the arms, possibly due to collisions between small cloudlets.

Paul Hodge (University of Washington) has studied by filter photography how the H-II regions are spread out around the nuclei of about a hundred spirals, using the Lick 120-inch reflector and the

Palomar 48-inch Schmidt. He finds a maximum about halfway out to the edge recorded on long-exposure photographs. His report was matched by Morton Roberts (National Radio Astronomy Observatory), who reported similar rings in the 21-cm H-I (neutral-hydrogen) radio emission in eight galaxies, including M 31, M 33, M 101, and several others studied by Courtès and Hodge. Except in M 31, the H-I rings are *well outside* the H-II rings, and there is little or no H-I radio emission from the strong H-II regions. Roberts thinks that there are H-I rings in over ninety other galaxies he has studied, most of them too small to be resolved by the 300-foot radio telescope at the National Radio Astronomy Observatory.

Figure 95 shows his plot of the 21-cm radio emission from M 31 (the Sb-type Andromeda galaxy). Units are "brightness temperature times km/sec" because Roberts summed the 21-cm radio emission (measured in degrees Kelvin by the radio telescope at each radio frequency near 1460 MHz, the 21-cm emission of hydrogen) from hydrogen in M 31 moving at all radial velocities, and therefore Doppler-shifted. The ring stands out as a circular ridge, foreshortened by the tilt of M 31. This

FIG. 95. An isometric (three-dimensional) plot of the total radio emission by neutral hydrogen (near wavelength 21 cm) in the Sb-type galaxy M 31. The peaks and valleys are shown by east-west curves across the galaxy, and by isophotes (lines of constant height above the plane) along the sides of the "ridge." The total volume of the "mountain" represents the total amount of interstellar atomic hydrogen (H I) in M 31. (Courtesy Morton Roberts)

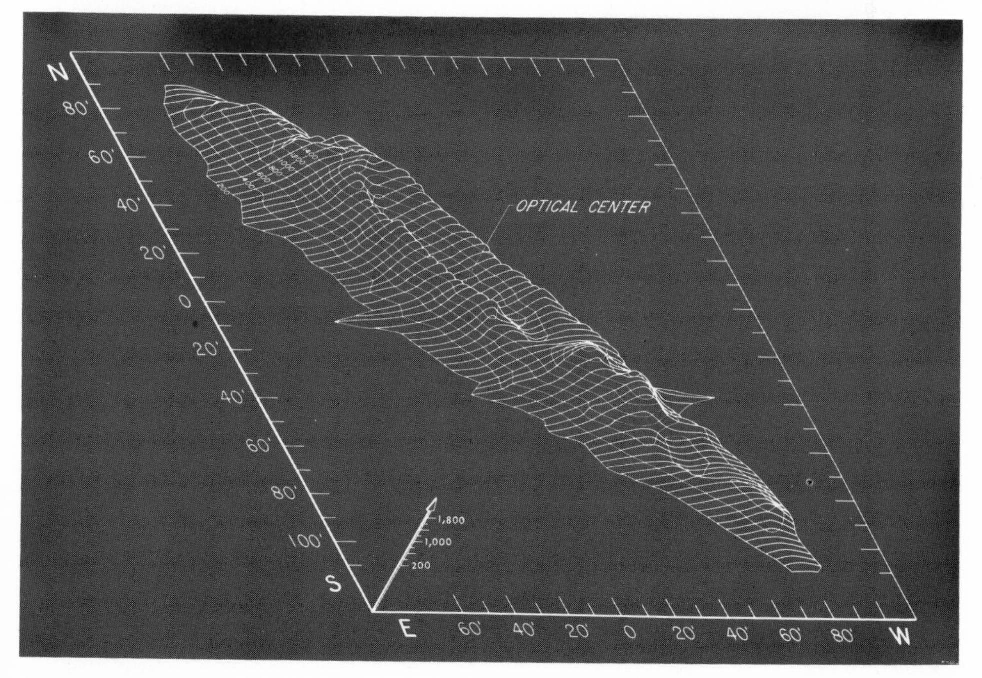

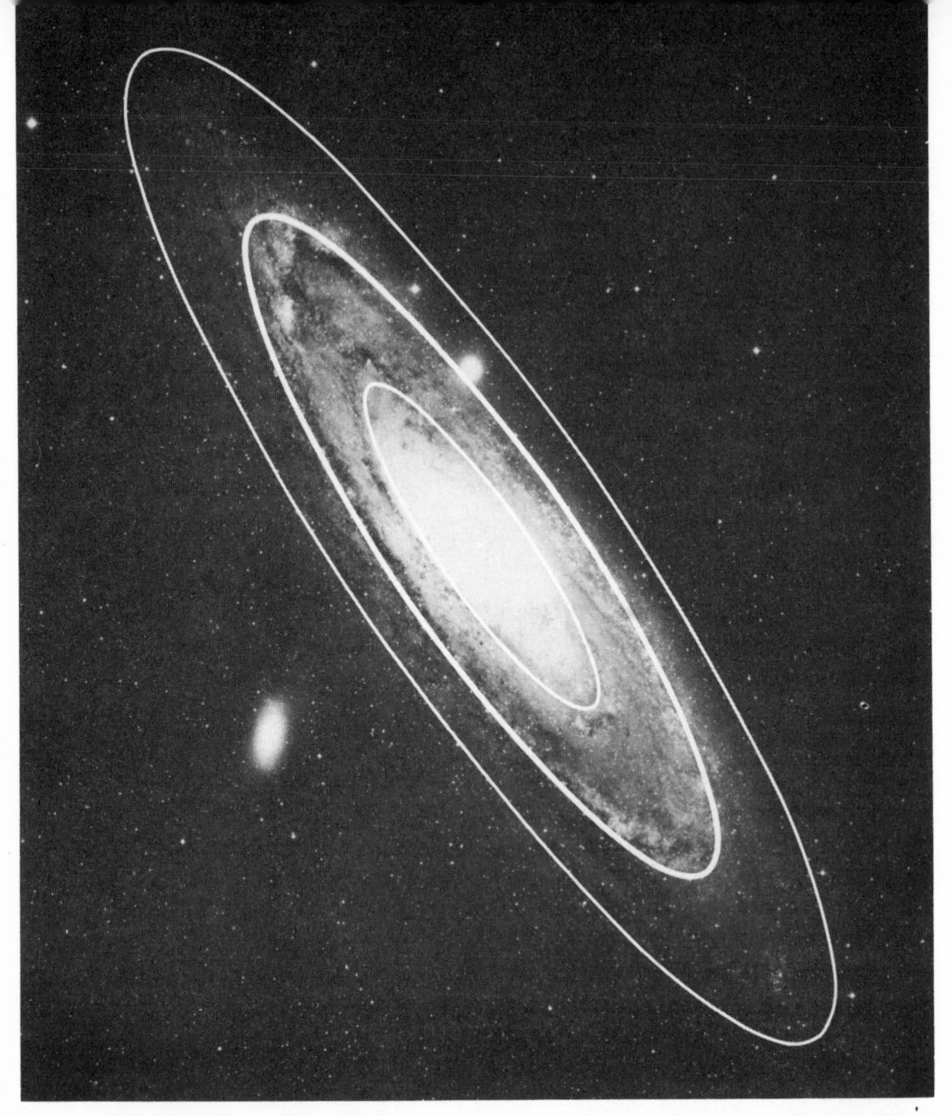

FIG. 96. Photograph of M 31 in Andromeda with the H-I emission ring plotted on it. The heavy line is the center of the "ridge" shown in Figure 92, and the other two ellipses show the smoothed inner and outer edges of the ring, more or less superimposed on the arms. (Courtesy Morton Roberts)

ring is plotted on a photograph of M 31 in Figure 96, coinciding more or less with the H-II regions and arms. Figures 97 and 98 show the results for M 33 and M 101, both Sc type, where the H-I ring is outside the H-II ring and arms.

These results pose two basic questions about spirals: (1) Why do these rings of gas around the nucleus differ in structure from the familiar

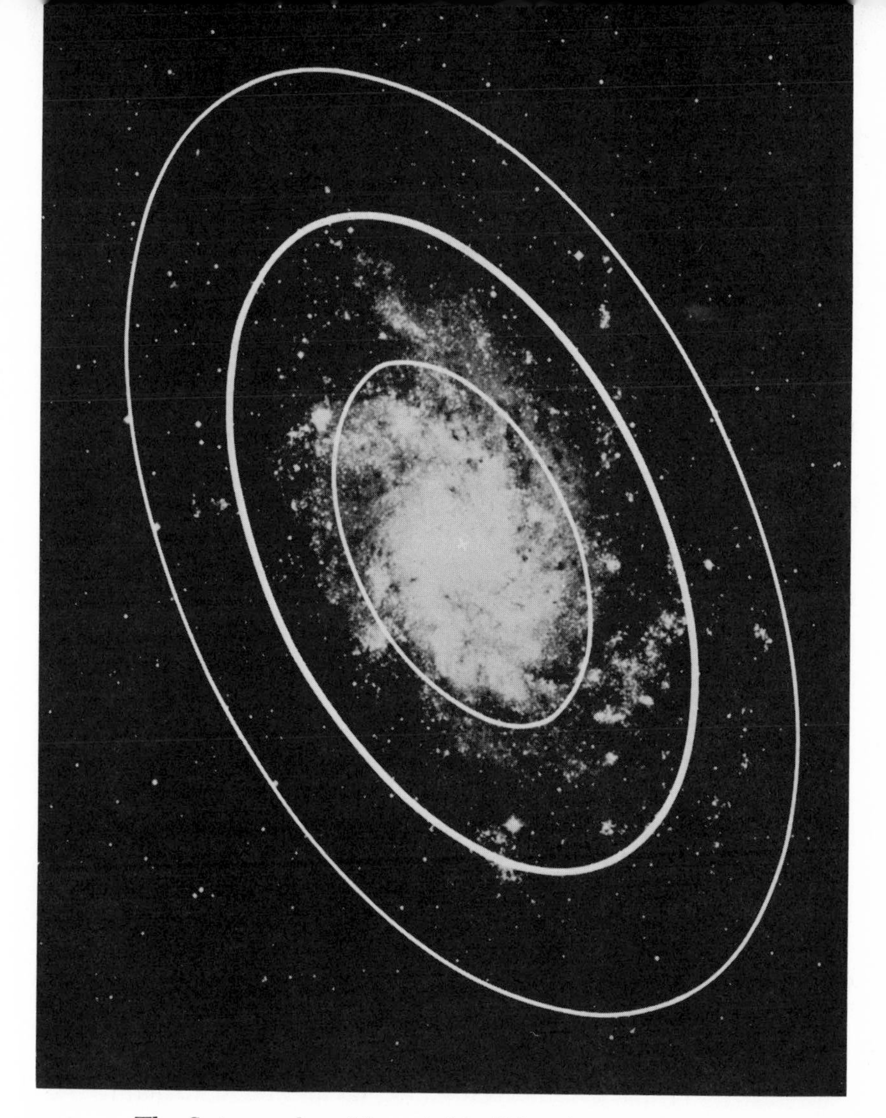

FIG. 97. The Sc-type galaxy M 33 (white-light photograph). The H-I ring measured by Roberts, plotted as in Figure 96, is farther out than the arms and most of the hydrogen-alpha emission shown in Figures 93 and 94. (Courtesy Morton Roberts)

arms? and (2) Why are the H-I rings not centered on the H-II rings? (Under conditions in interstellar space, we expect hydrogen atoms where-ever there are hydrogen ions; in fact, the hydrogen-alpha emission line results from electrons recombining with hydrogen ions to form the atoms.)

The first question is further compounded by F. Bertola (Asiago, Italy), who has studied the luminous efficiency (light radiated per gram mass of material) in fifteen galaxies. The inverse ratio, mass/light, has

FIG. 98. The Sc-type galaxy M 101. The H-I ring, plotted as in Figures 96 and 97, is again centered farther out from the nucleus than the spiral arms. (Courtesy Morton Roberts)

long been known to be large for E galaxies and small for spirals; in 1958 I found that the average E has sixty times the mass/light ratio of the average spiral. In Prague, Bertola reported that within each spiral the mass/light ratio decreases from one ring to the next, outward from the nucleus.

Another type of ring was observed by P. E. Wild (Switzerland), who finds circular dark "holes" in nearby galaxies. In many cases, supernovae have appeared at the bright edges of these "holes," and Wild thinks that such a bright edge about 10,000 light years in diameter is caused by a shock wave from an earlier supernova in the center of the hole. The shock wave might trigger new supernovae as it sweeps outward.

T. Schmidt-Kaler (Bonn, Germany) proposed a similar idea on a smaller scale in our own Milky Way Galaxy. On the Palomar Sky Survey prints he has found over a thousand rings of stars that he believes cannot be due to chance (because they are spread around the sky differently from individual stars). The average ring contains fifty to a hundred blue stars, probably young, and is not quite circular (one axis being about 70 per cent larger than the other). Assuming that these rings are all about the same true size, he reasons that those of smaller angular diameter are farther from us, and plots their distribution near the plane of the Galaxy. Presumably these rings indicate that "bubbles" of gas existed in our Galaxy several million years ago when conditions were right on the "bubble surface" for star formation, but few of his listeners were convinced.

The late Bertil Lindblad of Stockholm, a former president of the IAU and of Commission 28 (Galaxies), had developed a "density-wave" theory of spiral arms on the premise that stars, gas, and dust are moving through the arms in their orbits around the nucleus but are slowed down as they pass the center of an arm. That is, the arm is a region where the stars are closer together, and the interstellar gas somewhat compressed. In Prague, C. C. Lin (MIT) reported his mathematical analysis of the motions of gas under purely gravitational forces between all the masses of gas in one rotating disk-shaped galaxy 100 billion times more massive than the sun [see p. 106]. When he adjusted the mass distribution to match that of our Milky Way Galaxy, he found that a density pattern of two spiral arms must form, rotating in the same direction as the gas but at about half the speed. These results are confirmed by Mitsuaki Fujimoto and K. H. Prendergast at Columbia University. It is possible for the outer ends of the arms to curve either way (leading or trailing) but the trailing pattern is much more probable. The barred-spiral structure (SB) is *not* predicted by Lin's analysis, and this may indicate a fundamental difference between the S type and the SB. Details such as the pitch angle of the arms (5° to 10° outward from circular rings), and the gas rotation speeds, agree with radio observations of H-I arms in the Milky Way. New stars are expected to form in the compressed gas, then move out ahead of the arm, and this explains the young blue stars ahead of the H-I arms and the migration of older stars away from the arms which Bengt Strömgren found while he was at Princeton University.

If a magnetic field is added to this model of a rotating gas, Lin points out that it will not be "wound up" to match the spiral form, since it would be carried along by the rotating gas. However, S. Pikelner (Moscow, U.S.S.R.) finds the magnetic field to be along the arms of most spirals and attributes the higher gas density in the arms to a condensation wave in this magnetic field. At higher gas density the "tube" of magnetic field is broader, and this explains the thick spiral arms in Sa galaxies (thinner in Sc) as well as the thickening of arms near the nucleus of any spiral.

Radio observations show the gas density between arms to be about one quarter as large as that in the arms. Also, gas in the arms is clotted in large clouds or condensed into stars, whereas between the arms it is spread out uniformly. This stability of the interarm gas can be due to its higher temperature, and Pikelner assumes that it is heated by cosmic rays. If cosmic rays are produced by type-II supernovae (the very hot, long-lasting ones), it may be that supernovae "control" the interstellar gas temperature; when the gas gets cool enough to coagulate into stars, supernovae are formed and warm the gas to stable uniformity. On the average, there is only one supernova explosion in each galaxy every three hundred years, according to Fritz Zwicky (Cal Tech), but this provides enormous energy for heating the interstellar gas clouds—about 10^{49} ergs—enough to heat all the interstellar hydrogen within 10,000 light years to 10,000°K.

The Milky Way System

From our position within the Milky Way Galaxy it might be supposed that we would know everything about its structure and internal motions. Actually, the picture is by no means clear in detail, although the size, rotation, and halo around the nucleus are well established. Diagrams like Figure 99 show spiral arms outlined by 21-cm emission of interstellar hydrogen (H I), but these arms do not match the concentrations of luminous blue stars. Part of the trouble may be that stars are formed in the arms but then move out ahead of the arms, as

FIG. 99. Spiral pattern of interstellar hydrogen in the Milky Way as revealed by 21-cm radio observations. The accuracy of this plot, which combines measurements by Dutch and Australian radio astronomers, has recently been questioned. (Leiden Observatory diagram)

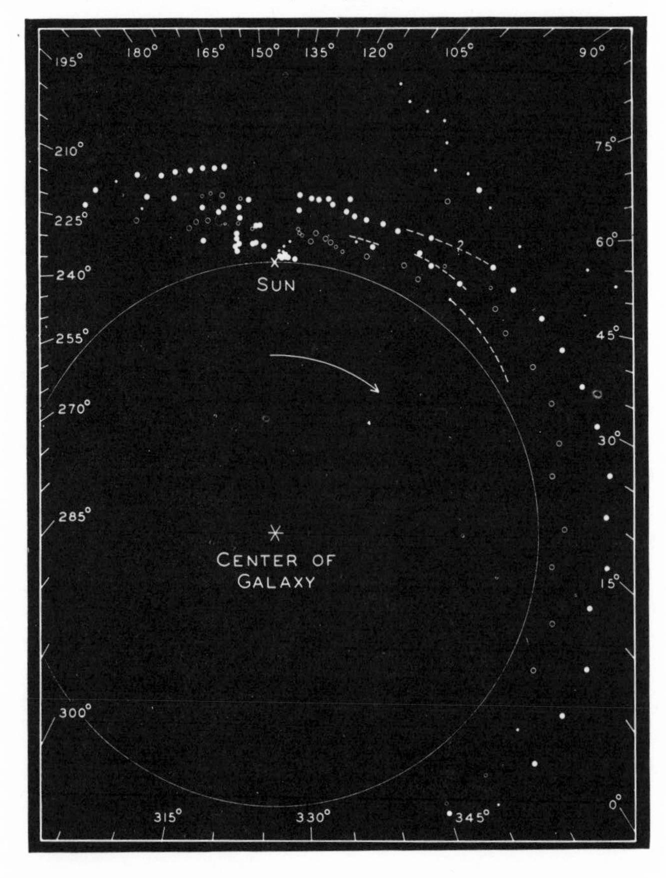

predicted by Lindblad and Lin. Another difficulty is that the plotted location of H-I emission in Figure 99 depends upon the manner in which the gas is assumed to rotate (in circular orbits around the nucleus, 33,000 light years from us in the direction of Sagittarius). Dutch astronomers, led by Jan Oort (who first worked out galactic rotation from stellar motions in 1925), have assumed that the gas moves with the stars, and that both gas and stars are concentrated in spiral arms.

Adrian Blaauw (Leiden) summarized recent radio studies which show that interstellar hydrogen near the sun forms a circular ring around the nucleus of the Milky Way Galaxy. In regions closer to the nucleus, the arm pattern spirals inward, inclined at about 15° to circles around the nucleus, but the gas seems to be streaming outward at 10 to 20 km/sec. Many different studies of stellar motions, H-II regions, and jets, or "spurs," of H-I clouds, complicate this picture. It may well be said that we are too close to get a good overall view of our own Galaxy, and that we cannot trace the spiral arms reliably, but the H-I ring outside the spiral pattern seems to resemble Roberts' measures on other Sc-type galaxies (see Figs. 97 and 98).

The history of a star, once formed, is now well established, but there is less known about the exchange of gas between stars and the interstellar medium in a series of steps called the Oort cycle. George Field (Berkeley) presented recent results on the coalescence of interstellar cloudlets to form larger clouds which can pull in more interstellar material, heat up, cool off, and condense into a star. A new hot star lights up the remaining interstellar gas as an H-II region. At the end of its life, the star finally blows some fraction of its mass back into interstellar space, where more stars can form.

Field has worked out the mechanics of coalescence, assuming that cloudlets stick together when they collide (a consequence of the magnetic fields now known to extend throughout the Galaxy). Observations of hundreds of H-II regions show many small clouds, each of mass (m) about ten times the sun's mass, and a dozen large groups (over 300 solar masses each). Some interstellar clouds are over 10,000 times as massive as the smallest cloudlets, but the numbers get smaller for larger masses (proportional to $m^{-1.5}$). The random collisions between these clouds lead theoretically to a rate of star formation that depends on the square of the interstellar density, a formula assumed by Maarten Schmidt in his studies at Mount Wilson ten years ago. Field finds that star formation from the Milky Way cloudlets would be about half complete in 8 billion years if none of the stars blew up.

S. Pikelner (Moscow) pointed out that the magnetic fields in the

Galaxy would be modified by the moving cloudlets in such a way as to flatten the disk of interstellar gas in the Milky Way and to reduce the motions of newly formed stars perpendicular to the disk. He expects that detailed study of magnetic-field effects will show a more rapid growth of large gas clouds, and consequently a larger increase in the proportion of giant stars than Field predicted.

The evidence for magnetic fields that might cause all this in spiral arms of the Milky Way or other galaxies was summarized by R. D. Davies (Manchester, England). The most direct observational effect is synchrotron radiation from electrons and ions spiraling in a magnetic field, but this only occurs in fairly intense radio sources like the Crab nebula, radio galaxies, and quasars. The general magnetic field in interstellar regions is indicated by the slight (5 per cent) polarization of starlight passing through clouds of dust particles lined up by the field. This was first detected in the Milky Way by J. S. Hall and W. A. Hiltner, and has recently been measured in other galaxies by Hall and Aina Elvius (Lowell Observatory). Another, similar effect takes place in clouds of ionized gas, where a magnetic field causes a change in polarized radio waves passing through. This "Faraday effect" shows only the sum of all magnetic fields in the line of sight, without a measure of their directions, and only if the radio waves are initially polarized.

Within 600 parsecs (2000 light years) of the sun the magnetic field is about 10^{-5} gauss along the spiral arms in the plane of the Milky Way and south of the plane, but is reversed north of the plane. More locally, there are irregularities in the magnetic-field strength and direction over regions that are 1 to 100 light years across. The puzzling thing about this is that gas clouds can continue their motions; even these small magnetic fields exert enough force on moving ions to compress the gas and hold it back. H. Alfvén, the Swedish expert on magnetohydrodynamics, expects electric currents to be set up in the ionized gas clouds, and these may produce luminous filamentary nebulae like the ones observed in M 82 by Elvius at Lowell Observatory. It was agreed that much more work is needed to map out the interstellar magnetic fields in the Milky Way and in other galaxies.

Closely allied to interstellar polarization is the reddening and dimming of starlight as it passes through tenuous dust clouds. The fine dust particles reduce the light by scattering, and cut out more blue light of short wavelength (λ) than red light of larger λ. This dimming takes place at both ends of the light path from an external galaxy to us—due to dust clouds both in the Milky Way and in the other galaxy—and is expected to be roughly proportional to $1/\lambda$, although recent measurements by H. L. Johnson (University of Arizona) and others show deviations from this "reddening law." Clues about the nature of the dust particles may be derived from measurements of the dimming at different wavelengths. Particles of ice and of graphite (carbon), and mixtures of these two, have been studied by J. M. Greenberg (Rensselaer) with this in mind. He suspects that the broad interstellar absorption band at 4430 A is caused by dust particles, but cannot yet identify the chemical responsible.

Colors and Spectra of Galaxies

It has long been known that the overall colors of galaxies (after correction for the reddening by interstellar material along the line of sight) are blue in Irr, Sc, and SBc types, yellow in Sa and SBa types, and red in E types. This fact, and the analysis of spectra, are consistent with the interpretation that E-type galaxies and the nuclei of spirals are composed mostly of old, Population-II stars, while spiral arms are composed mostly of young, Population-I stars. The spectra of most spirals also show emission lines of hydrogen ($H\alpha$ strongest), oxygen ions ("forbidden" lines of O II and O III), nitrogen ions, and neon ions, just like nearby H-II regions. In the past, most of the interest in these lines and the absorption lines (ionized calcium strongest) has been for the purpose of measuring radial velocities by the Doppler shift—for internal motions [see p. 222], or for refining the Hubble law [see pp. 23 and 135], or for using that law to obtain distances. I have measured the strengths of the emission lines in galaxy spectra with the idea of finding the mix of stars (blue giants, red giants, yellow, red, and white dwarfs) that will provide sufficient ultraviolet light to create H-II regions and match the colors and spectra of various types of galaxies [see p. 118].

Recently several other efforts have been made to understand both emission lines and absorption lines in the spectra of galaxies. D. E. Osterbrock and R. A. R. Parker (University of Wisconsin) worked on the peculiar galaxies with strong broad emission lines first studied by Carl Seyfert when he was at McDonald Observatory in 1938. From measurements of the line intensities in NGC 1068, Osterbrock and Parker obtained the density and temperature (over 8000°K) of the interstellar electron gas which they conclude must be heated by collisions of high-velocity cloudlets. G. R. Burbidge and E. M. Burbidge (San Diego, California) studied the intensities of $H\alpha$ and the forbidden red line of ionized nitrogen in spectra of normal galaxies and found systematic differences reflecting different densities and temperatures in Irr-, S-, and a few E-type galaxies. W. W. Morgan (Yerkes Observatory) finds that there is little or no $H\alpha$ emission (no H-II regions) in spiral arms that are nearly circular.

For the past six years Hyron Spinrad (first in Pasadena, then in Berkeley, California) has been studying absorption-line intensities in galaxy spectra in an attempt to measure the mix of giant and dwarf stars. In Prague he reported recent results of accurate intensity measures in the spectra of the nuclei of M 31 and M 81. About half of these measures (at wavelengths between 3800 A in the ultraviolet to 10,700 A in the infrared) were in absorption lines. He and B. J. Taylor matched these intensities by adding together the spectra of various types of stars, first getting the combined color about right, then the absorption-line intensities. This showed that a large fraction of the stars in the nuclei of both galaxies are red dwarfs and infrared stars with low luminous efficiency, or high mass/luminosity. This ratio "M/L" amounted to 34 for M 31's and 51 for M 81's nucleus (consistent with Kinman's recent studies at Lick, and approaching the value of 90 that I found for E-type galaxies in 1960). Spinrad's final fit is good to about 5 per cent, but predicts too weak TiO bands and Na (sodium) line. Very high metal abundance must be assumed, showing that the stars in a galaxy's nucleus are probably second generation, containing heavy elements "brewed" by nuclear reactions in earlier giant stars. Along this same line, J. S. Mathis (Wisconsin) has found that the interstellar helium content in three galaxies is about the same as in the Milky Way.

Discussion revealed that refinements are necessary in Spinrad's work to take account of interstellar absorption (strengthening some absorption lines, and reddening the color) and white dwarfs (which might increase M/L still further). Also, the small number of hot blue stars may not be sufficient to account for emission lines.

W. W. Morgan reported similar efforts to classify galaxy spectra, each as a mix of three components: "Orion-type" blue stars, "amorphous" Population-II stars, and "intermediate" types. A spectrum of M 31 can easily be classified as amorphous near the nucleus and intermediate in the arms. Morgan has classified many spectra in this way, including composite spectra of globular clusters and the Sagittarius star cloud (which is similar to the M 31 nucleus).

More quantitative measurements of intensities in galaxy spectra (though in less detail than Spinrad's) are being carried out by several astronomers, included E. Vandekerkhove (Belgium), G. de Vaucouleurs (Texas), and the writer. In Prague I reported on the use of a cascaded image tube for this purpose (see Fig. 100). This image tube (designed and built by Merle Tuve and W. Kent Ford, Jr., at the Carnegie Institution, Washington, D.C.) simply strengthens the image to be photographed. It produces an image five or ten times brighter than the original on a phosphor screen (like a small TV tube), and this can be

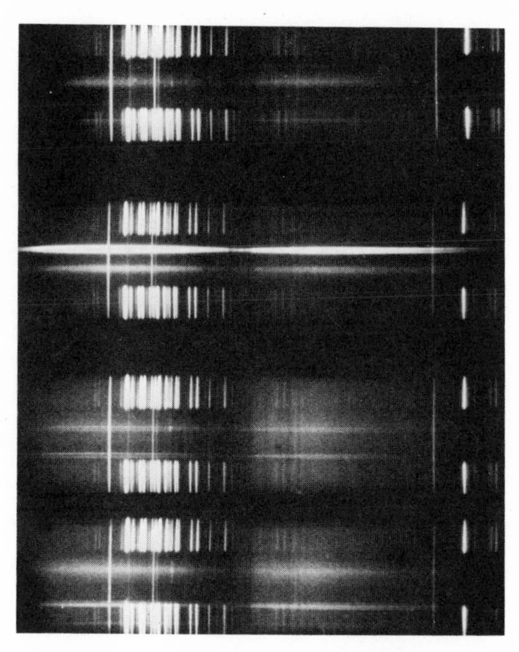

FIG. 100. Spectra of three galaxies (NGC 1796, 5039, and 4835) and a star, obtained with a Carnegie image tube in the spectrograph at the Newtonian focus of the Córdoba 61-inch telescope. The long vertical lines are atmospheric. Crowded comparison lines are from neon; yellow is at the left, and the region from center to right is second-order blue to yellow. (Courtesy Córdoba Observatory)

photographed in a fifth to a tenth of the normal exposure time, making the 61-inch reflector at Córdoba, Argentina, about as "fast" as a 200-inch telescope. [Córdoba's southern latitude makes possible observations of galaxies in the southern skies not visible from Palomar or other northern observatories.] Moreover, the spectrum on the phosphor screen of the image tube is all of one color (greenish white), which simplifies the calculation of accurate intensities from spectrograms like those shown in Figure 100.

When intensity curves are plotted for the spectra of over 150 galaxies already observed in this way, we will have evidence of the stellar mixes in various types of galaxies (Irr, S, SB, and E). Similar data are already available from **UBV** color measurements by a number of astronomers, including D. B. Wood (Berkeley), who has estimated star mixes. H. L. Johnson (Arizona) has extended these measures into the the infrared for ten galaxies, and finds that galaxies must contain many infrared stars (confirming Spinrad's results).

Masses and Luminosities

It is natural to think that photographs like Figures 69, 71, 75, 92, and 93 show the actual distributions of material in galaxies, although this interpretation rests on an incorrect assumption that the luminous efficiency is everywhere the same. In fact, the ratio of mass to luminosity (M/L) differs widely between spirals and E types, and even between the nucleus and arms in one spiral. Current research interest centers on three broad questions: How is the material spread out in galaxies of various types? What fractions of galaxies have masses and luminosities of various sizes? and Why does the ratio M/L differ from one location to another? All three are closely associated with the problem of star formation, or the conditions under which nonluminous matter becomes luminous.

Masses of galaxies, and the density distributions of mass within galaxies, are derived from measured motions and three assumptions: (1) that these motions are caused by gravitation, (2) that they are symmetrical, and (3) that they have "settled down" to long-term stability. The rotations of single galaxies, the orbits of galaxies in pairs, and the random motions of individual galaxies in clusters, all lead to mass estimates which I recently collected for the plots shown in Figure 101. This had to be done with care, since the calculated mass (and luminosity) in every case depends on the distance of the object; in most cases the distance (D) was derived from Hubble's law of red shifts, D = v/H, but different astronomers used different values for the constant, H, now thought to be about 100 km/sec per million parsecs. Aside from the

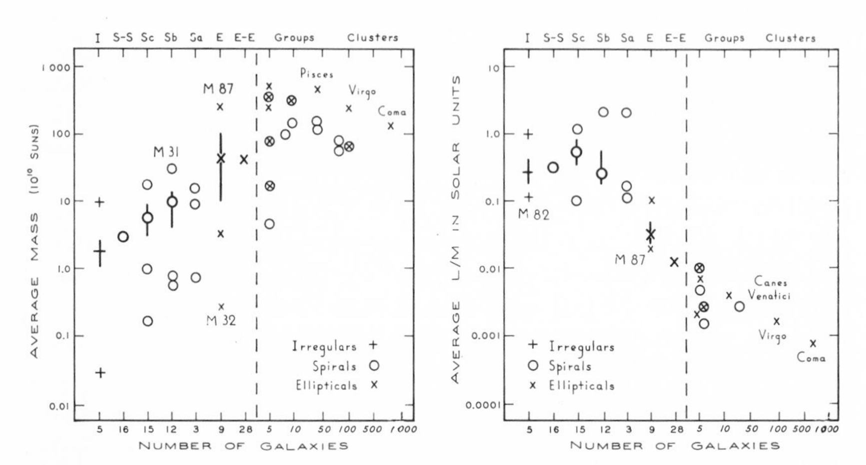

FIG. 101. Estimates of masses and the luminous efficiency (L/M) for galaxies: 45 single, 88 in pairs, 6 groups, 4 clusters. On the left, mass (M) is plotted on the vertical logarithmic scale for various types of galaxies in groups of different sizes. On the right is a similar plot of L/M in solar units. To the left of the dashed lines the number of galaxies indicates how many single irregulars (I), double spirals (S-S), single Sc's, etc., have been averaged in the mean value (dark symbol) for each type. To the right of the dashed line, the numbers in italics indicate the approximate number of galaxies in the groups and clusters plotted. (From Smithsonian Astrophysical Observatory *Special Report* 195)

three assumptions above, there are other uncertainties and errors in the observations, so that most of these masses (M) and values of L/M may be half to twice the estimated amounts plotted in Figure 101. Recent additions by the Burbidges (San Diego), F. Bertola (Italy), and R. Duflot (France) have not been plotted in Figure 101 but would not change it significantly.

The range in values of M (Fig. 101) is very large—from less than 10 billion solar masses to 500 times larger than that. The range in L/M is even greater, and this has led many astronomers to doubt the large mass estimates. Of course, the assumptions (1), (2), and (3) could be wrong; the most questionable is (3) when applied to the stability of groups and clusters of galaxies. However, there is no reason to expect all galaxies to be of nearly the same mass. Stars are known to have masses as small as 0.05 sun and as large as 60 or 70 suns. The luminous efficiency of individual stars ranges from less a thousandth of the sun's to a thousand times the sun's, but one might expect the average L/M for the billions of stars in a galaxy to fall in a narrower range.

The average L/M (or M/L) in galaxies has been studied by many astronomers, including H. J. Rood (Wesleyan University), who has measured accurate luminosities for 315 galaxies in the Coma cluster. When all measures are combined, as in Figure 101, an interesting feature shows up. This was first noted by A. Poveda (Mexico) in a different form, and more recently by I. D. Karachenzev (Soviet Armenia) and the writer. It is the trend of L/M in Figure 101 from high values for spirals to lower values for E types, lower yet for groups, and smallest of all for clusters of E-type galaxies. J. P. Pskovsky (Moscow) points out that L/M increases with increasing M among spirals and Irr types, and S. Huang (Northwestern University) theorizes that the high luminous efficiency of these types is due to their high content of neutral hydrogen (H I). Erik Holmberg (Sweden) had previously concluded that higher initial density of matter caused stars to form earlier in E-type galaxies, thus accounting for their old, Population-II stars and small L/M. The H-I, H-II, and blue-star content of spirals and Irr types Holmberg explains as a result of their low density and consequent slow rate of star formation, which is still going on. In 1964 I did some calculations roughly confirming this idea, and more recently L. Gratton (Italy) has repeated them more accurately. On the other hand, Ambartsumian (Soviet Armenia) thinks that some E types are increasing in luminosity, so that L/M would increase with age—just opposite to Holmberg's conclusion.

Studies of the differences between spirals and E types have been hampered by lack of data on the latter. Rotation is difficult to detect in their spectra. Ivan King (Berkeley) has recently made detailed studies of the luminosity in different parts of E-type galaxies, and with R. Minkowski he has measured rotation in a few, confirming the high values of M/L (about 50). J. L. Sersic (Córdoba, Argentina) concludes from his observations that NGC 6438 is in the process of splitting into two parts. He has collected data on sizes and masses of E types, and finds evidence suggesting that smaller ones are fragments of previous large E-type galaxies, but he has no explanation of the enormous energy required to blow a galaxy apart [see pp. 166–167].

Radio astronomers have recently observed rotations in many galaxies, using the Doppler shift in the 21-cm H-I radio radiation, and these have been used to estimate masses (including six plotted in Fig. 101). M. S. Roberts (NRAO) found that he could measure the 21-cm radiation much farther from the nucleus of NGC 925 than optical spectra can be measured—but his rotational velocities fit nicely onto the optical ones. At Prague, J. Heidmann (Paris Observatory) reported on his observations of the width of the 21-cm emission line in fifty-five small galaxies, mostly of SB type, with the large radio telescope at Nancay, France. From these measures he derives approximate masses (which he admits may be three times too large or too small) based also on optical measures of size and tilt. His mass estimates for Irr and SBc types are smaller than for SBa types, and the L/M values are about 0.10 (as in Fig. 101, where SB-type galaxies and normal spirals are averaged together).

The proportions of galaxies that have various luminosities were studied by Edwin Hubble at Mount Wilson in the thirties. A plot of the percentage of galaxies having each luminosity (or absolute magnitude) is called "the luminosity function," which Hubble found difficult to derive from observations because the brightness (apparent magnitude) of each galaxy we see combines both its luminosity and its distance. Errors in the distances used to calculate the luminosities can introduce large uncertainties in calculated L. However, the galaxies in a cluster are all at about the same distance, so their relative luminosities are the same as their relative observed brightnesses. George Abell (University of California at Los Angeles) has studied the luminosity functions of galaxies in clusters for the past ten years, and reported his latest findings to the IAU. Most of the cluster distances are well known from red-shift measures and the Hubble law, so Abell's curves can be plotted for true luminosities (expressed as absolute magnitudes).

The brightest galaxies in clusters have absolute magnitude about —23.5 (luminosity about 2000 billion suns) and there are more and more galaxies of lower luminosity. For some reason, there is a peak at absolute magnitude −20.5 (120 billion suns). After a small decrease, the luminosity function increases for lower luminosities; that is, there are more and more fainter galaxies, to the limit observable on Abell's photographs made with the Palomar Schmidt telescope. He finds the same curve for each of half a dozen clusters. Detailed inspection of the photographs shows that the excess galaxies at luminosity 120 billion suns are all E type.

Hubble thought that the luminosity function peaked at absolute magnitude −14 (40 million suns) and dropped off nearly to zero for galaxies less luminous than 20 million suns, but Fritz Zwicky (Cal Tech) has long maintained that the curve keeps on rising, so that there are always more galaxies of lower luminosity. H. C. Arp (Mount Wilson) has recently found a pair of small compact galaxies of absolute magnitude −12.7 (luminosity 10 million suns), not much larger than a globular cluster. This raises the disturbing possibility that Zwicky is right, that galaxies are not a well-bounded class, and that the smaller ones merge with objects the size of star clusters in the Milky Way.

Clusters of Galaxies

There are similar difficulties in defining clusters of galaxies. Many galaxies occur in pairs, others in small groups of five to ten, and in the large clusters containing hundreds of members there are internal pairs and groups. Abell, Karachenzev, and others feel that there are also groups of large clusters, although A. G. Wilson (California) claims to have evidence of an even spacing, the clusters being like tennis balls packed closely in a carton. The observational basis for a good statistical study was provided in 1955 by the Palomar Sky Survey and by the similar survey under C. D. Shane at Lick Observatory. George Abell catalogued clusters of various sizes and richness from the Palomar Atlas; C. D. Shane and C. A. Wirtanen counted galaxies in 10′ squares on the Lick photographs, and analyzed these counts with J. Neyman and E. Scott (Berkeley) for "clumpiness." (Part of the clumping is due to nearby interstellar dust clouds in the Milky Way, which obscure galaxies behind them.) In a recent study of all these data, T. Kiang (London Observatory) confirms Abell's result that clustering occurs on all scales, although Fritz Zwicky (Cal Tech) concludes from his own data that there is no clustering beyond the largest ones, which are about 160 million light years apart.

The difficulties of work on clusters are illustrated by three recent studies: G. Omer, T. Page and A. G. Wilson at Cal Tech counted galaxies in the region of the large Coma cluster and determined that the cluster's true diameter is about 3° rather than 5° or larger, as previously thought. It has about eight hundred member galaxies on the Palomar Atlas plates, but they are mixed up with many background galaxies and a few foreground galaxies. If the luminosity function follows the pattern set by Zwicky and Abell, there are many more member galaxies too faint to be seen on 48-inch Schmidt photographs.

B. E. Markarian (Soviet Armenia) finds "chains" of about ten galaxies each that he considers were formed together, and H. C. Arp finds peculiar galaxies halfway between pairs of quasars. Many astronomers consider these small groupings to be the result of chance and of the vast numbers of galaxies on the Palomar Atlas photographs. However, B. A. Vorontsov-Velyaminov and his co-workers in Moscow have catalogued peculiar and interacting galaxies and find that more than half of all galaxies have faint companions.

The pairs (S-S and E-E), groups, and clusters noted in Figure 101 are generally accepted because they stand out from the background, but some of them can be criticized as chance groupings, and no one can claim that we have identified all their members. Some astronomers agree with I. D. Karachenzev, whose recent study concludes that the giant clusters in Coma and Virgo are unstable, and will disperse during the next few billion years. (This vitiates their average galaxy masses plotted in Fig. 101.) Others feel that the large clusters are stable and that their excessive mass is due to as yet unseen matter. For instance, Vera Rubin (Carnegie Institution) reports 230 blue objects in the Virgo cluster, each of which is probably a compact galaxy. It is possible that nonrotating galaxies collapse under their own gravitational force and become so dense that space curvature prevents our seeing their light. Highly ionized intergalactic gas is another possible form of unseen matter in clusters of galaxies and even in the vast spaces between them. Its effect on the spectra of quasars is under study, as noted below. Ordinary dust is ruled out by the fact that its reddening and dimming of distant galaxies is not observed, but large chunks of solid material (like meteoroids) could be present and invisible.

Part of the interest in unseen material comes from the cosmologists, many of whose relativistic models of the universe seem to require 100 times greater density of material than provided by galaxies of normal mass (about 10^{11} solar masses, or 2×10^{44} grams each), spread out uniformly (about 10 million light years apart) in the numbers we see. This observed density is about 10^{-31} gm/cm^3, whereas the cosmologists prefer about 10^{-29} gm/cm^3. Recently J. P. Pskovsky (Moscow) re-estimated the observed density, using the large cluster masses plotted in Figure 101, and got 7×10^{30} gm/cm^3, much closer to theoretical models.

X-Ray Emission

Before turning to radio galaxies and quasars, I must mention the discussion of recent work on X rays from astronomical sources [see p. 212]. This region of the spectrum (from less than 1 A- to 30 A-wavelength) was first observed only five years ago from rockets shot above the atmosphere (which is opaque to X rays), and has yielded surprising results—such as the fact that some galaxies and quasars are emitting strong X rays. The first job, of course, is to establish just how the X rays are emitted; then the measured X-ray emission and spectra of various sources can be used to learn more about the content, structure, and history of galaxies.

In a special discussion at Prague, Herbert Friedman summarized his work at the Naval Research Lab, Washington, D.C., the work supervised by R. Giacconi at American Science and Engineering in Cambridge, Massachusetts, and that supervised by P. Fisher at the Lockheed Lab in California. The NRL group has concentrated on surveys to detect X-ray sources over as much of the sky as possible, and Giacconi's group has studied individual sources in more detail. The early X-ray "telescopes" were ionization counters behind slits or honeycomb baffles that limited the "X-ray view" to a region of the sky about eight degrees square. A large region of the sky was scanned as the rocket rotated during its four or five minutes above the atmosphere. Most of the three dozen X-ray sources are near the Milky Way and are probably nearby objects. The Crab nebula is one, and H. M. Johnson (Lockheed Lab) reported another identification to the IAU—an old nova in Scorpius. He noted the interesting fact that two of the X-ray sources are also radio sources.

Figure 102 shows the difficulty of identifying X-ray sources. The map at the left is a plot of Friedman's most recent scan across part of the Virgo cluster of E galaxies made with a $1° \times 8°$ field of view from an Aerobee rocket on May 17, 1967. The scan's X-ray counts, plotted on the right, show the extragalactic X-ray sources (established by these and previous observations): the quasar 3C 273 and the E-type radio galaxy M 87 with two jets (the radio source Virgo A). Three other, new sources of somewhat lower X-ray intensity are not yet identified at declinations 4°.7, 7°.0, and 10°.3. Each count on the intensity scale is an X ray of wavelength between 1 A and 10 A. Knowing the distances of 3C 273 and M 87 (1.6 billion and 32 million light years, from their red shifts), Friedman and E. T. Byram (also of NRL) calculate that these galaxies radiate 7.3×10^{45} and 1.5×10^{43} ergs per second in

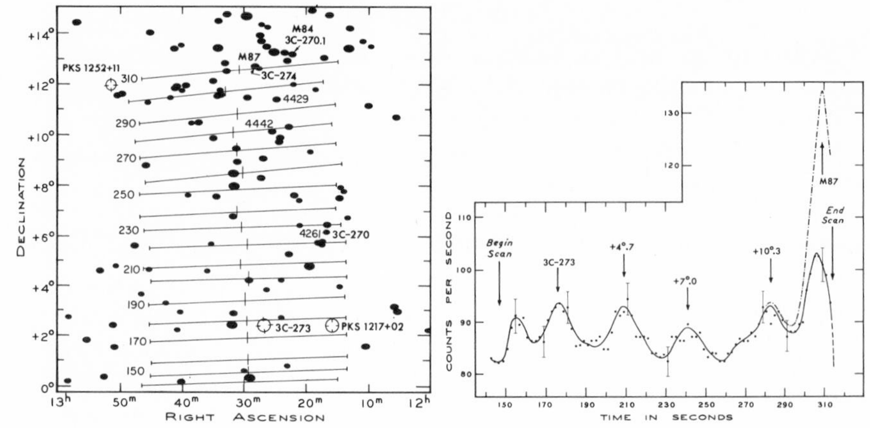

FIG. 102. Evidence for X-ray sources. The map on the left shows elliptical galaxies in Virgo (dark spots) and three radio sources (open circles). The lines show successive positions of the X-ray-scanning slit as the NRL rocket rotated northward for 170 seconds (times given near left edge) above the atmosphere on May 17, 1967. In the chart on the right, the recorded X-ray counting rates are plotted against time; five X-ray sources are shown. After 280 seconds from launch, the counting rate is corrected for increasing atmospheric absorption of 1- to 10-angstrom X rays as the rocket fell back toward the earth (dashed curve). (Courtesy U.S. Naval Research Laboratory)

X rays (1 A to 10 A)—about seventy times their total radio outputs, and about equal to their optical outputs. Most of this energy output seems to result from the synchrotron process (high-speed electrons and ions spiraling in a magnetic field), although it is difficult to explain where these speeding charged particles come from, and how new ones are provided to replace the energy radiated.

At Prague, R. Giacconi discussed other X-ray sources in the Milky Way, many of them in Scorpio and Cygnus. By using different counters and different thicknesses of Mylar windows as X-ray filters, he, George Clark, and H. Bradt (MIT), and L. Peterson (San Diego) measured X rays at various wavelengths (energies) from about 0.2 A to 21 A (60,000 to 600 electron volts). They find four different types of spectra, one with an absorption line. All the spectra decrease toward high energies (E), roughly proportional to $1/E$, as expected for synchrotron radiation. (Because X-ray quanta are larger at shorter wavelengths, the counting rate is proportional to $1/E^2$.) The galaxy spectra are a bit steeper than this, but no major "break" in the spectrum curves has yet been found.

During April 1967 a British group led by K. A. Pounds (Leicester University) made the first X-ray scans of the southern Milky Way from a Skylark rocket flight in Australia. The group reports one very strong source in Centaurus and ten others in Vela and Taurus (aside from the Crab nebula). It also noticed that the strong Scorpius X-ray source seems to have changed (although American observations show no variability larger than about 20 per cent).

W. H. Tucker (Cornell) discussed the possible ways that X rays might be produced, and concluded that most of them come from the synchrotron process in a magnetic field, possibly augmented by radiation from clouds of very hot gas. Near the source, interstellar material would be heated by X rays and should show several absorption edges at wavelengths longer than 30 A (not yet observed). V. B. Zeldovich (Moscow) proposed that gas falling into a very dense neutron star would hit the surface at speeds over 100,000 km/sec and generate X rays as they are generated in a doctor's X-ray tube. One difficulty is that the remaining gas and the surface of the neutron star would be heated to 25 million degrees. It was then suggested that a larger collapsed mass might have sealed itself from our view by the relativistic effect on space curvature, in which case X rays can still be produced by gas atoms sucked toward the collapsed mass and colliding on the far side. If this were true, X rays may show the presence of "invisible matter."

There was also discussion of the X-ray background between sources in scans like those used for Figure 102. If it is due to more distant X-ray sources, like the quasar 3C 273, there must be five quasars for each square degree of the sky. Evidence tending to favor this explanation is that the spectrum of the X-ray background matches the X-ray spectra of galaxies, and that the background seems to be about equal all over the sky, without any peak in the Milky Way (expected if it were of local origin). Another possibility is that there is an intergalactic gas consisting of ionized hydrogen at a temperature of about 1,000,000°K.

This X-ray background should not be confused with the 3° background radiation in radio wavelengths of a few centimeters discovered by A. A. Penzias and R. W. Wilson at Bell Telephone Labs in 1963. The latter is explained as the remnants of radiation that filled the whole universe nearly 10 billion years ago, shortly after the "Big Bang" of evolutionary cosmology. Although it then corresponded to a very high temperature, this radiation was "cooled" by expansion of the universe until it now corresponds to only 3° above absolute zero. Recent radio observations at wavelengths of a few centimeters confirm the 3° spectrum, and George Field (Berkeley) has shown that its effect on interstellar CN molecules, observed in spectra of distant stars, also confirms the 3° temperature at a wavelength of 2.7 millimeters.

Radio Galaxies and Quasars

Radio observations in the early fifties revealed that many giant elliptical galaxies are strong sources of radio emission; these galaxies are now called radio galaxies. The 21-cm emission line of atomic hydrogen (H I) is a separate phenomenon. A radio galaxy has a continuous radio spectrum with energy decreasing toward higher frequencies (shorter wavelengths), usually in the manner expected from the synchrotron process (as first shown by the Russian, Shklovsky, in 1950). Some peculiar E-type galaxies (such as M 87 with its "jet" that confirmed the synchrotron process) were found to have radio emission. Hundreds of radio sources catalogued at Cambridge, England (the smaller ones known by their numbers in the *Third Cambridge Catalogue*—"3C"), were also identified with nebulae in the Milky Way, but many remained unidentified with optical objects until Maarten Schmidt at Cal Tech recognized late in 1962 the first quasi-stellar radio source, 3C 273, as a fairly faint, slightly fuzzy star with a large red shift ($\Delta\lambda/\lambda = 0.16$). By 1964 there were eight more identified "quasars," and by now there are several hundred (103 with measured red shifts). Because the radio quasars were found to be optically bluish, Allan Sandage (Mount Wilson) began a systematic optical search for hazy faint blue stars in regions of the sky far from the Milky Way. With Willem Luyten (Minnesota), G. Haro (Mexico), T. D. Kinman (Lick), and others, he established a new class of "blue stellar objects," or "radio-quiet quasars," which also have large red shifts. Harlan Smith (Texas), J. B. Oke, T. Matthews, A. R. Sandage, and others found that many of the quasars and one galaxy are optically variable.

As larger radio telescopes with better resolution were brought into action, it was found that quasar radio emission comes from very small regions, usually two on either side of the optical object, and that the radio emission also varies. A great deal of interest was generated in the extreme conditions implied by the small size, the large red shifts (up to 2.1), and the peculiar spectra. It was (and is) generally assumed that the large red shifts are "cosmological" and obey Hubble's law, so that the quasars are at billions of light years' distance and have very large luminosity (10^{46} ergs/sec, or 10^{13} times the sun). Since no explanation was readily available for such large energy sources and their rapid fluctuations, several astrophysicists accepted the suggestion of James Terrell (Los Alamos) that quasars could be much nearer ("local"), much smaller, and much less luminous, their large Doppler shifts being due to some other cause.

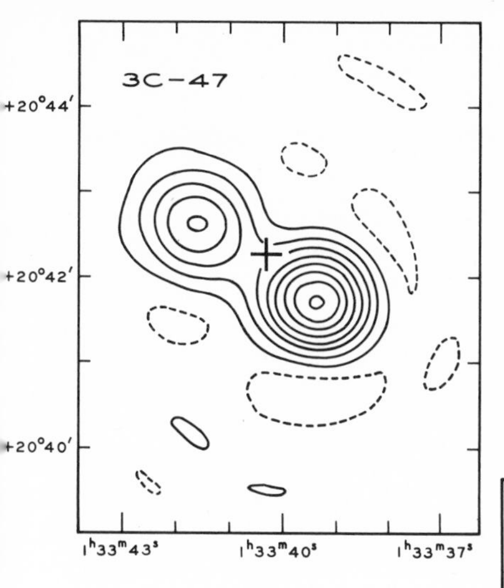

FIG. 103. Isophotes of the measured radio emission from 3C 47. Location of the optical quasar is shown by the cross about halfway between the two peaks, which are about 40″ apart. (Courtesy of Sir Martin Ryle)

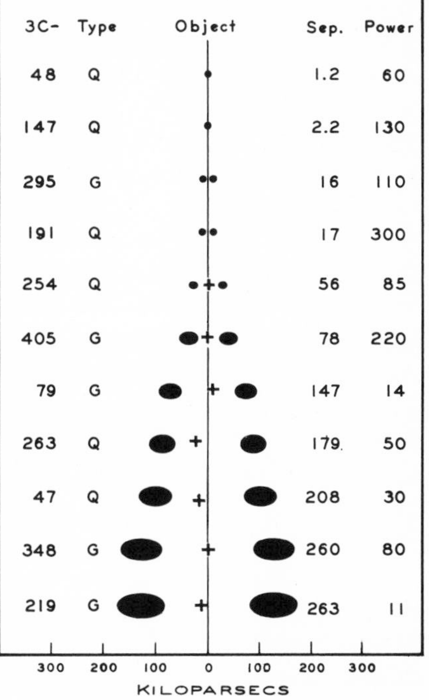

FIG. 104. Separations and radio-power outputs of eleven double radio sources identified with optical quasars (Q) or galaxies (G) for which red shifts have been measured. The separations are given in kiloparsecs, and the radio power in units of 10^{25} watts per steradian per cycle-per-second, assuming "cosmological" distances derived from Hubble's law. The black ellipses show the size and separation of the radio clouds, and the crosses show the location of the optical object. (Courtesy of Sir Martin Ryle)

This background, and more, was summarized by Allan R. Sandage and Sir Martin Ryle (Cambridge) at a special and very popular "evening discourse" in Prague on August 28. Much of what they had to say refers to the controversy over "local" and "cosmological" distances of the quasars, and part of this was repeated in a one-sided debate on August 30. Ryle has observed radio sources for fifteen years (he was knighted for this work) and he leads the Cambridge group of radio astronomers. He based his arguments on recent radio observations of high resolution and sensitivity. The new long-base-line radio interferometers have resolutions better than 0.01 second of arc, and the new large "dishes," "bowls," and antenna arrays can detect radio sources of "brightness" less than 0.01 flux unit, or 10^{-28} watt per square meter per cycle per second.

The improved resolution allows accurate measures of the separations of two centers of radio emission in each quasar as well as the wider separations of double radio sources associated with 60 per cent of the radio galaxies. Figure 103 is a plot of the radio emission from 3C 47, showing two components about 20″ either side of the optical image, which has red shift $\Delta\lambda/\lambda = 0.425$. Using Hubble's law, the distance of 3C 47 is 3.4 billion light years and the radio-cloud separation is 208 kiloparsecs (680,000 light years). This separation, and ten others, are shown in Figure 104, together with the radio power outputs (P_R in units of 10^{25} watts per steradian per cps) computed from the Hubble-law distance[1] and the measured radio flux (F_R in units of 10^{-26} watts per square meter per cps), according to the formula $P_R = D^2F_R$. (P_R is analogous to intrinsic luminosity in optical measurements.) If a smaller distance (D) were used, on the "local" quasar hypothesis, the six quasars would have separations and radio powers much smaller than the five radio galaxies plotted in Figure 104 (a point previously noted by L. M. Ozernoi in Moscow). Since few doubt the large distance of radio galaxies, Ryle argues that quasars are also at "cosmological" distances. (He also uses the measured separations in a theoretical model of quasars, described below.)

A similar argument, stressed by Sandage, is based on calculated radio surface brightnesses (B_R) of galaxies and quasars by David Heeschen,

[1] Computing the separations is more complicated than indicated here. The Hubble law of red shifts gives $D = v/H = c\ \Delta\lambda/\lambda\ H$ correctly for small red shift ($\Delta\lambda/\lambda$ less than 0.1). But for red shift greater than 0.2 the relativistic formula (p. 185) must be used instead of $v = c\ \Delta\lambda/\lambda$, and the distance is so large that the curvature of space-time changes the geometry. Although the separation (S) can be computed from the angular separation (a) in seconds of arc (Fig. 100) and the red shift, the formula is not the simple one, $S = D \sin a = c\ \Delta\lambda \sin a/\lambda\ H$, which is correct for "flat" Euclidean space. Various cosmological theories give different curvature, therefore different formulae for S. In his studies Ryle uses the Einstein–de Sitter cosmology according to which

$$D = cz\ (2 + z)/(1+z)(2 + 2z + z^2)H$$

where z stands for the red shift $\Delta\lambda/\lambda$, and H is the Hubble constant, 100 km/sec per megaparsec. For the radio power output (P_R), one expects in flat space that $P_R = D^2f_R$, where f_R is the measured radio brightness or flux density. But in the Einstein–de Sitter cosmology it is

$$P_R = f_Rc^2z^2(2 + z)^2(1 + z)^2/(2 + 2z + z^2)H^2$$

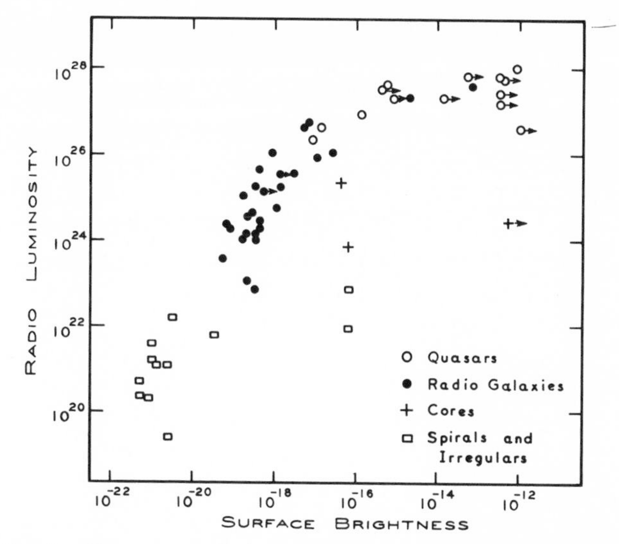

FIG. 105. Total radio output (P_R) at about 1400 megacycles plotted against radio surface brightness (B_R) of quasars, radio galaxies, spirals, and "cores" (nuclei). (Arrows indicate plotted values that are lower limits to B_R.) Neither quantity has been corrected for red shift. If quasars were local, their values of B_R would be the same, but P_R would be millions of times smaller, and the quasars would not match the radio-galaxy curve, as they do here. The three cores and two peculiar spirals (NGC 1068 and M 82) do not fit the general trend. (From *The Astrophysical Journal*)

using radio fluxes measured with the 300-foot dish at the National Radio Astronomy Observatory, Green Bank, West Virginia. Heeschen also calculated the radio luminosity, or total radio power output, P_R. Figure 105 is a plot of P_R versus B_R (at 1400 megacycles frequency) showing that the quasars fall on the same curve as the radio galaxies if their red shifts and distances are "cosmological."

The large power outputs (totaling over 10^{61} ergs during the estimated lifetime of a quasar—about the same as for a radio galaxy) are expected to cause quasars to expand. Ryle points out that quasars probably evolve into radio galaxies, and that the production of such high power output is well explained by the synchrotron process. No one understands fully where the high-speed electrons come from, and what produced the magnetic field, but synchrotron emission matches the radio spectra of both quasars and radio galaxies. At higher frequencies (f), the radio flux decreases as $f^{-0.7}$; in the lower frequencies it has a cut-off. The "cut-off frequency" is higher for quasars than radio galaxies, implying that quasars are smaller and more concentrated, as observed.

Sandage blames the "local hypothesis" on the observed rapid light variations in quasars. Some of these variations occur in a few days and imply (since no physical effect can travel faster than light) that the light source is only a few light days (0.01 light year) in radius. At distances of a billion light years or more, these sources would have angular diameters less than 0.003 second of arc, but this is not inconsistent with observations showing that the variable light source is starlike, and the diameters of radio-emitting quasar clouds are less than $0''.02$. Such a small synchrotron model is difficult because of its high density, but L. Woltjer (Columbia) has shown it possible if the magnetic field has the proper pattern—a pattern that fits the observed polarization. The density cannot be large enough to cause appreciable gravitational red shift, since the measured red shifts of several emission lines are the same as for absorption lines in each of dozens of quasar spectra. If the red shifts were gravitational, they would differ systematically, and in any case, the high-gravitational field of such a collapsed mass would prohibit low-density gas clouds where the observed emission lines must originate.

The absorption lines in twenty quasar spectra have a somewhat smaller red shift, indicating that clouds of cool gas are moving outward from each quasar—and that each is losing mass. This is consistent with the ejection of gas clouds to produce the double radio sources, and implies that the quasars used to be more massive than they are now—probably at least 10^{10} solar masses. Sandage is convinced that all this evidence hangs together, showing that quasars were previously galaxy-sized objects similar to Morgan's N-type, or Zwicky's compact galaxies, or the Seyfert type with peculiar nuclei. He stressed the precision of Hubble's law ($V_r = HD$) for clusters of galaxies out to red shifts of 0.5 or so, including many radio galaxies.

Ryle reported to the IAU some very interesting results of recent Cambridge radio surveys which indicate a peak in quasar explosions some 8 or 9 billion years ago. Assuming that their radio outputs (P_R) are all about the same, the numbers of quasars at various distances can be computed from the numbers observed with radio brightness greater than 10, 5, 2, 1, 0.5, 0.2 flux units, and so on, down to 0.01 flux unit. The new observations confirm his earlier view that there are more quasars per unit volume at distances corresponding to red shift $\Delta\lambda/\lambda = 2$ than there are nearby, and also show that this "population density" of quasars drops off rapidly at distances corresponding to $\Delta\lambda/\lambda = 3$. Of course, we are "seeing" these distant objects as they were 8 or 9 billion years ago, and Ryle follows Maarten Schmidt's conclusion that most of them "exploded" then—very few of them earlier. The population density extrapolated outward to greater distances (earlier times) is still adequate to account for the unresolved radio background. (This is not the 3°K radiation mentioned earlier, but the lower-frequency radio emission of more distant quasars and radio galaxies.)

Geoffrey R. Burbidge (San Diego) had the last word in rebuttal against "cosmological" quasar distances. He claimed that most of the "proofs" given by Ryle, Sandage, Woltjer, and others are inconclusive, and that it would be premature to accept large distance estimates before the theoretical models of quasars are worked out in more detail. For instance, it may be that their red shifts are *partly* gravitational and *partly* due to peculiar motions. He stressed the apparent "lineups" of quasars with H. Arp's distorted galaxies, which are certainly not billions of light years distant.

Another puzzling observational fact is that the red shift of absorption lines differs from the red shift of emission lines in the spectra of several quasars. These absorption-line red shifts are all about 1.95, and Burbidge suggests that this preferred value indicates a new phenomenon, different from cosmological Doppler shifts. However, S. Shklovsky (Moscow) offered a cosmological explanation which is not yet worked out in full detail. He assumes that the 1.95 red-shifted absorption lines are due to uncondensed gas clouds in galaxies between us and the more distant quasars with emission-line red shifts of 2.0 or 2.1. It is possible that the expansion of the universe was not uniform in the distant past; as we look back in time 7 or 8 billion years, the Hubble law of expansion may be found to change toward the "static universe" first proposed by Einstein in 1918.

If Shklovsky is right, the history of the universe does not lead back to a Big Bang 10 billion years ago but leads to a "pause" in the expansion, a pause that ended about 8 billion years ago. During the pause, the expansion of the universe was less than 1 per cent of its value today. (That is, the Hubble constant was about 0.5 km/sec per megaparsec.) This means that there could be a large extent of time before 8 billion years ago, and that the age of a galaxy could be much more than 10 billion years. The pause might explain the large apparent "population density" of faint radio sources and the background radiation mentioned by Ryle. Shklovsky claims that it also accounts for the present average density of matter (10^{-30} gm/cm^3 without assuming "unseen masses"). His preliminary calculations based on Einstein's field equations yield a radius of 15 billion light years, a deceleration parameter (q_0) of -1.3 and a small positive cosmological constant ($\Lambda = 4.3 \times 10^{-56}/\text{cm}^2$).

The energy source, variability, spectrum, and structure of quasars were discussed extensively at Prague. Many aspects have already been discussed in *Sky and Telescope* (September 1967—see p. 289), and I will only summarize them briefly. The variations are generally smallest in low-frequency radio-power output, and larger in higher frequencies (doubling of the output occurs at wavelengths of 2 or 3 cm). Optical variations and spectra give some evidence of two components, the blue continuum varying more than the emission lines. Some of the variations are extremely abrupt (output decreasing by 70 per cent in twelve hours), and all of them are irregular in nature.

Because of their large red shifts, the optical spectra of quasars show lines normally in the unfamiliar far ultraviolet, down to the hydrogen Lyman-alpha line at 1216 A. The lines are broad, indicating large internal motions, and their strengths indicate extreme conditions, although the abundance of chemical elements seems to match ours in the Milky Way.

There is strong evidence that much of a quasar's light and radio output comes from very small regions, the optical ones in different locations from the radio ones. Some of this evidence is based on the rapid light variations; some is based on recent high-resolution radio measurements with special interferometers [see p. 203]. With a base line of over a thousand miles, Canadian astronomers have shown that eight quasars emit most of their radio output from regions smaller than 0″.005. Similar interferometer work is starting at NRAO in cooperation with Berkeley and Jodrell Bank (England). Tape recordings of the radio flux from one source are taken at each radio telescope, together with a standard (accurate) radio frequency signal, and later compared. Signals can be compared directly over shorter base lines, such as the 1-mile one at Cambridge with a resolution of 25″. Even simpler is the "scintillation effect," caused by interplanetary clouds of ionized gas—the "solar wind" near the earth. The effect is similar to poor seeing in an optical telescope; if the flux received from a radio source "twinkles" like a star, the source is less than about 0.3 seconds of arc in diameter—as is the case for many quasars.

The compact double radio sources illustrated in Figures 103 and 104 led Sir Martin Ryle to consider a simple model of their structure, illustrated in Figure 106. He assumes that a sudden release of energy (explosion) in the optical object ejected two gas clouds moving outward at equal speeds (v) in opposite directions. The speeds are large enough that our view (from the bottom of Fig. 106) shows the closer gas cloud at a later time (t_2) than the more distant one (t_1), and therefore at a different distance from the explosion. This explains why the radio-emitting clouds are not equidistant from the optical images in Figures

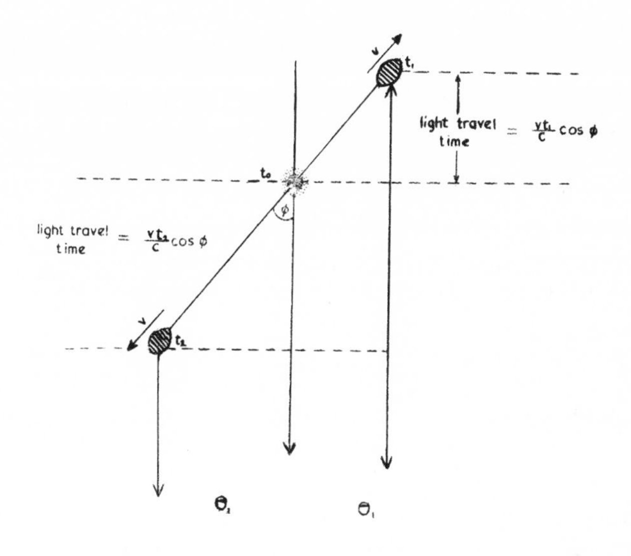

$$t_1\left(1+\frac{v\cos\phi}{c}\right) = t_2\left(1-\frac{v\cos\phi}{c}\right) = t_0$$

FIG. 106. Simple model of a compact double radio source. Two gas clouds moving at equal and opposite speeds (v) were ejected at time t_0 from the galaxy in the center. Due to the different distances from us, we see the lower gas cloud at a later time (t_2) and at larger angular distance (θ_2) from the galaxy than the upper one (t_1 and θ_1). (Courtesy Sir Martin Ryle)

103 and 104, and also why they are not of equal radio-power output (P_R)—evidently P_R changes with time.

From the study of more than twenty such double sources, Ryle can plot P_R versus time after the explosion. He finds that P_R increases by a factor of 10 during the first 100 years, then remains roughly constant for 10,000 years, then decreases by a factor of 1000 in the next million years. Measures of flux at longer radio wavelengths show that P_R rises more slowly, probably because of self-absorption at these longer wavelengths. Between 30,000 and 500,000 years after an explosion the high values of P_R are expected to resemble the radio characteristics of quasars; later these objects become more like radio galaxies, and finally become too weak to detect by their radio radiation. The velocities of gas-cloud ejection are close to the velocity of light except for the Seyfert galaxies, like NGC 1068, where the energy release must have been much slower.

Ryle's recent observations of four components of 3C 84 (the galaxy NGC 1275) show that a stream of ionized gas is now being ejected, and effects of this "jet" are apparent on two other galaxies, 5 million light years away. If this jet has been running at the same power for 10 million years, it must have expended 4×10^{60} ergs, about the total energy output of a quasar.

The differences between power outputs of quasars, and the variations in power output of one, lead Ryle to favor a series of supernova explosions as the energy source for ejecting gas clouds, an idea first proposed by G. R. Burbidge and recently modified by W. H. McCrea (Sussex, England). Other ideas discussed in Prague include the destructive interaction between matter and antimatter, the creation of matter in limited regions, unstable "plasma clouds" (ionized gas in a magnetic field), and the general concept of energy released through gravitational collapse. Hannes Alfvén and his co-workers in Sweden calculate that annihilation of matter by antimatter (atoms composed of positrons, negative protons, and other antiparticles) can produce the radio spectra observed from quasars. Alfvén therefore thinks that quasars are concentrated blobs of one kind of matter surrounded by the other kind. In sharp contrast, W. H. McCrea suggested three years ago that the continual creation of matter postulated in Steady-State cosmology might occur at a rate proportional to the local density, thus adding greatly to the condensed mass in a galaxy's nucleus or in a dense quasar.

Unstable plasma clouds (similar to solar flares, but much larger) might be formed by the gravitational contraction of a large (galaxy-size) gas cloud in a magnetic field. Variations of density, electric charge, and magnetic-field strength in moving plasma can carry large amounts of energy. One idea, discussed by F. D. Kahn (Manchester, England), is that a small region can be filled with an excess of such energy, which is suddenly released as light and radio emission. L. M. Ozernoi (Moscow) prefers a large stable configuration that he calls a "magnetoid." His statistical analysis of quasar light variations shows that they cannot be due to random supernova flashes, so he thinks that a quasar is a single condensed gas cloud with mass equivalent to about 1 billion suns, and a strong ringlike (toroidal) magnetic field. The pulsations and radiation of this "magnetoid" can last for a million years and account for the variations in quasar luminosity, the radio spectrum, and the ejection of material in two opposite directions.

Obviously, there are many theoretical interpretations of the newly discovered quasars. Astronomical observations made over the past three years have narrowed down the possible explanations, and seem to establish cosmological sizes and distances. Sandage estimates on this basis that there are over 100,000 quasars brighter than magnitude 19.7 (visible on Palomar Sky Atlas plates), and many more of them must be studied before a satisfactory model can be accepted by physicists and astronomers.

Although a large majority of the astronomers at Prague were convinced that quasars are at large "cosmological" distances, there are still some doubts. Six months later, T. A. Matthews (formerly at Mount Wilson Observatory) announced evidence that may rule out distances as large as even one billion light years.—TLP

Structural Changes in a Quasar

(*Sky and Telescope*, March 1968)

In the northeastern part of the constellation Coma Berenices is the quasar 3C 287, which is making astronomical news because it may help solve the vexing question of whether these objects are extremely distant or not.

Astronomers first learned of 3C 287 in 1959 as a fairly strong cosmic radio source of very small angular extent. In 1964 a two-color photograph taken with the 200-inch reflector on Palomar Mountain showed in the source's position a very blue, starlike image of magnitude 18. Later that year Maarten Schmidt recorded the optical spectrum with the 200-inch Palomar telescope, measuring three emission lines from which he deduced a very large red shift, $z = 1.055$. (By definition, z is the increase in wavelength divided by rest wavelength.) This value is typical for a quasar.

And now Thomas A. Matthews, University of Maryland, points out that major structural changes appear to have occurred within 3C 287 during the last eighteen years. The evidence consists of five plates taken since 1950, whose interpretation requires great care because of differences in the telescopes and emulsions that were used.

In the course of the Sky Survey with the 48-inch Schmidt at Palomar Observatory in May 1950, blue and red plates were obtained of this region of the sky. Matthews finds that the red plate has an image that resembles a double star incipiently resolved, its northern component being about two magnitudes brighter than the companion some three seconds to the south. In addition, the peculiarly shaped extensions to the south appear much redder than any normal galaxy or star.

On the blue Sky Survey plate, only the northern component is seen, probably because the companion is below the plate limit. But both objects appear on a 200-inch reflector plate taken in blue light by Allan Sandage in 1964. They resemble a double star, without visible nebulosity.

Next come two plates taken by Matthews in the course of a program of high-resolution photographs of identified radio sources. The first was with the 100-inch reflector on April 25–26, 1965, using Eastman IIa-D (yellow-sensitive) emulsion, the second with the 48-inch Schmidt on March 17–18, 1966, upon the same emulsion. In each case the exposure time was as long as possible without fogging by the light of the sky.

As seen in the enlargements reproduced in Figure 107, the 1965 100-inch plate shows the two concentrations noted by Sandage, separated by 3½ seconds of arc. A narrow bridge joins them, its surface brightness comparable to the southern component. Both components' images are starlike, but no extended faint nebulosity is seen.

Strikingly different is the 48-inch plate of 1966. In place of the narrow bridge, a broad area of nebulosity extends from the northern concentration just to the position of the southern one. Careful measurements demonstrate that the nebulosity had definitely increased in brightness relative to neighboring stars and galaxies. Within eleven months the nebulosity had either brightened over a distance of not less than two seconds of arc or had expanded by this amount. If the change had

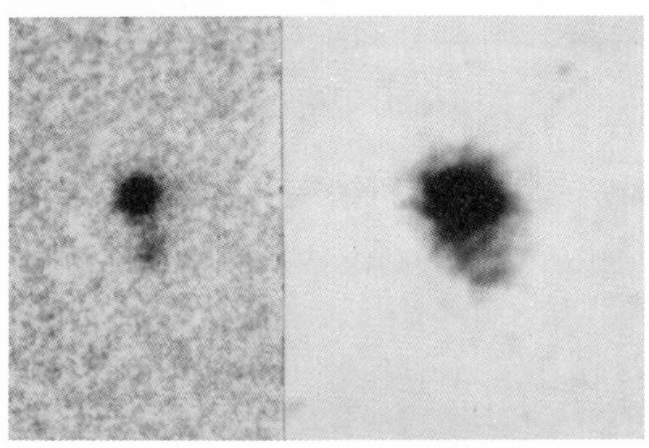

FIG. 107. The negative on the left is enlarged from a plate taken with the 100-inch telescope at Mount Wilson on April 26, 1965; the one to the right is enlarged to the same scale from a plate taken with the 48-inch Schmidt at Palomar on March 18, 1966. Both are on Eastman IIa-D emulsion. After allowing for the difference in telescopes and atmospheric "seeing" (different sizes of star images), there seems to be a change in the gray area that took place in 3C 287 during the intervening eleven months. (Courtesy of T. A. Matthews, University of Maryland)

proceeded at the speed of light (the fastest possible) it could move a little less than 1 light year in eleven months, and since we see this as 2″, the distance to 3C 287 would be only about 100,000 light years. This quasi-stellar radio source would be even nearer to us if the speed were less than that of light. Even if the time span of the change were taken to be sixteen years (from the 1950 Sky Survey plates), the distance could be no more than 1,600,000 light years—much less than the 2 billion light years derived from the red shift in the spectrum of 3C 287.

One proponent of the "local" theory of quasi-stellar objects is James Terrell, Los Alamos Scientific Laboratory. In 1966 he suggested that the large red shifts of quasars are a result of high-speed ejection from the nucleus of our Galaxy. On the assumption that 3C 287 is 100,000 light years away, it furnishes a test of the Terrell hypothesis.

Thus, if 3C 287 is traveling through space directly away from the galactic nucleus, at the speed implied by its red shift, its position on the sky would have changed by 6.4 seconds of arc between 1950 and 1966. But measurements of the photographs showed that the proper motion, if any, amounted to less than one second during the sixteen-year interval. In this respect, 3C 287 does not support the Terrell hypothesis, even though its relative nearness seems to be indicated by the changes in the nebulosity. . . .

The disproof of Terrell's hypothesis is based on the fact that we on earth are located over 30,000 light years from the nucleus of our Galaxy. If 3C 287 had been shot out of that nucleus with a speed of 185,000 km/sec (about 62 per cent of the velocity of light, from the formula on page 185 and the measured red shift of 1.055), it would reach 100,000 light years' distance in about 160,000 years (the time since the assumed explosion in our Galaxy's nucleus), and it would have sideways motion (across our line of sight) of almost 60,000 km/sec, or a proper motion of 0″.4 per year across our sky, which is not observed. At the maximum possible distance consistent with Matthews' estimates (1.6 million light years), the proper motion would be 0″.025 per year—barely detectable on photographs like Figure 107—and the assumed explosion would have taken place over 3 million years ago.

Matthews' evidence that 3C 287 is nearby, and these two calculations, highlight the fact stressed by G. R. Burbidge in August 1967 (see p. 249) that several different theoretical models of quasars can fit the present observational data. (Sandage would claim that the observations of moving nebulosity are doubtful when based on photographs like those in Figure 107.) Suppose, with Burbidge, that 95 per cent of the measured red shift of 3C 287 were not Doppler shift but due to some other effect. Then the radial velocity would be only 15,000 km/sec, and if the distance is 100,000 light years, the proper motion predicted by Terrell would be only 0″.1 per year—barely detectable.—TLP

The Problem of 3C-287

GEORGE S. MUMFORD

(*Sky and Telescope*, September 1968)

The quasar 3C-287 has been identified on photographs as an eighteenth-magnitude blue starlike object, with a much fainter red companion 3.3 seconds of arc away. In December 1967, Thomas A. Matthews (University of Maryland) reported to the American Astronomical Society that these two components were linked by a nebulous bridge, which became much brighter in 1966 (see p. 254). His evidence consisted of five photographs taken since 1950, three of them with the 48-inch Palomar Schmidt, and one each with the 200-inch and 100-inch reflectors.

This change, and even the existence of the bridge, have now been denied by J. Kristian and P. V. Peach, Mount Wilson and Palomar Observatories, in *Astrophysical Journal Letters* for June [1968]. None of four 200-inch plates taken in 1964, 1967, and 1968 shows any trace of nebulosity between or around the components. Of the dozen 48-inch plates of 3C-287, which have lower resolution and smaller scale, a partial blending of the images of the two components is present on some. This, Kristian and Peach believe, can be explained by a combination of seeing, photographic, and instrumental effects. However, they have no photographs between May 1964, and February 1967, an interval during which Matthews recorded the bridge.

The issue has far-reaching implications. Proof that a quasar can undergo observable structural changes within a few years would have meant that it is a small object inside our Galaxy, rather than at the enormous cosmic distance corresponding to its spectral red shift.

The confusion about quasars is typical of new discoveries in the sciences; the first observations tend to be inaccurate, and the fit with current theory may require basic changes (such as the red shift not

being a Doppler shift) or new hypotheses (such as an explosion in the Galaxy's nucleus). The statistical evidence shown in Figures 103, 104, and 105 is more convincing to the editors than that in Figure 107, so we favor the large (cosmological) distances of quasars.—TLP

7

Cosmology

For centuries, speculation about the universe has provided men a basis for philosophy and religion. The foregoing chapters have shown some of the benefits of such cosmological studies in fitting a wide variety of observational facts into a self-consistent whole. Three thousand years ago there were far fewer facts, and an intelligent, well-informed man could exercise his preference among several sketchy cosmologies; today there are many more facts (most of them summarized in broad theories, such as Newton's theory of gravitation, Einstein's general relativity theory, the theory of quantum mechanics, a nuclear theory), and a good deal more thought is required to check a cosmology for consistency. Whereas the ancients were preoccupied with a limited region (such as the eastern Mediterranean), the sun, moon, nearby stars, and the gods' supervision, modern cosmology is more concerned with the astronomers' universe—the domain of galaxies and quasars. In fact, modern cosmology avoids emphasis on our local surroundings, since the earth is most certainly an "odd-ball" planet; the solar system, nearby stars, the Milky Way Galaxy, or even our local group of galaxies may be unique, or not representative of the whole universe.

"Cosmic quantitites"—or concepts, or qualities—are ones that apply (or may apply) to the whole universe; among these quantities are the abundances of chemical elements, the average density of matter, the background (3° K) radiation, gravitation, and other basic properties of matter. An important cosmic quantity is long duration in time. Any one star is a "transient object"; galaxies are longer lived; and matter endures forever (with a small fraction convertible to energy). Nevertheless, evolutionary change is part of the "Big-Bang" cosmology (see p. 136), which assumes some "beginning" of the universe. Other cosmic attributes that can be checked observationally are uniformity and isotropy. That is, we can count galaxies at different distances, and in different directions, *to learn whether or not they are evenly distributed (see p. 135). Most astronomers consider that the distribution is isotropic—the same in all directions.*

The concept of a Big-Bang beginning of the universe may be related to the origin of cosmic rays, as speculated in the thirties.—TLP

Mysterious Messengers from beyond the Solar System[1]

VIKTOR F. HESS

(*The Sky*, November 1937)

Twenty-three years have passed since cosmic rays were first discovered, but the general public did not pay much attention to them until Professor Piccard started his famous excursions into the stratosphere in 1931. Only then did the world begin to realize that scientists of all nationalities had opened an entirely new and interesting field of research by penetrating into the problems of cosmic rays, their origin, and their peculiar properties.

The most striking characteristic of cosmic rays is their enormous penetrating force. We know that they pass through six hundred feet of water as easily as through two hundred feet of solid marble or sixty feet of lead. In order to protect ourselves against their influence, we must seek shelter in the depths of a cavern covered by at least three hundred feet of rock. . . .

As soon as cosmic rays were discovered, scientists wondered about their origin. Although even today this problem has not yet been completely solved, we have made considerable progress insofar as we know for certain that these rays result from enormous cataclysms in the universe, from the composition and destruction of atoms, somewhere in the remotest distance of the skies, outside our solar system.

How did science come across cosmic rays, and how did we learn about their peculiar qualities?

Radioactive substances have a peculiar effect on air, transforming it

[1] This important article by Professor Hess first appeared in *The World Observer*.

from an electric isolator into a conductor. This effect is called ionization. In 1903 Ernest Rutherford of Manchester University and J. C. McLennan detected ionization within a receptacle which was protected by metal sheets against the so-called alpha and beta rays emanating from radium. This was attributed to the so-called gamma rays, supposed to come from the ground.

If this were their source, the ionization effect should have diminished in higher regions above the surface of the earth. Since radioactive substances contained in the walls of ordinary towers proved to have a disturbing influence, Professor Theodor Wulf, a German Jesuit physicist, undertook experiments on top of the all-steel Eiffel Tower . . . and established the theory that another influence, unknown until then, was the cause. In 1911 Professor D. Pancini of Rome came to similar conclusions after making measurements on and below the sea.

Professor Albert Gockel, a German physicist, was the first investigator who took to the balloon in order to clear up the riddles of ionization. In 1910 and 1911 he performed several ascensions, and was much surprised to find an increase of ionization above several hundred feet.

At this time I achieved a method of measuring the exact penetrating force of gamma rays and could also prove that the effects of radioactive emanations in the air were almost negligible. I then started a systematic study by using balloons that carried hermetically sealed receptacles to altitudes of 1000 to 15,000 feet above sea level. At first, ionization diminished, but as soon as my balloons attained the height of 3000 feet, a remarkable increase was observed. At 15,000 feet, ionization was several times stronger than on the ground, and this, at last, was a conclusive proof of the long-suspected cosmic origin of the rays causing this extraordinary ionization.

In order to establish whether these rays, which we now could legitimately call "cosmic," were in any way dependent on the sun, I repeated my experiments during an almost total eclipse. The results indicated that they were not dependent on the sun.

The outbreak of war [World War I] interrupted the further progress of my investigations.

Since then, several very interesting theories with regard to the origin of cosmic rays have been brought forward by such eminent scientists as W. H. Nernst, R. A. Millikan, and Sir James Jeans. We all believe that cosmic rays come from the outside universe, but we have not been able to establish clearly whether the processes causing them take place in the stars, in the nebulae, or in interstellar space.

The so-called latitude effect, registered for the first time by the Dutch scientist J. Clay, has proved the corpuscular nature of cosmic rays. . . .

During the summer of 1931 I was able to build an observatory, specially designed for the study of cosmic rays, on top of the Hafelekar, a mountain range towering above Innsbruck, at an altitude of 7000 feet. There, my assistants helped me with continuous registration of cosmic rays. We could determine their relation to the position of the sun and the different stars.

It seems as though the cosmic rays reach the earth in homogeneous quantities, from all directions in the universe. This leads to the conclusion that they come neither from the Milky Way nor from any visible nebula. We believe that their origin is to be sought within some nebulous mass many millions of light years away. . . .

In connection with cosmic rays, Carl D. Anderson has discovered the positron, a new element of atomic structure. Recently, we have some reason to hope that our studies will lead to an experimental proof for the existence of another element of atomic structure, called the neutrino, which is supposed to exist, but for which no tangible evidence has yet been found. This particle has dimensions equal to those of the electron, but has no electric charge. . . .

EDITOR'S NOTE [1937]

A recent newspaper article quoted Arthur H. Compton, University of Chicago, Nobel Prize winner, and P. Y. Cheu, of the National Tsing Hua University, Peiping, as saying that an explosion billions of years ago which shattered the universe may have been responsible for the cosmic rays now bombarding the earth. These scientists said that the space-explosion theory was looming larger with our increased knowledge of the rays. . . .

According to Big-Bang cosmology (see p. 136), an explosion 10 billion years ago started galaxies on their outward journeys, but the explosion's million-degree flash of radiation has since cooled to 3°K, and must be observed in radio wavelengths; the cosmic rays are now known to be high-speed atoms, or ions (mostly protons).

The Belgian Abbé Lemaître calculated the early Big-Bang history of the universe in an effort to explain cosmic rays. George Gamow (then at Georgetown University) used the ultra-high temperatures in the first few hours after the Big Bang to explain the present abundances of chemical elements in terms of nuclear reactions that took place then (see Volume 7 of this series, Stars and Clouds of the Milky Way*). The radiation temperature and matter density both decreased with time—a* consequence of the expansion of the universe and of general relativity theory (described in Volume 1 of this series, Wanderers in the Sky).

Evolution of the whole universe, according to this theory (relativistic cosmology), depends on values chosen for three or four parameters which can be adjusted to fit our present view of the galaxies—the constants in the law of red shifts, the average density of matter, and changes in these quantities with distance. In fact, the set of "evolutionary models" of the universe include some that did not start from a "big bang" t_0 years ago, some that are infinite in volume ("open" models), and some that are finite ("closed" models, because the gravitational effect of matter causes a "curvature of space-time," which limits the total volume of space). In each of these possible "world models" the curvature changes with time, except for a static model, first worked out by A. Einstein and W. de Sitter (a Hollander) before Hubble's discovery of the recession of galaxies (see p. 23). In effect, Hubble's law of red shifts reveals the decreasing curvature, or "expanding radius," of the universe. By choosing a specific value for the "cosmological constant," Einstein and de Sitter were able to avoid predicting an expansion, which the theory would otherwise do.

All these relativistic, evolutionary cosmologies are based on a general assumption that our view of the universe from the Milky Way Galaxy is the same as that from any other galaxy (see Fig. 10). This "cosmological principle" avoids the assumption that we are in the center—or in any other preferred position. It means that there can be no center and no "edge." Extending this to the fourth dimension, time, three British astrophysicists (Bondi, Gold, and Hoyle; see p. 137) assumed there should be no beginning, no end, and no change (on the average) with time. The resulting Steady-State cosmology must include creation of matter (to keep the density constant in spite of the observed expansion), so it is sometimes called the Continuous-Creation theory.

Recently, modifications of both the Steady-State and the Big-Bang cosmology have been proposed.—TLP

Continual Creation

GEORGE S. MUMFORD

(Sky and Telescope, February 1965)

The Steady-State cosmology advocated by the English scientists H. Bondi and F. Hoyle considers the large-scale properties of the universe as not changing in space or time. Thus the average density of matter in the universe should be invariable. But this can be reconciled with the observed recession of galaxies only by postulating that new matter is continually being formed. The current version of the Steady-State cosmology considers that this new matter is hydrogen atoms, spontaneously formed at a rate that is everywhere and at all times the same.

W. H. McCrea of the University of London has pointed to a difficulty in steady-state cosmology. To account for the formation of galaxies requires the existence of about a hundred times as much mass in the form of gas between the galaxies as in the galaxies themselves. Observationally, there is yet no positive evidence for the widespread occurrence of intergalactic matter.

The seeming contradiction can be removed, McCrea suggests, by changing the assumption that the formation of new matter is everywhere uniform. Instead, he proposes that it is produced only from existing matter. Normally, all material is in galaxies. Thus the creation of new matter simply causes galaxies to grow to a certain limiting size. Occasionally a fragment may break off, the embryo of a new galaxy.

This new theory does not require the existence of any significant amount of intergalactic material, nor does it require its absence. One of the problems encountered in earlier Steady-State theory was how to create condensations in uniformly distributed intergalactic matter to start galaxy formation. McCrea's proposal avoids this difficulty, and calls for less departure from conventional thinking in a number of other ways. He presented his theory in *Monthly Notices* of the Royal Astronomical Society for October 1964.

Relativity and Solar Evolution

(*Sky and Telescope,* November 1965)

Four years ago C. Brans and R. H. Dicke at Princeton proposed a modification of the theory of general relativity whereby the familiar universal constant of gravitation (G) would depend on the distribution of matter in space. Since then, a number of physicists and astronomers have looked for possible observational tests of this revised theory. It should slightly alter the motions of solar-system bodies, and recently Dicke pointed out that the minor planet Icarus affords a sensitive test [since it passes close to the sun in its highly elliptical orbit].

When the Brans-Dicke theory is applied to the universe as a whole, it predicts a series of cosmological models that are analogous to those deduced from general relativity. The chief difference between the two series is that in the Brans-Dicke models the gravitational constant has been decreasing steadily ever since the early history of the universe. The possibility of such a progressive change in G had already been considered in 1937 by P. Dirac [in England] and in 1948 by E. Teller [then at the University of Chicago]. Unfortunately, there is little hope of discriminating between the two series of models by means of current cosmological observations (red shifts or counts of galaxies), according to R. C. Roeder and P. R. Demarque, University of Toronto.

They have therefore studied the evolution of individual stars as a possible test. The luminosity of a star is extremely sensitive to changes in the gravitational constant, being approximately proportional to G^8. During the time the sun has been evolving, the decline in G has amounted to 5 or 10 per cent, in the Brans-Dicke theory.

The two Toronto scientists have calculated the evolutionary changes in the sun's luminosity during the last 4.5 billion years, each of their several calculations assuming a different rate of change in the gravitational constant. By making slight changes in the chemical composition of the solar model, they found that the present sun could have evolved in each case.

Roeder and Demarque therefore conclude that there is no conflict between the solar data and Dicke's theory. In particular, the solar data permit a rate of change in G proportional to t^n where n is any fraction between $-1/14$ and $-1/8$ (time being counted from the beginning of the expansion of the universe). . . .

Antimatter May Matter

GEORGE S. MUMFORD

(*Sky and Telescope*, May 1966)

An alternative to the Big-Bang and Steady-State theories of the universe was proposed about a decade ago by the Swedish physicist O. Klein, who suggested that the universe initially consisted of an extremely dilute cloud of protons and electrons. Part of this cloud contracted under gravitation and later formed the galaxies.

In Klein's theory, the initial cloud was a mixture of equal amounts of *particles* and *antiparticles*. The possibility that the universe contains matter and antimatter in equal quantities had been familiar to scientists since soon after the discovery of the positron (positive electron) more than thirty years ago. This belief was strengthened by the discovery of the antiproton. In fact, it now appears that for every kind of elementary particle there exists a corresponding kind of antiparticle.

Recently, in *Reviews of Modern Physics* [for October 1965], the well-known Swedish astrophycist H. Alfvén presented an updated version of Klein's cosmology. According to Alfvén, the universe originated in a tenuous *ambiplasma* that contained protons and antiprotons, electrons and positrons. This cloud began to contract under gravitation. However, as the density within it increased, so did the chances that particles of matter (protons and electrons) would collide with those of antimatter (antiprotons and positrons), with the resultant annihilation of both. Because of this conversion of mass to energy, radiation pressure would increase, and the contraction would stop, to become the expansion that is now observed.

If the original ambiplasma contained equal amounts of matter and antimatter, three kinds of regions should result: one consisting purely of matter, another containing only antimatter, and a buffer zone of ambiplasma separating the other two. Regions of the same content would attract each other, while repelling those of opposite content. Thus larger and larger agglomerations of matter and antimatter would grow and coalesce.

When water is sprinkled on a very hot griddle it does not vaporize instantly, but small droplets persist briefly, each insulated from the hot surface by a *Leidenfrost layer* of water vapor. Something analogous may be expected at the interface between matter and antimatter. Alfvén suggests that the ambiplasma zone will be very hot because of annihilations taking place in it, and thus the regions of opposite content will be held apart.

Where in the universe would the antimatter be concentrated? We know that the earth consists of matter, as must the moon, since the collision of a Ranger spacecraft with it does not cause a violent nuclear explosion. Likewise, the sun is composed of matter, since particles in the solar wind that reach the earth during magnetic storms and auroral displays cause no conspicuous annihilation phenomena. The other planets doubtless consist of matter also.

Yet in the universe as a whole, the possibility cannot be excluded that half of the celestial objects consist of antimatter. The nature of cosmic rays, however, seems to indicate that there is no significant amount of antimatter between us and nearby stars in the Milky Way.

As for the stars themselves, the nuclear properties of matter and antimatter are similar, so a star built of antimatter would emit energy derived from thermonuclear processes. "Antiatoms," each consisting of a nucleus of antiprotons and antineutrons surrounded by a swarm of positrons, would cause spectral lines at precisely the same wavelengths as ordinary atoms do.

There would be one difference however. The magnetic or electric fields produced by Zeeman or Stark effects [see pp. 323, 324] in antimatter would have the direction of field opposite to that in ordinary matter. The trouble here is that we don't know *a priori* the direction of any magnetic field in space.

Radio observations may provide a clue. Alfvén points out that an ambiplasma would be a strong emitter at radio wavelengths (long), but not at gamma-ray wavelengths (very short). Thus all cosmic radio sources could derive their energy from annihilation, although other explanations are possible.

The energy of quasi-stellar sources (quasars) seems too great to be explained by nuclear reactions, and for some of them total annihilation of matter and antimatter may be the only possible source. Even a supernova explosion may result from the collision of an antimatter star and one of matter.

On a larger scale, it is conceivable that every galaxy we see and all members of our metagalaxy consist only of matter, while antimatter resides in other metagalaxies. In this view, our metagalaxy condensed until the disruptive forces became too great and it began to fly apart. On this hypothesis, it is not the entire universe that is expanding, but just our metagalaxy. . . .

Not all the many proposed cosmologies have been described here; there are many more possibilities, some mentioned in the next four summaries of special meetings on relativistic astrophysics.—TLP

Dallas Conference on Super Radio Sources

LOUIS C. GREEN

(*Sky and Telescope*, February 1964)

In mid-December [1963] an exciting international symposium on gravitational collapse as related to the problem of ultra-high-energy radio sources was held in Dallas, Texas. The gathering of some four hundred astronomers and relativity experts was sponsored by the Southwest Center for Advanced Studies, the University of Texas, and Yeshiva University, with the support of a number of government agencies.

In recent years astronomers have felt an increasing need for new ideas concerning possible sources of the energy that maintains the intense radio radiation we receive from certain extragalactic objects. The older thought that such a source represented the collision of two galaxies has dropped out of favor, for it has become apparent that in the best-examined cases no second galaxy is involved. Indeed, among the well-known objects, only for NGC 1275 in Perseus is collision still a possibility.

Furthermore, the collision of two galaxies is none too effective in releasing energy. Even in the dense parts of a galaxy, the stars are small compared with their distances apart, and the star clouds of colliding systems should pass through each other. Energy release would be significant only when gas and dust clouds in one galaxy collide with those of another.

It has been known for some time that energies of the order of 10^{57} to 10^{61} ergs are needed to account for the radiation that we receive here on earth from many of the radio sources. The recent recognition of several very remote, very strong, small-diameter radio sources [quasars] suggests that the upper limit of 10^{61} ergs may be too conservative. Jesse L. Greenstein, California Institute of Technology, described the peculiarities of their optical spectra.

Such a source is perhaps ten to thirty times as luminous as the brightest elliptical galaxy. For example, 3C 273 in Virgo, which is not the intrinsically most luminous of these quasi-stellar sources, has an apparent photographic magnitude of 13 and a spectral red shift corresponding to a velocity of recession of 47,400 kilometers per second. This indicates a distance of about 1.6 billion light years[1] and an intrinsic brightness of 3C 273 equal to a trillion (10^{12}) suns! Since our sun radiates 3.8×10^{33} ergs per second, this amazing object must release about 4×10^{45} ergs each second.

How long has this tremendous outpouring of radiation been going on? We do not know with certainty for any of the quasi-stellar objects, but presumably the outflow has continued since some catastrophic event—some great explosion in the quasar. Assuming that each quasar expanded at the velocity of light to its present size, calculations yield ages larger

[1] This distance was computed from the Hubble law: Velocity of recession increases 100 kilometers per second for each million parsecs distance. [See formula on p. 185.]

than 10^6 years. In this time 3C 273 would have emitted at least 10^{59} ergs—an enormous amount of energy.

From the relative amounts of optical and radio energy, we believe these sources are producing synchrotron radiation, which was first clearly recognized in the case of the Crab nebula and has since been suggested for a number of other cosmic radio emitters. It occurs on earth in the radiation of those large particle accelerators, the synchrotrons, which physicists use so widely. "Relativistic electrons" are required; that is, electrons with velocities close to that of light, moving through magnetic fields.

In astronomical cases, it is difficult to see how such negatively charged electrons can be separated from positively charged ions in any large numbers. The most likely ion is the proton (hydrogen nucleus). The mechanism that accelerates the electrons will accelerate an equal number of protons, and the latter, in view of their much greater masses, will take the major portion of the available energy. Furthermore, the electrons are much more subject to energy losses, so that in the end only 1 per cent of the total energy goes to them. Thus if we wish the electrons to be able to radiate 10^{59} ergs, we must search for some process capable of supplying 10^{61} ergs to the cosmic body as a whole.

As a start in our search, we ask what energy would result from the complete annihilation of all the matter in our own Milky Way Galaxy. Very roughly, the Galaxy's total mass is 4×10^{44} grams. From the Einstein equation $E = mc^2$ we find that 4×10^{65} ergs is roughly the energy (E) obtainable from converting the total mass (m) of the Galaxy.

No known nuclear process or series of processes is capable of converting mass into energy in this wholesale fashion. In fact, if hydrogen is changed into iron, the most efficient possible series of reactions, we could transmute only 0.01 of the mass. In actual practice, the fraction is likely to be much smaller. Furthermore, unless the temperatures are considerably higher than in the centers of main-sequence stars (such as the sun), the rate of energy release is about a thousand times too slow.

In view of these facts, scientists are reviewing other energy sources than nuclear reactions—in particular, rapid gravitational collapse. In 1854 H. Helmholtz, the German physicist, suggested this process for the origin of the sun's energy, but it has long since been found inadequate to keep the sun shining at its present rate for more than 15 million years, and nuclear processes were recognized as far more satisfactory. What then is new about gravitational contraction to make us reconsider it now?

Helmholtz pictured the sun as contracting from a uniform rarefied

FIG. 108. William A. Fowler, physicist at Kellogg Radiation Laboratory of California Institute of Technology. (Photograph by Al Mitchell, courtesy of Graduate Research Center of the Southwest)

medium of very large, virtually infinite radius to its present size. Instead, let us consider as one possibility a dispersed body of perhaps 100 million times the sun's mass, which might evolve to reach a very small radius.

Fred Hoyle of Cambridge University and William Fowler of California Institute of Technology discussed these possibilities. . . .

As the original giant body begins to contract, with an accompanying increase in its rate of rotation, the crucial question is whether or not the magnetic lines of force are "frozen" into the material. If, as seems likely, magnetic field and material move together, an early breakup into many fragments is impossible. . . .

In the next stage, further contraction and hastening rotation of the outer region lead to an ever more flattened system, which breaks up into spherical bodies, each containing about 100 solar masses. High temperatures will develop in these large masses (or be already present as a result of earlier contraction)—temperatures high enough to modify substantially the ordinary carbon-nitrogen energy generation cycle. This is a relatively leisurely process in the solar interior, but here, after going slowly at first, it explodes in a "whoosh." The time scale for hydrogen burning becomes about a week, and one after another the subcondensations of 100 solar masses go through the whoosh. Hoyle suggests that the tremendous energy released in this rather catastrophic process would be observable as a quasi-stellar radio galaxy. There might be 10,000 of these subcondensations in one system.

Irregularity in the occurrence of the whooshes may be related to brightness fluctuations (such as those charted in Fig. 109) reported by Harlan J. Smith, University of Texas. . . .

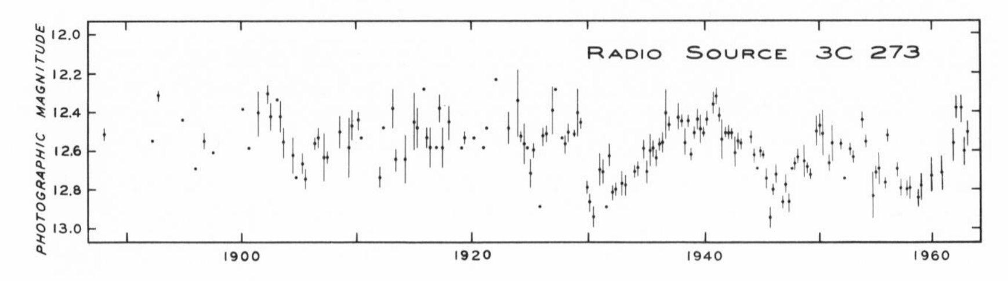

FIG. 109. Brightness variations of the quasar 3C 273 during the last seventy-five years, estimated by Harlan Smith and Dorrit Hoffleit from old photographs. (Yale University chart)

If little rotation has been retained, the contraction of the central region would continue. A dramatic possibility is that the object goes on contracting until it develops an "event horizon," becoming unobservable to the outer world except by the static gravitational field it leaves behind. According to general relativity theory, light reaching us from a strong gravitational field should be reddened, indicating some loss of energy. If the gravitational field is that of a star, the degree of reddening depends on the mass of the star divided by its radius. As the radius shrinks, the reddening increases, and so does the energy lost by the light. Eventually the radius could become so small that the light could escape only at the cost of losing all its energy. If there were no energy left, there would be no light either. At such a time, the body is said to have withdrawn inside an event horizon.

If rotation does not become fast enough to stop contraction before the event horizon is reached, indefinite further contraction takes place, with "crushing of the rotation." In the usual cosmological theories, this course of events leads to the final contraction of the body into a point —a singularity—but Hoyle emphasized that in his Steady-State cosmology the singularity is not reached.

With somewhat greater rotation, gravitational waves would be more important, and might promote fission of the immensely massive and dense primary body. Hoyle suggested that the separating fragments might be somehow analogous to the multiple structure of certain radio sources. He noted a suggestion by the Russian astronomer V. L. Ginzburg that in some quasi-stellar radio sources there must be a high density of photons as well as of electrons, and that by collisions with electrons, the photons could gain substantial energies—the inverse Compton effect[2]—so that these sources could emit intense X rays as well as visible light. . . .

[2] Radiant energy can be converted to kinetic energy of electrons when a photon passes an electron, an effect discovered by A. H. Compton at the University of Chicago in 1923. In the "inverse Compton effect," electron kinetic energy is transferred to a passing photon, increasing its frequency.—TLP

E. Margaret Burbidge, University of California at San Diego, considered several alternative explanations. The interaction of matter and antimatter could annihilate mass, converting it to energy. But how are matter and antimatter to be kept apart until the right moment of interaction? Energy might be gained from turbulent motions in the early evolution of a galaxy, but ellipsoidal galaxies, which are much stronger radio emitters than the spirals, are far too smoothly structured for turbulence to be important.

Winding up a magnetic field in which gas is intermeshed would cause the lines of force to break and produce a flare, but this does not fit well with the slow rotations and lack of gas in elliptical galaxies. As for supernova explosions, since 10^8 of them would be needed, how could they all reach the same evolutionary stage simultaneously? It is difficult to explain how a trigger mechanism would work. . . .

On the final afternoon of the conference, F. J. Dyson (Columbia University) pointed out that the basic difficulty with collapse as a source of the energy for radio galaxies is that its time scale is much too short. Little energy is released until the density becomes very high, and then a great deal is produced in a very short time, of the order of a day. E. L. Schucking of the University of Texas emphasized that cosmic rays, X rays, gamma rays, and the contents of intergalactic space had been largely omitted from the symposium discussions. P. G. Bergmann of Yeshiva University warned nonrelativists that all the models discussed were special cases of very high symmetry, because of the mathematical difficulties of more general treatments. He pointed out that at some future date theories that are both relativistic and quantized will give us new ways of looking at the universe in the small; whether they will be important in the large no one knows. With this session a most interesting conference came to an end.

Relativistic Astrophysics

LOUIS C. GREEN

(*Sky and Telescope*, March and April 1965)

Exciting new observations and theories from the farthest frontiers of astronomy were reported at the Second Texas Symposium on Relativistic Astrophysics, held at Austin, December 15–19, 1964, under the sponsorship of the University of Texas and the Southwest Center for Advanced Studies. . . .

Geoffrey Burbidge of the University of California at San Diego spoke on extragalactic sources of high-energy radiation. We have long been aware of the presence in the universe of a large flux, or flow, of nucleons with relativistic energies. Nucleons are the protons and neutrons of which atomic nuclei are composed. Relativistic energies mean that their velocities are so great that the ideas of special relativity apply, such as the increase in mass of a fast-moving object [as described in Vol. 1 of this series, *Wanderers in the Sky*]. These relativistic nucleons—and their combinations in the form of atomic nuclei—are the principal component of the energetic particles which make up cosmic rays, first detected fifty years ago but incompletely understood at that time.

By contrast, recognition of high-energy cosmic electrons and positrons is much more recent. After World War II it was found that much of the radio radiation we receive from the universe is *nonthermal* [different from that predicted by Planck's law for a hot body].

In many cases of nonthermal radio radiation, the mechanism is the motion of relativistic electrons in the magnetic fields which occur in many astronomical objects. Observational identification of this synchrotron radiation (so-called because it occurs in terrestrial synchrotron accelerators) is aided by its visible component, which is distinctly bluish and partially polarized.

Much of the nonthermal radio radiation that reaches the earth comes from a large, roughly spherical region that is more or less concentric with our Galaxy. This region is called the galactic halo; it is regarded as similar in size and shape to the radio halos detected in other galaxies. These may be rapidly changing features; they radiate their energy, and their matter tends to fall into the equatorial plane of each galaxy after about 10 million to 1 billion years.

A galaxy may be able to re-create its halo several times, through what appear to be colossal explosions in its central parts. The number of known halos and their short lifetimes strongly suggest such regeneration, and recently Allan Sandage and Roger Lynds have shown that an event of this kind is now occurring in the galaxy Messier 82 [see Fig. 73]. The explosion carries matter out in all directions, but that part moving in the main plane of the galaxy is quickly slowed by collisions with the material already there. Above and below the plane, the explosion products expand into regions where there is little matter or radiation and where magnetic fields are weak. Some heavy atomic particles may escape to form a source of cosmic radiation.

Collisions between atoms of the expanding gas and those already present will generate other nucleons, mesons, electrons, gamma rays, and neutrinos. (*Mesons* are particles intermediate in mass between the heavy nucleons and the light electrons.)

In this mixture of products, both the electrons and positrons will spiral around the magnetic lines of force, yielding synchrotron radiation. They will also interact with the accompanying light, upgrading it by the inverse Compton effect into gamma radiation and X radiation.

Burbidge suggested that the X-ray source observed in the direction of the center of our Galaxy [see p. 213] might be the result of an explosion such as we see taking place in M 82, or it might be due to the infall of sufficiently hot intergalactic matter. For synchrotron radiation, the energy of the particles must be greater than the energy of the magnetic field itself.

An electron flux decays through radiation much more rapidly than a proton flux; thus, if protons are present at all, their flux will dominate, by a factor of perhaps 100. For radio galaxies the electron flux can be calculated at 10^{59} ergs; thus the proton flux would be about 10^{61}. Quasi-stellar sources need somewhat less energy because they are smaller.

For several additional reasons, Burbidge believed that more energy might be needed, and he suggested that to account for a radio galaxy, a source of 10^{64} ergs was required—about the energy obtained by the total annihilation of 10^{10} solar masses! The size of the problem is clear. He felt we should look to Hoyle and Fowler's suggestion of the gravitational collapse of very large masses or to the more drastic hypothesis (by John Wheeler at Princeton) that extreme gravitational forces might crush the nucleons out of existence.

Thomas A. Matthews of California Institute of Technology summarized present knowledge of identified extragalactic radio sources. He began with preliminary results of a systematic attempt to identify the brighter radio sources on the Palomar Sky Survey 48-inch Schmidt plates, covering the sky north of declination $-30°$. Images down to a magnitude of about 18 can be recognized as galaxies or as starlike, and some of the characteristics of the galaxies can be determined.

Of the 119 sources examined, ninety-five are optically identified with a high degree of certainty. About two thirds of these are of Morgan's type D [see Figs. 28 and 110], galaxies that have extensive outer envelopes, some of them being supergiant systems or E (elliptical) galaxies. About 7 per cent are "dumbbell" galaxies or normal spiral and irregular systems. About 26 per cent are quasi-stellar objects (quasars), many of them very blue (Morgan's N-type galaxies), with nearly starlike images on the 48-inch Schmidt plates (as shown on Fig. 110). . . .

According to Allan Sandage of Mount Wilson and Palomar Observa-

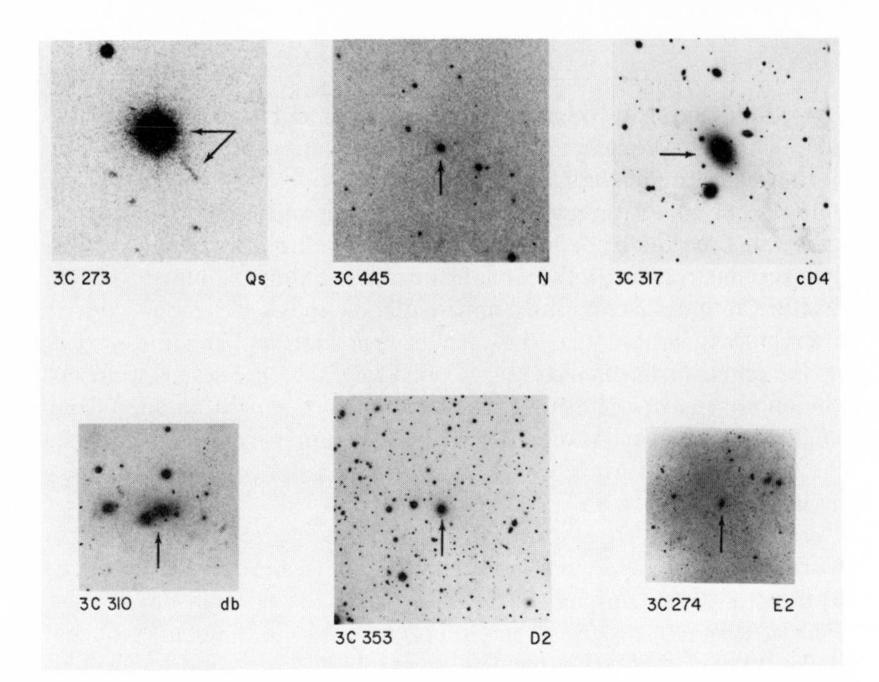

FIG. 110. Six negative photographs of cosmic radio sources, all printed to the same linear scale. In the first picture, the distance from the center of the quasar 3C 273 to the tip of its jet is 50,000 parsecs. Each of the photographs (all but one taken with the Hale 200-inch reflector) has Morgan's morphological type (see Fig. 28) printed under the lower right corner. The last picture, of 3C 274 (the radio galaxy M 87), is from a Palomar 48-inch-Schmidt plate. (Mount Wilson and Palomar Observatories photographs)

tories, about thirty-four definite or probable QSSs [quasi-stellar sources, or quasars] had been announced by the middle of December 1964, with another five as yet unpublished. The QSSs show a marked ultraviolet excess in their light. . . .

The quasi-stellar sources have a deep significance in cosmology because they are the most distant identified objects in the universe. Do their red shifts and apparent brightnesses fit a linear velocity-distance relation, or does that relation become nonlinear at the greatest distances? Sandage showed the latest plot of red shift against magnitude, commenting that there are still not enough measured points for a trustworthy determination of any deceleration. Different cosmological models predict various values of the deceleration parameter. In particular, when the so-called cosmological constant is assumed to be zero, the deceleration parameter is positive for the expanding-universe models of general relativity, which implies the expansion is slowing down as time goes on. On the other hand, the Steady-State theory of cosmology predicts a value of —1 for the parameter.

English astronomer H. P. Palmer discussed the angular sizes of radio sources. His observations were made with the 250-foot Jodrell Bank telescope and another dish, located some eighty-two miles to the east, the two forming a radio interferometer working at a wavelength of 73 cm. Neither an optical nor a radio interferometer gives an image of what it is examining, but it can tell the angular separation of two point sources, or the approximate diameter of a circular disk [see p. 204]. Because of the approximations, Palmer emphasized that it is safer to talk about *angular scale* rather than angular diameter.

Of some twenty-five QSSs that he observed, only three had angular scales over 20 seconds of arc, while fourteen were two seconds or less; five were unresolved. In contrast, among the sources identified as galaxies twenty-five out of thirty-five were found with angular scales larger than 25 seconds. These facts strongly suggest that the QSSs are more remote than the galaxies.

Maarten Schmidt of Mount Wilson and Palomar began his discussion of QSSs by describing them as starlike objects associated with radio sources. From 1 to 10 per cent of the radiation may originate in jets. In their optical spectra, nine out of ten show broad emission lines, 20 to 50 angstroms wide (see Figs. 112 and 113), but no absorption lines. The emission lines often do not stand out against the continuum, and are harder to measure than those in radio galaxies. In six other cases there are no identified lines in the QSS spectra.

In principle, there are three possible explanations of the large red shifts of the QSSs: they could be Doppler shifts arising from the rapid motions of nearby objects; they might be gravitational red shifts, such as occur when radiation originates in a very strong gravitational field;

FIG. 111. Maarten Schmidt of Mount Wilson and Palomar Observatories, who discovered the large red shifts of quasars. (Photograph by Al Mitchell, courtesy of Graduate Research Center of the Southwest)

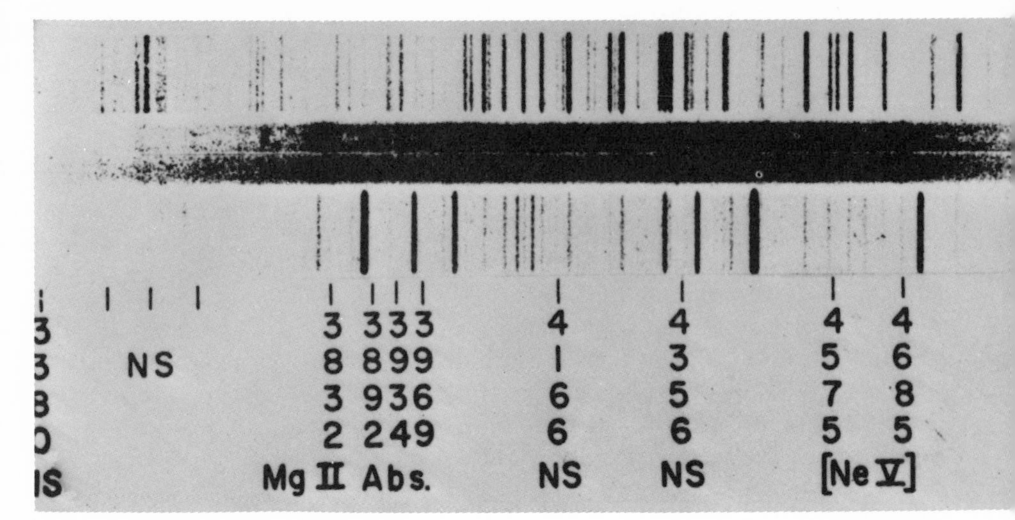

FIG. 112. Two spectra of 3C 48, a quasi-stellar object, bordered top and bottom by laboratory spectra. In this negative reproduction, emission lines appear dark. The only labeled lines belonging to 3C 48 are the broad fuzzy ones of ionized magnesium (at 3832 A) and neon (4575 A and 4685 A). All the features labeled "NS" or "Abs." originate in the earth's atmosphere. These are Palomar spectrograms taken with the 200-inch telescope. (From *The Astrophysical Journal*)

or they could be cosmological red shifts, due to the general expansion of the field of galaxies.

The first of these hypotheses runs into many difficulties. Careful measurements by W. H. Jeffreys [then at Wesleyan and Yale Universities] have shown that the proper motion of 3C 273 is less than 0.002 second of arc per year, which for a nearby object would imply very little transverse motion despite an enormous line-of-sight velocity. Furthermore, we observe no blue shifts. Why should all these objects be moving away from us? There are other arguments—none of them conclusive—that also make the interpretation of the red shifts as ordinary motions rather unattractive.

If we assume the red shifts are gravitational, we also find ourselves in difficulty, because the spectra show so-called forbidden lines, indicating that these spectra originate in an extremely rarefied gas [see p. 318]. . . .

The usually accepted explanation is that the QSS red shifts are cosmological in origin. The red shifts then yield distances. From these and the observed energy fluxes we can estimate minimum QSS diameters of the order of three to thirty light years. In these densely packed volumes, the electron scattering of radiation is large, and photons diffuse slowly through. This makes it hard to understand the short-period light varia-

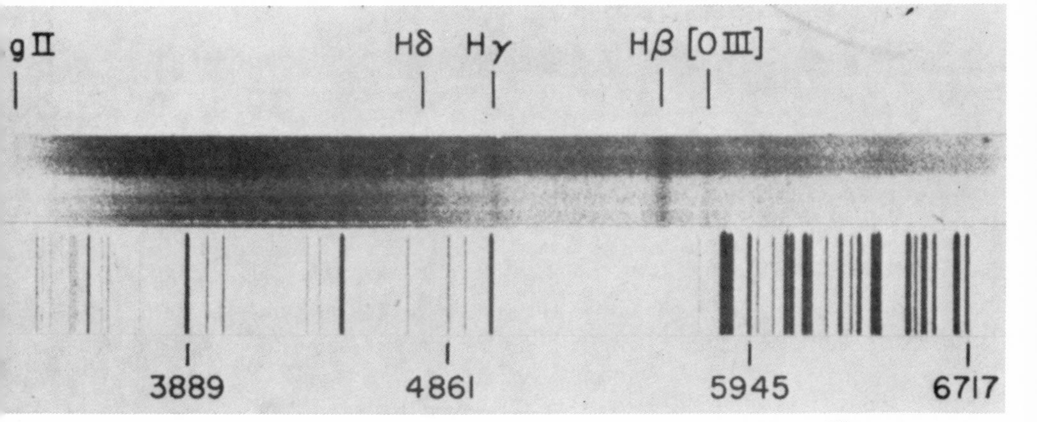

FIG. 113. This Palomar spectrogram of 3C 273 shows recognizable lines of hydrogen (Hγ, Hβ) and ionized oxygen [O III]. These features are shifted far to the red (right) of their normal positions, indicating that the source is receding from us at 47,000 km/sec. (From *The Astrophysical Journal*)

tions. Schmidt pictures a nucleus surrounded by a spherical body of *filaments*. Such a structure increases the volume, while greatly reducing the opacity encountered by the photons in working their way out. Outside the filaments, Schmidt proposed, could be a shell, perhaps 1500 to 6000 light years in diameter, from which the radio radiation comes.

Bruno Rossi (Massachusetts Institute of Technology) discussed the cosmic rays of highest energy—very fast-moving particles from the depths of space [see p. 259]. He pointed out that all our information about cosmic ray energies greater than 10^{13} electron-volts comes from secondary or tertiary showers, not from individual primaries. (When the very energetic primary cosmic ray particles strike our atmosphere, they collide with atoms and molecules, generating whole showers of secondary particles. These in turn collide with atmospheric atoms and molecules, generating tertiaries, so that in the end the progeny of a single cosmic ray primary may spray over hundreds or thousands of square yards of the earth's surface.)

Some ten great cosmic ray showers have now been observed for which the energies of the primary particles must have been greater than 10^{19} electron-volts (e.V.). These highest-energy primaries tend to be protons rather than heavier nuclei. If, as has been suggested, the lower-energy cosmic rays come from our Galaxy while the higher-energy ones are extragalactic, then there should be a change in the slope of the curve obtained when the number of particles having less than some particular energy is plotted against that energy. There appears to be such an inflection in the curve between 10^{15} and 10^{18} e.V. . . .

Peter Meyer (University of Chicago) discussed cosmic-ray positrons. It has been known for a decade that the primary cosmic-ray flux contains electrons, but only recently has it been shown that some of these are positrons—that is, positively charged electrons. There are two possible sources of electrons that may be important astrophysically; the first produces positive and negative electrons in equal numbers when a cosmic ray proton collides violently with the nucleus of a hydrogen atom of the interstellar gas. Such a collision could result in the formation of two nucleons and a positive, negative, or neutral pion (pi-meson). A positive or negative pion decays by way of a muon (mu-meson) to neutrinos and positive or negative electrons, respectively. The neutral pion decays very rapidly into two positrons and two electrons.

The second source is the nuclear reactions that occur in a supernova. In this case, the negative electrons overwhelmingly outnumber the positrons. Supernovae occur frequently enough to supply the flux of negative electrons in cosmic rays, but the presence of a small positron component suggests that the first process also operates. . . .

Possibilities of detecting cosmic neutrinos were considered by Frederick Reines (Case Institute of Technology). He pointed out that there are four kinds of such particles: the neutrino and antineutrino associated with reactions involving a nucleus and an electron or positron; and the neutrino and antineutrino associated with the decay of those particles of intermediate mass called muons. Hereafter, we shall use the term neutrino for the first kind (electron neutrino), unless otherwise specified. All four kinds of neutrinos occur among cosmic ray secondaries. . . .

Perhaps the two most important reactions for detecting neutrinos are these: antineutrino and proton yield positive electron and neutron; neutrino and neutron yield negative electron and proton. The probability that either reaction will take place in any given case is extremely small, but increases with increasing energy. From theory, we know that the sun and stars produce a neutrino flux, so far undetected. (Only antineutrinos have been observed, and these only near laboratory reactors. Even a high extraterrestrial flux of neutrinos and antineutrinos could exist unrecognized.) . . .

Why should one attempt neutrino astronomy experiments with their fantastically low yields? Some answers to this question were given by J. H. Bahcall (California Institute of Technology). First, neutrinos offer the only way of directly observing the deep interiors of the sun and stars. The neutrino energy spectrum should tell us what elements are there and what reactions are running. Moreover, if neutrinos are to be used in cosmological studies, we must first be familiar with local contributions to the general flux.

To illustrate his first argument, Bahcall turned to the proton-proton chains of reactions—the most important source of the sun's energy. They yield neutrinos with a continuous spectrum of energies, and also neutrinos with tightly bunched energies; that is, an emission-line spectrum of neutrinos. A plot of such a neutrino energy spectrum would somewhat resemble the microphotometer tracing of the optical spectrum of a star with emission rather than absorption lines. In the case of the sun, at least, the neutrino energy spectrum should tell us what reactions are occurring and at what temperatures.

It would also be desirable to study the neutrino flux from strong radio sources, which should tell us something about the origin of the radio radiation. If certain reactions, which we can write on paper and which are not inconsistent with present knowledge, do in fact occur in nature, it should be possible to detect these radio sources' neutrino fluxes.

Since neutrinos interact with matter so very weakly, they can travel enormous distances without undergoing reactions. A short calculation suggests that this distance may be as great as 10^{20} times the radius of the universe as computed from theories of the expanding universe. Therefore, may not neutrinos approach us from every direction, with an arbitrarily great flux? This situation is like that underlying the famous Olbers paradox regarding starlight [see p. 33].

But there is evidence that the neutrino flux is not so very great. Consider the reaction of a neutrino and neutron to give a proton and a negative electron. If the neutrino flux were practically limitless, every neutron would have found a neutrino with which to interact, and only protons and electrons would exist in large numbers. Since in fact neutrons somewhat outnumber protons in the universe, the neutrino flux cannot be infinitely great. . . .

R. H. Dicke (Princeton University) described his ingenious repetition of an experiment performed in Budapest, Hungary, by R. Eotvös in 1906. The purpose of this experiment is to determine how precisely the gravitational mass of a body (in the law of gravitation) is equal to its mass in Newton's laws of motion—its inertial mass.

The exact equality of these masses was assumed by Newton and is commonly taken for granted by engineers, physicists, and astronomers in their daily work. General relativity offers an explanation for the equality. It is therefore important to ascertain whether experiments can detect any difference between these two kinds of mass. Dicke's work (described in *International Science and Technology* for May 1964) has established their equality to within one part in 10^{11}; they may in fact be precisely equal.

W. A. Fowler (California Institute of Technology) considered a possible model of the quasi-stellar radio sources. This supposes that at the center of a QSS there is a massive object undergoing relativistic collapse. Around this central body is a spherical shell, perhaps 10 light years in radius, of material at a temperature of 10^5 degrees; this shell contributes the continuous spectrum that is observed optically. Outside it is a second shell, ten times as extensive and a tenth as hot; from this region come the emission lines in the optical spectrum. Still farther out, in a shell 1000 light years in radius, move the fast electrons that are responsible for the radio radiation.

Fowler went on to envisage successive cycles of energy release in the central region. No radio radiation would come from the first cycle, but in time this would begin. A flat energy spectrum might result from the repeated injection of electrons into the outer regions. The light variations of such an object as 3C 273 might then be interpreted as the relaxation oscillations of the shells after a cycle of energy release had occurred. "After you strike it, you let it ring." It is not clear whether an impulse is needed every ten or twelve years—the cycle length of 3C 273's light variations—or whether only an occasional impulse is needed to make the QSS continue to oscillate with such a period.

Observational Aspects of Cosmology

LOUIS C. GREEN

(*Sky and Telescope*, April 1966)

On December 15, 16, and 17 [1965], at Miami Beach, Florida, a conference on Observational Aspects of Cosmology was held under the auspices of the University of Miami. Leading scientists from several countries presented results of recent radio and optical observations, and discussed their possible theoretical interpretations. This conference was in the same tradition as the two Texas symposia on relativistic astrophysics, held in Dallas in 1963 and at Austin in 1964. . . .

Allan Sandage (Mount Wilson Observatory) spoke briefly on the test for deceleration of the expansion of the universe. This consists in plotting the measured red shifts of galaxies against their apparent magnitudes (after appropriate corrections). A changing expansion rate would reveal itself by a departure from linearity in this plot. Sandage pointed out that this test cannot be satisfactorily applied by plotting individual nonradio galaxies; their dispersion in intrinsic luminosity is too great.

When the red-shift–magnitude curve is drawn for the ten brightest members in each of various clusters of galaxies, the scatter is reduced but is still great. If, however, only the brightest galaxy in each cluster is plotted, the scatter in the red-shift–magnitude curve is much improved. Radio galaxies, which on the average are optically as luminous as the brightest cluster members (absolute magnitude −21.6), have very little scatter, despite the fact that their radio fluxes may differ by a factor of 100,000, and they fall on the same straight line as the brightest members. The quasars, with an average absolute magnitude of −24.7, fall on a parallel line (see Fig. 114).

Cosmologists prefer to talk in terms of a red-shift–magnitude relation instead of a velocity-distance relation because red shift and magnitude are observed quantities. To obtain velocity from red shift, we must make use of some particular cosmological model. For small red shifts (as of stars in our Galaxy), all models give the same familiar result, the simple Doppler relation ($\Delta\lambda/\lambda = v/c$). For red shifts of 1.0 or more, such as have been recently observed, the simple relation fails, and another [the equation on p. 185, or one more complicated] must be used. . . .

Sandage commented that he saw no hope of deciding among various cosmological models by studies of the increase in numbers of galaxies

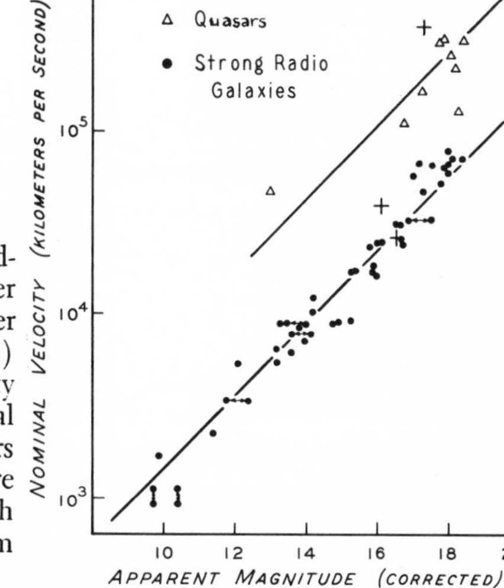

FIG. 114. In Allan Sandage's red-shift–magnitude diagram, the upper line represents quasars; the lower line, giant galaxies. Red shift (z) has been multiplied by the velocity of light (c) to give "nominal radial velocity (cz). Note that the quasars are about three magnitudes more luminous than giant galaxies with the same red shift. (Adapted from *The Astrophysical Journal*)

to fainter and fainter magnitudes. The principal difficulty is the strong tendency of galaxies to cluster. . . .

Different cosmological models predict different variations of apparent angular diameter with distance, but the effects to be measured are small, and the exact size of the soft-edged photographic or radio image of a galaxy is quite elusive. Thus it may be easier to approach the problem in terms of angular separations of galaxies in clusters.

Maarten Schmidt (Cal Tech) remarked on the four ultraviolet spectral lines that have been mainly used in red-shift determinations of quasars and other very remote objects. These lines, whose unshifted wavelengths are 2798, 1909, 1550, and 1215 angstroms, are due to singly ionized magnesium, doubly ionized carbon, triply ionized carbon, and neutral hydrogen, respectively. In addition, quasars often show strong "forbidden" lines, which can arise only in a large volume of very low-density gas. These last have played an important role in the argument that the quasars cannot be very compact nearby objects in our own Galaxy.

By last December [1965], spectra had been observed for thirty-four quasars. Of these, twenty-three show several lines and therefore can yield reliable red shifts, eight others show only one line each, and three show no lines at all. A number of red shifts are larger than 1.0 and two exceed 2.0. The intensity of the continuous spectrum tends to fall off noticeably toward 1200 angstroms (unshifted wavelength). T. D. Kinman (Lick Observatory) reported that in one quasar, PHL 938, both the Lyman-alpha emission line and the carbon emission line at 1550 A have absorption lines in their violet wings. . . .

The magnesium line (unshifted wavelength 2798 angstroms) in 3C 345 has been studied by Margaret and Geoffrey Burbidge (University of California at San Diego). Their wavelength measurements of the components of this line show radial velocities differing by as much as 3000 kilometers per second from the nominal recessional velocity of 130,710 kilometers per second. Both E. J. Wampler (Lick Observatory) and the Burbidges detected an apparent absorption feature on the short-wavelength side of the Lyman-alpha emission line at 1216 angstroms. One of the suggested explanations attributes it to intergalactic hydrogen.

For the majority of several hundred discrete sources, made up of radio galaxies and quasars, the radio spectra follow a simple law: flux density is proportional to some power, α, of the frequency. [See p. 211.] This was pointed out by K. I. Kellermann of the National Radio Astronomy Observatory in West Virginia, who discussed a model in which electrons are injected into a magnetic field in a series of short bursts. On this basis, the radio spectrum of each source can be divided into three regions. At long wavelengths the spectrum will be unaffected by radiation losses, and the spectral index will be about —0.25. At intermediate wavelengths the radiation losses will begin to affect the spectrum shortly after an outburst, and the spectral index on the average will be larger in this region, $\alpha = -0.75$. At high frequencies radiation losses will be even more important.

If observations are made immediately after an outburst, even the high-frequency region of the spectrum will not have had time to decay, and the spectrum throughout will be characterized by $\alpha = -0.25$. Later the intermediate and high-frequency regions will steepen, and the latter region may have a spectral index of —1.3. The rapid changes actually observed in quasars suggest that the time between bursts may be only a few years.

P. Véron of Mount Wilson and Palomar has found the very interesting result that radio galaxies increase in number with distance (decreasing flux) at the rate which one would expect for objects uniformly distributed in space, but the quasars increase more rapidly. He attributes this result to some evolutionary effect in the latter. . . .

Y. Ne'eman of Tel-Aviv University emphasized that the expansion of the universe might not be proceeding at the same rate at all points. He suggested that the massive core of a quasar might be lagging behind the original expansion.

A particularly interesting group of papers dealt with the general background radiation at centimeter radio wavelengths [see p. 243]. Last

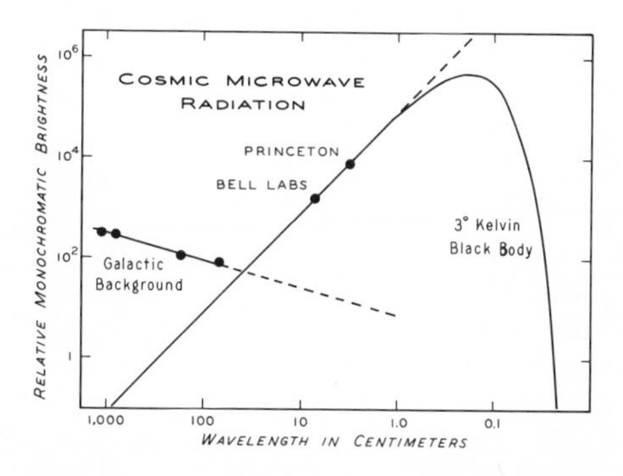

FIG. 115. The peaked curve shows the expected intensity of the cosmic black-body radiation if its temperature is 3°K, as the Princeton and Bell Laboratories data suggest. Test observations are possible only in a narrow "window" of wavelengths, since above 20 cm the Milky Way's radio noise masks the cosmic effect; below 1 cm our atmosphere is too opaque. (Chart courtesy P. G. Roll and D. T. Wilkinson)

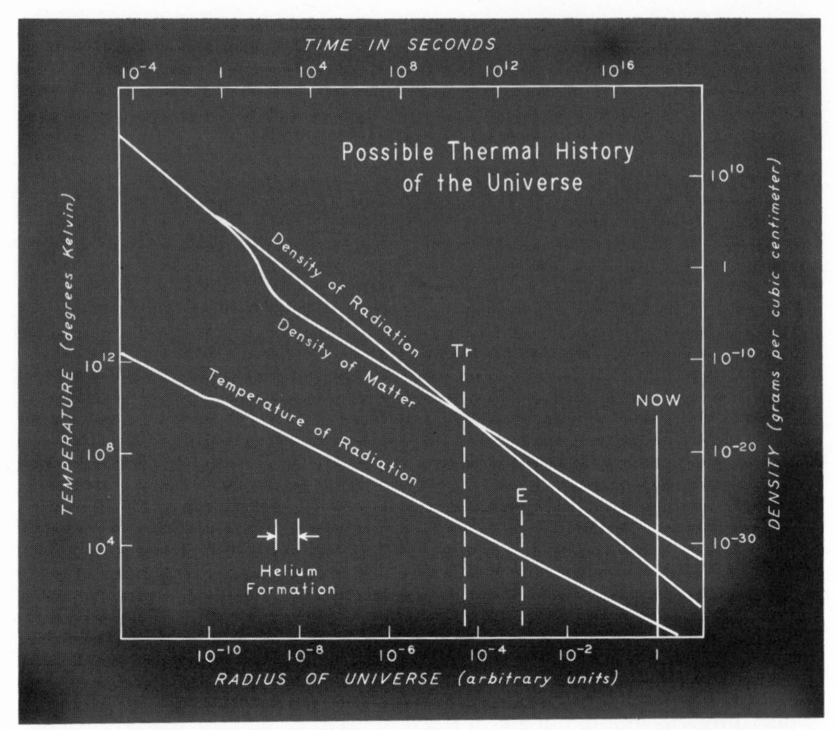

FIG. 116. As the primordial "fireball" faded, the densities (right-hand scale) of radiation and matter decreased, along with the thermal radiation temperature (left-hand scale) of the universe, while the radius (bottom scale) grew. The dashed line *Tr* marks the transition from a radiation-filled to a matter-filled universe. *E* indicates the time when thermal equilibrium between matter and radiation was lost. All scales are logarithmic. (Adapted from a diagram by R. H. Dicke and his co-workers at Princeton University in *The Astrophysical Journal*)

year A. A. Penzias and R. W. Wilson (Bell Telephone Laboratories, Holmdel, New Jersey) announced that in their attempts to track down every contribution to the radio noise picked up by their horn-reflector antenna, a certain small amount was unexplained. Within the accuracy of their observations, this contribution was isotropic (the same from all directions), unpolarized, and showed no seasonal variation. At 7.3-cm wavelength, the intensity of observed radiation corresponded to a black-body temperature of 3° after corrections (see Fig. 115).

At the same time four Princeton physicists, R. H. Dicke, P. J. E. Peebles, P. G. Roll, and D. T. Wilkinson published a short paper on "Cosmic Black-Body Radiation." In an expanding universe, at any past time all the mass and radiation must have been contained in a smaller volume. If we call the "radius" of the universe *R*, the average density of matter varies as R^{-3}, while the temperature of the radiation varies as R^{-1}. However *radiation density* changes as R^{-4}, which is more rapid than the change in the density of matter (see Fig. 116).

In the far past the radiation density was dominant. At the epoch when the temperature was 10^{10} degrees Kelvin, an equilibrium that was characteristic of this temperature existed between matter and radiation. If the radiation at some time in the past had the characteristics of black-body radiation, during the expansion of the universe it would retain such character, but at successively lower temperatures. Thus if today we observe a faint isotropic black-body radiation, it may well be the remnant of the intense radiation flux of the universe in its "fireball" stage.

Do the intensities at various wavelengths establish that the observed residual radiation is of the black-body kind? In addition to the measurements by Penzias and Wilson at 7.3 centimeters, Roll and Wilkinson have found 3.0° at 3.2 centimeters, and G. Field of the University of California has inferred 3.0° at 0.25 centimeter from the rotational structure of the interstellar absorption bands of cyanogen. It is too early to be sure whether the background radiation is black body or not, and experimenters are seeking data at other wavelengths.

Significant cosmological implications would follow from so low a temperature as 3°K for cosmic black-body radiation. Peebles has studied its bearing on the question of how much helium existed in the early history of the universe, before any helium could be made in stellar interiors. As the original fireball cooled, there came a time when neutrons and protons could form deuterium, and the deuterium could burn to tritium, and this to helium. (Owing to the great numbers of neutrons in the fireball, the reactions were different from the familiar ones of hydrogen-burning inside a star.)

The amount of this primeval helium depends upon the density of matter at the time these reactions became possible. We can calculate this density if we can ascertain from observation what the primeval abundance of helium was. The temperature at which helium formation takes place is virtually independent of density. From the temperature then and from the present temperature of the cosmic black-body radiation, we can tell how much the universe has expanded. Next, from the density at the time of helium formation and the known expansion, we can find the present density of matter in the universe.

From observations, J. Oort has estimated that the average density now of all matter is 7×10^{-31} gram per cubic centimeter. A rather similar result, 3×10^{-32}, was calculated by Dicke and his colleagues on the assumption that 25 per cent of primeval matter was helium. . . .

In another paper, N. J. Woolf (University of Texas) pointed out that the abundance of helium in primeval matter may have been well below 25 per cent, based on the chemical abundance in the stars of two globular clusters. The differences in the abundance of heavy elements among Population-II stars implies the former existence of a still earlier generation of "Population-III" stars, mostly massive objects that aged rapidly. These could even have been formed from a medium containing no primeval helium. Woolf has suggested that at the present time the primitive helium abundance might still be deduced from low-mass, slowly aging Population-III stars, such as may exist in dwarf galaxies and intergalactic globular clusters.

Cosmologists are concerned with the amount of intergalactic matter because it may make an important contribution to the average density of matter in the universe. At Mount Wilson and Palomar Observatories J. Gunn and B. Peterson have observed the optical spectrum of the quasar 3C 9, measuring the depression (absorption) in the continuum on the short-wavelength side of Lyman alpha. From this, they have deduced the intergalactic density of hydrogen to be about 10^{-34} gram per cubic centimeter. A similar result has been obtained by J. A. Koehler from observations of 21-cm absorption in the radio galaxy Fornax A with the 210-foot radio telescope at Parkes, New South Wales. Of course, this density refers only to neutral hydrogen. If intergalactic hydrogen is mostly ionized, its total amount may be considerably greater.

One also expects absorption lines from the gas within clusters of galaxies. Neutral hydrogen has in fact been observed in absorption in the Virgo cluster by Koehler, working at 21-cm wavelength. . . .

It is possible that membership in a cluster affects a galaxy in some way so that it is more likely to be a strong radio source. This might be due to the unseen mass (intergalactic material) in clusters of galaxies, or to the manner in which such clusters were formed. However, a careful study shows that cluster membership has little or nothing to do with the radio luminosity of E galaxies.—TLP

Elliptical Galaxies as Radio Sources

(*Sky and Telescope*, November 1968)

The brightest member of a cluster of galaxies is generally a giant elliptical galaxy and commonly a strong radio source. Does this mean that among the ellipticals the more luminous ones are generally the stronger radio emitters? Or is being a dominant galaxy in a cluster the important reason for strong radio emission?

To decide between these possibilities, 191 elliptical galaxies that are not members of clusters have been surveyed for radio emission at Owens Valley Radio Observatory. These systems of Hubble types E and So were examined at a frequency of 2640 megaHertz (wavelength 11.4 centimeters).

The work was reported by D. H. Rogstad and R. D. Ekers of California Institute of Technology. The observations were made in 1967 and 1968 with Caltech's twin 90-foot paraboloidal antennas, spaced 100 feet apart (east to west) to form a high-resolution interferometer. Three sets of 20-minute runs were made on each object in the survey.

Only galaxies with known spectral red shifts were included, in order that their distances and their optical and radio luminosities could be determined. The observations plotted in Figure 117 imply that an elliptical system must have a photographic absolute magnitude brighter than -20 to be a *strong* radio source, that is, with a radio energy output of more than 1.7×10^{41} ergs per second.

This is a necessary but not a sufficient condition, for in fact only six percent (three out of 48) of the ellipticals that are optically that luminous are also strong radio sources. Including other workers' data raises this proportion to 7.7 percent.

As the chart shows, quasars and many radio galaxies are intrinsically stronger sources than the ellipticals covered by this survey. In addition to the few intense radio ellipticals, several E and So systems were found to be weaker emitters. Even so, they had radio luminosities some 100 times greater than those of spiral galaxies that emit detectable radio energy.

To compare the results for these noncluster ellipticals with those in

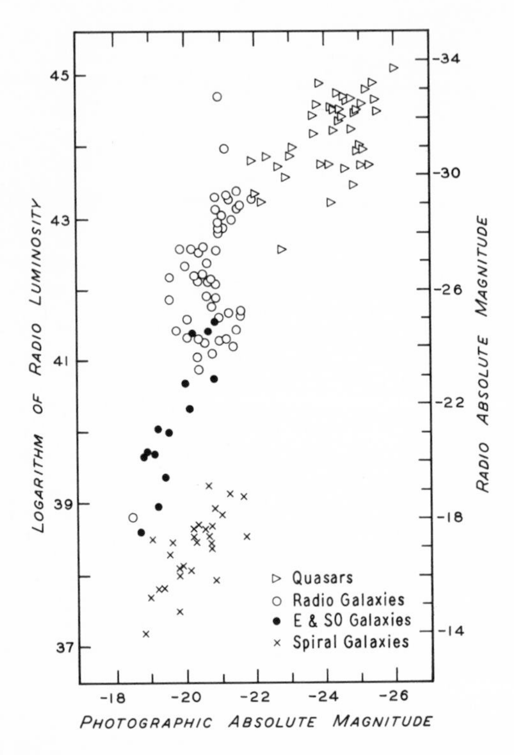

FIG. 117. Luminosity values for elliptical galaxies found to be strong radio sources in the Cal Tech survey are plotted as black dots. Note that they are intrinsically fainter (on both scales) than the quasars and radio galaxies (courtesy of Allan Sandage) but have larger radio luminosity than spirals (courtesy of M. L. De Jong).

clusters of galaxies, the authors turned to 1965 observations with the Owens Valley interferometer by E. B. Fomalont and Rogstad. Some 111 clusters had been surveyed at a frequency of 1445 megaHertz. Now their fields were all reexamined on the Palomar Sky Survey prints, and identifications of individual galaxies could be made if they were close to the measured positions of radio sources.

After certain assumptions to put the 1965 results on the same statistical basis as the noncluster work, it was found that the probability of a cluster elliptical brighter than magnitude −20.0 being a strong radio source was 6.3 percent—very nearly the same as the noncluster result. It therefore seems possible to account for the radio emission from clusters by the probability of individual giant E galaxies being strong radio sources. It is also found that a cluster's principal radio source is not necessarily the brightest member nor is it always centrally located in the cluster.

The Fourth "Texas" Symposium

LOUIS C. GREEN

(*Sky and Telescope*, August and September 1967)

Relativistic astrophysics is the branch of modern astronomy that deals with the large-scale universe in its extreme aspects—distances, energies, and time spans so enormous that they are described by Einsteinian rather than Newtonian physics. It is the science of quasars, radio galaxies, and the past and future of the universe.

At Dallas, Texas, in 1963, physicists and astronomers first gathered for a symposium on relativistic astrophysics, then again at Austin in 1964, and at Miami in 1965. The most recent meeting was in New York City, January 23–27, 1967, where the hosts were the Belfer Graduate School of Science of Yeshiva University and the Goddard Institute for Space Studies. The broad scope of "Texas" gatherings is indicated by the range of topics discussed: quasars, the universal background radiation, primordial galaxies, cosmic rays, neutrinos, and observational cosmology.

It became strikingly evident early in the symposium that our knowledge of these subjects is much more advanced than it was only a year ago.

Opening the meeting, Maarten Schmidt of the California Institute of Technology discussed the optical spectra and red shifts of quasars. He defined a quasar as a blue, starlike object with a large red shift and, in many cases, variable light. More than 150 of these sources have been recognized on photographs, and red shifts have been measured for more than ninety.

The red shift is defined as $z = \Delta\lambda/\lambda$, where λ is the laboratory wavelength of an emission or absorption line in the spectrum, and $\Delta\lambda$ is the difference between the observed and the laboratory wavelength. A few years ago the largest z known was 0.46, for a very distant optically observed galaxy. Today, for quasars, we have become accustomed to red shifts as high as 2.2.

Students often ask how it is possible to have red shifts larger than 1. Would this not indicate recession from us at a speed greater than that of light? In Newtonian physics it would, for the classical relation for Doppler shift is $z = v/c$, where v is the object's velocity, and c is the velocity of light.

In Einstein's relativity theories, both general and special, this equation holds approximately if v is not too large a fraction of c. But as v increases, the approximation becomes poorer and poorer. The formula as given by special relativity [see p. 185] furnishes the following equivalents:

z	0.10	0.50	1.00	4.00
v/c	0.095	0.385	0.60	0.925

Only if z is infinite does v equal the velocity of light. In general relativity, different relations between z and v are possible, and some of these may become quite complicated, but they are always such that observable zs imply velocities less than c.

The strongest emission lines in the spectra of quasars are those of ionized magnesium (Mg II) originating at 2798 A, triply ionized carbon (C IV) at 1549 A, and the hydrogen Lyman-alpha line at 1216 A. Of course, these lines are not ordinarily observable from the earth's surface, since our atmosphere absorbs wavelengths shorter than 3000 A. But the large red shifts mentioned above bring them into the accessible part of the spectrum.

Quasar lines are enormously wide, extending 30 A to 80 A at their original wavelengths and becoming broader still when shifted into the observable region. From the further fact that a number of them come from high levels of ionization, and some are "forbidden," we can conclude that this portion of the radiation originates in rarefied hot gas: density about 10^7 particles per cubic centimeter, temperature around 30,000°K.

Some quasars have absorption lines as well as emission. For example, in the source 3C 270.1 (with $z = 1.52$), the C-III line at 1909 A appears in emission, but the C-IV line at 1549 A is in absorption. 3C 191 has some fifteen absorption lines, including Lyman alpha. The elements producing ten of these fifteen lines have been identified by applying a red shift slightly smaller than the 1.95 indicated by the object's emission lines.

In the quasar Parkes 1116 +12, the absorption spectrum yields a z of 1.95 and the emission lines 2.0. Parkes 0237 −23, with the largest red shift so far found, has $z = 2.22$ for emission (see Fig. 118).

Recently Schmidt and Jesse Greenstein at Cal Tech showed that a number of absorption lines in Parkes 0237 −23 could be identified by

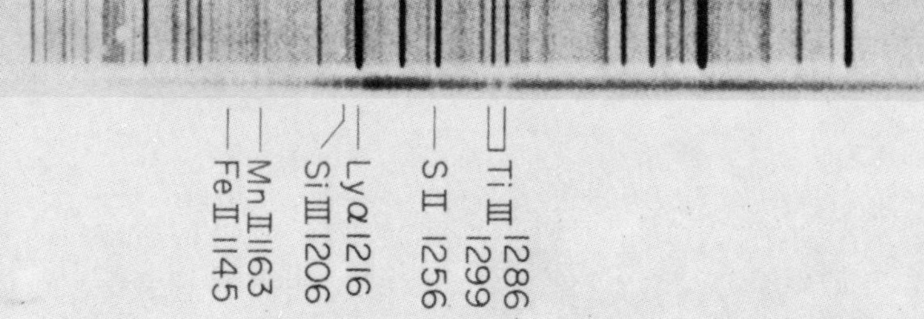

FIG. 118. A spectrogram of Parkes 0237—23 taken with the 200-inch telescope at Palomar Observatory. The labels here identify heavily red-shifted *absorption* lines by element and by normal wavelength in the ultraviolet. The strong comparison line above Lyman alpha (Lyα) is at wavelength 3888 A. (From *The Astrophysical Journal*)

using a red shift of 1.96 for some, 2.20 for the others. It therefore appears more probable at present that the similarity to the white dwarf spectrum is nothing more than a matter of chance.

In an attempt to establish the percentage of quasars of each absolute magnitude, Schmidt chose all cases from the third Cambridge (3C) catalogue for which there were both optical and radio data as well as measured red shifts. There were only sixteen objects in all. He used the Hubble law in calculating absolute luminosities, and found them grouped around the value 10^{10} suns, or 2×10^{43} ergs per second. However, the radio-power outputs are spread over a wider range, with numbers increasing toward the low-power end.

At this point, we note a relation to Allan Sandage's "interlopers," or radio-quiet quasi-stellar galaxies [see p. 188], which may be far more numerous than quasars. In one area of 40 square degrees, Sandage found eleven such objects, which were shown by their spectra to have large *z*s, up to 2.0. Compare this average of one in 4 square degrees with the one quasar in 250 square degrees to the limit of the 3C catalogue. To as faint as magnitude 19.7, there are probably at least 100,000 radio-quiet quasi-stellar galaxies in the whole sky. Thus, there may well be a hundred to a thousand such radio-quiet objects for each radio-emitting quasar. . . .

Astronomers have been considering four possible interpretations of the enormous red shifts of the quasars:

1. *Cosmological.* Velocities of recession resulting from the expansion of the universe.

2. *Gravitational.* The effect on light (and all electromagnetic radiation) when it leaves a strong gravitational field, thus not related to motion of the source.

3. *Local Hypothesis.* Velocities of ejection following an explosion of mammoth proportions in our Galaxy.

4. *Unknown Cause.* Part of all extragalactic red shifts may have an unknown cause, one which is effective for many radio sources, and particularly for the quasars.

At present, one of the principal difficulties with the cosmological explanation (1) arises from the rapid fluctuations in light of quasars, some occurring in intervals as short as one day. Unless there is a kind of synchronous oscillation of the object, all parts involved in the light changes must be close enough to one another for brightening to be triggered everywhere by the same signal; that is, such a rapidly varying quasar cannot be bigger than one light day (roughly 200 astronomical units), a small size for so luminous an object. On the other hand, it is only the fluctuating part of the source that needs to be so small. . . . The tremendous distances implied by the observed red shifts require sources of fantastically large energy, at least 10^{60} ergs. It is the need to explain such high energies that has inspired most theoretical studies of quasars.

If the gravitational hypothesis (2) is accepted, one must somehow obtain the low densities required for the observed forbidden lines to occur in quasar spectra, and at the same time explain the high densities required by gravitational red shifts. William Fowler of California Institute of Technology and Fred Hoyle of Cambridge, England, have suggested a low-density gas surrounding a large number of neutron stars.

A neutron star is a hypothetical one in which densities as high as 10^{14} to 10^{15} grams per cubic centimeter and temperatures around 10^{9}°K have caused most of the electrons and protons to react to form neutrons. Although the mass would be about equal to the sun's, the radius would be exceedingly small, perhaps only ten kilometers! Being directly proportional to the mass and inversely to the radius, the gravitational red shift of such a star would account roughly for the observed value of z for quasars.

If such neutron stars exist, is there some other way to detect them? At their very high surface temperatures (around 10 million degrees), most thermal radiation occurs at about 4-A wavelength, in the X-ray region of the spectrum. The recent discovery of discrete cosmic X-ray sources [see p. 212] gave hope that one or more might prove to be neutron stars.

However, what was regarded as an unusually favorable case, the X-ray source in the Crab nebula, has been shown by Naval Research Laboratory scientists to be much too large to be a neutron star. Furthermore, the spectra of this source and the one in Scorpius are too dissimilar from black-body radiation to be simple neutron stars.

If the red shifts are interpreted as local motions (3), as James Terrell of Los Alamos Scientific Laboratory suggests, the kinetic energy involved is also about 10^{61} ergs, and as difficult to explain as in (1). Also, their emission lines would require special explanation.

In addition, if an explosion 5 or 10 million years ago created our local quasars, then similar objects would be expected from similar explosions in neighboring galaxies. Some of these should be approaching us [since they would be ejected in all different directions—some of them toward us], with their spectra therefore blue-shifted, but none has been observed.

At Mount Wilson and Palomar Observatories, Halton C. Arp has found cosmic radio sources to be associated with peculiar galaxies like the one in Figure 119 [see pp. 222 and 249]. Quite commonly, two radio sources are found about equally spaced along an almost straight line through the peculiar galaxy, suggesting that the sources have been ejected from it and are at roughly the same distance from us. Therefore

FIG. 119. H. C. Arp has compiled a photographic *Atlas of Peculiar Galaxies* (Mount Wilson and Palomar Observatories, 1966). His No. 145 is shown here. The chart shows the positions of the only two radio sources in the revised 3C catalogue within an area 6° × 8° centered on the galaxy. (Chart from *The Astrophysical Journal*; photograph courtesy Mount Wilson and Palomar Observatories)

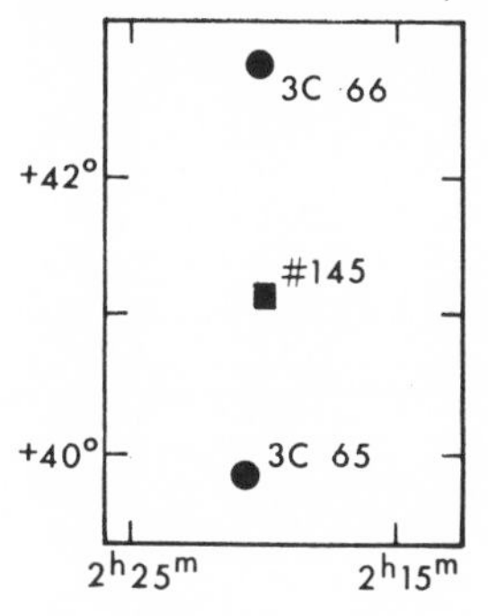

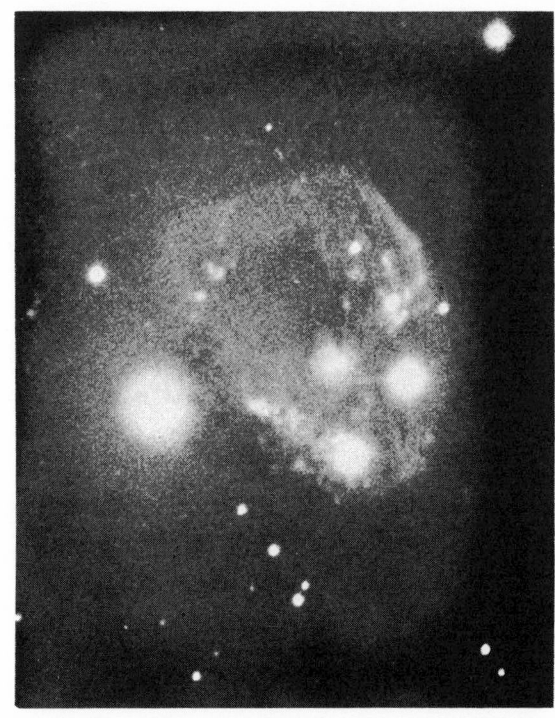

Arp proposes that any substantial differences in their observed red shifts cannot be cosmological in origin but have some as yet unknown cause (4).

Arp views the quasars as neither local nor extremely remote, and therefore not excessively bright. He places them at intermediate distances, with the luminosities of ordinary galaxies. . . .

Allan Sandage has investigated the dependability of the red shift as a distance indicator, using the data plotted in Figure 120 for the brightest members of galaxy clusters [see p. 238]. If these brightest cluster members all have exactly the same intrinsic luminosity, then 15 percent differences can occur in *z* for objects at the same distance. The same kind of analysis for radio galaxies yields a similar spread in *z*.

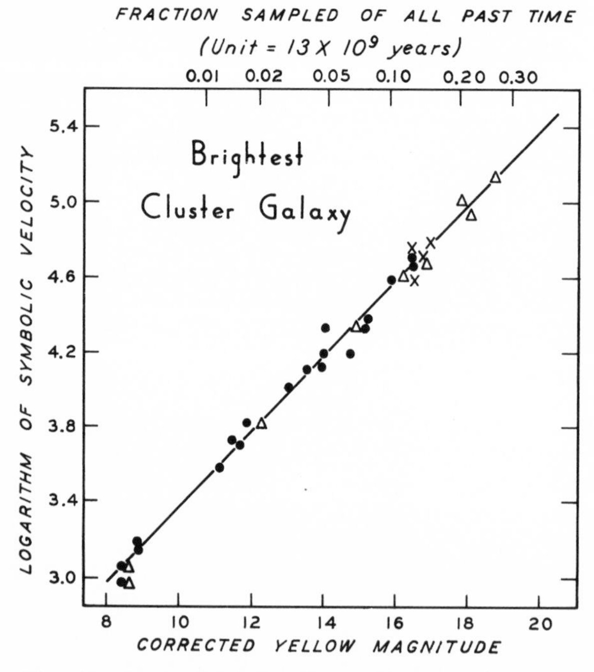

FIG. 120. Allan Sandage's "Hubble diagram" shows the close relation between red shift (z) and apparent magnitude for the brightest galaxy in each of thirty-two clusters. "Symbolic velocity" is *z* times the speed of light; "corrected magnitude" is that which an observer would see in yellow light if he were stationary with respect to a galaxy. Dots indicate photoelectric magnitudes by Sandage, triangles by W. Baum; crosses indicate photographic data. The line has the theoretical slope corresponding to a linear Hubble law of the expanding universe. (Mount Wilson and Palomar Observatories chart)

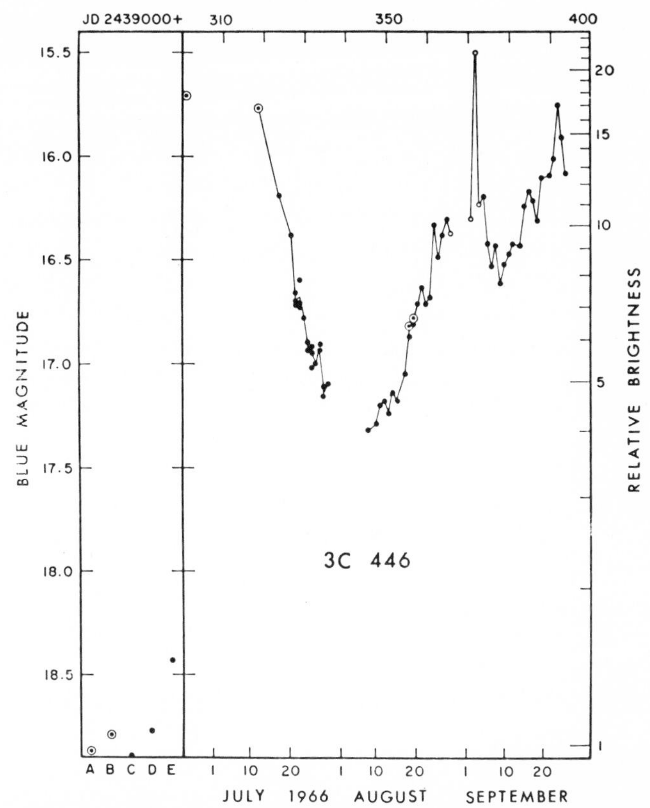

FIG. 121. Striking changes in photographic magnitude of 3C 446. At lower left the dates are: *A* and *B*, October 3 and 5, 1964; *C* and *D*, July 3 and 29, 1965; *E*, October 24, 1965. The Julian Day numbers given at the top allow astronomers to calculate intervals in days easily. (Lick Observatory chart by T. D. Kinman, E. Lamla, and C. A. Wirtanen)

On the other hand, the apparent angular diameters of the galaxies decrease with increasing remoteness just as predicted if z is a good indicator of distance, so we are inclined (but not forced) to attribute the scatter in Sandage's plot to a dispersion in luminosities.

Very shortly after the first quasars were discovered, it was found that at least some of them varied in light. The largest change that has so far been observed is that of the source 3C 446; Figure 121 shows how erratic such variations can be. These optical changes result primarily from fluctuations in the quasar's continuous radiation, the emission lines retaining the same strength but becoming much less prominent when the continuous background brightens. For example, during the three-magnitude increase in brightness of 3C 446 plotted in Figure 121, the C IV line at 1549 A decreased in equivalent width from 230 A to about 40 and showed no change in mean wavelength. Yet it was actually no fainter than before the brightening of the continuum.

Turning now to the radio spectrum, two years ago it was confirmed that quasars vary intrinsically. These variations tend to be more intense at the shorter wavelengths, but there is no good correlation between the variations in the radio spectrum and those observed in the optical region for the same quasar. . . .

The general background radiation of the universe is a low-intensity microwave radiation that bathes the earth from all directions and apparently pervades the whole universe [see pp. 283 and 293]. . . .

Evidence continues to accumulate for the thermal character of this general background radiation. By now, direct measurements have been made by five different groups of researchers, at wavelengths ranging from 1.5 to 20.7 centimeters. The observed intensities all fall close to the theoretical curve for black-body radiation at a temperature of roughly 3°K [see Fig. 112].

An indirect determination of the temperature corresponding to a wavelength of 2.6 millimeters has been made at the University of California by George B. Field and John L. Hitchcock. They measured the intensities of the first two rotational lines in one of the bands of the CN molecule (cyanogen) in the spectra of the stars Zeta Ophiuchi and Zeta Persei in the Milky Way. These two lines (at 3874 A and 3876 A) [shown in Fig. 122] are caused by absorption of starlight by interstellar cyanogen gas. They have the same upper energy level in the molecule but originate from different lower states. CN molecules are excited to these states by absorbing 2.6-mm radiation reaching the gas. Hence these measurements of optical CN absorption provide a value for the intensity

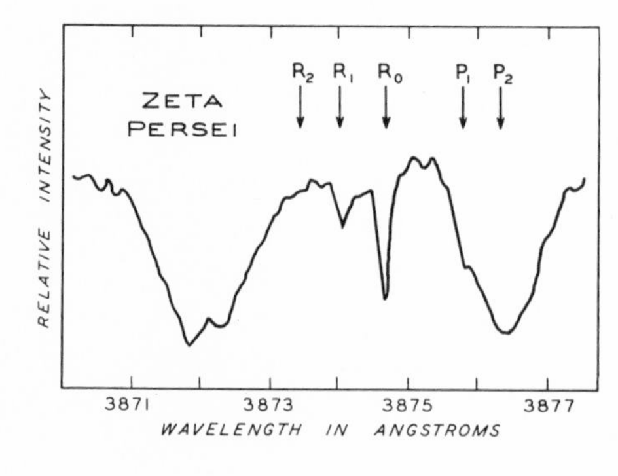

FIG. 122. Intensity tracing of a short part of Zeta Persei's spectrum showing very narrow interstellar cyanogen absorption lines, arising from the molecules' ground state at R_0 and from the first excited state at R_1 and P_1. The measured ratios of the strengths of the absorption lines indicate relative populations among the energy states of CN molecules in space, and these depend on the exciting background-radiation's "temperature." (Diagram from J. F. Clauser, NASA Goddard Institute for Space Studies)

of the cosmic background radiation at 2.6-mm wavelength, and it again falls on the 3° black-body curve.

It is difficult to conceive of any physical process, other than the cooling of a primordial fireball's radiation by the expansion of the universe, that would produce the present thermal (3°K) background. It should be isotropic, and R. B. Partridge and D. T. Wilkinson of Princeton have measured the microwave radiation from different parts of the celestial sphere to see if its intensity is accurately the same in all directions. No significant variation was detected. The Princeton workers find the differences between measurements along the celestial equator and at the north celestial pole to be no greater than 0.5 per cent of the background flux.

An important aspect of the cosmic black-body radiation is its relation to the abundances of the helium isotopes He^4 and He^3, and of deuterium (which is the hydrogen isotope H^2), as they were created during the very early stages of the universe.

Today we often think of the elements heavier than hydrogen as resulting from nuclear reactions in the interiors of stars. Fifteen years ago George Gamow suggested that all the elements had been formed in the early stages of the expansion of the universe as a whole [see p. 136]. This theory was abandoned when it was realized that the elements heavier than helium could not be formed in significant amounts unless the present-day mean density of matter overall were far above that required to yield a closed universe, 2×10^{-29} gram per cubic centimeter. This itself is far greater than the observed mean density of luminous material. The discovery of cosmic microwave radiation implies higher temperatures in the early phases of the universe, and makes generating heavy elements still more difficult for the theory.

On the other hand, in the hot, high-density phase [a few minutes after the "Big Bang"], H^2, He^3, He^4, and a little Li^7 and Be^7 would be formed (Fig. 123). Calculations by P. J. E. Peebles of Princeton, who used several values for the present density of matter in the universe and various rates for the early expansion, indicate that very substantial amounts of He^4 would have been formed. If the time scale of expansion is that predicted by the general theory of relativity, helium abundances would be about 26 to 28 per cent of the total mass. At time scales a tenth as long, the amount of helium becomes unacceptably high (over 60 per cent). Modifications of Einstein's cosmological equations, such as suggested by C. Brans and R. H. Dicke of Princeton in 1961 [see p. 263], might possibly shorten the early stages of expansion by a factor of a million. In such rapid expansion, insignificant quantities of deuterium and the helium isotopes would be produced.

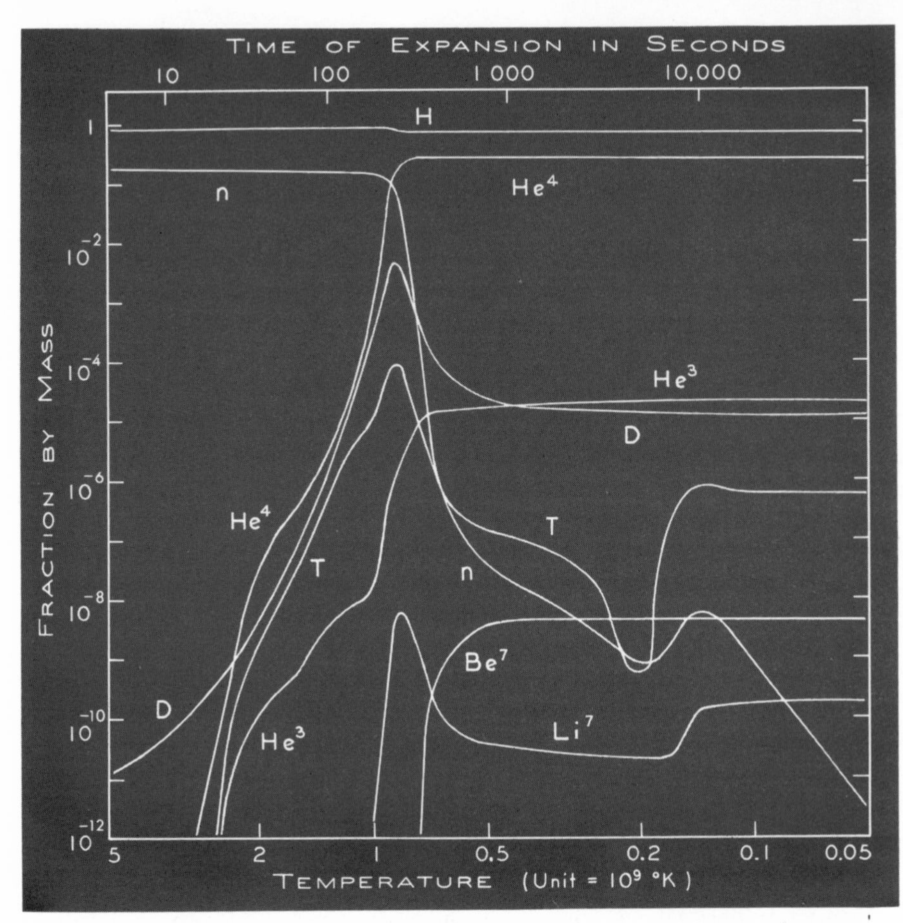

FIG. 123. Rapidly changing composition of the universe during the very early expansion of the primordial fireball, calculated by Robert V. Wagoner, William A. Fowler, and Fred Hoyle. During this time interval, as the fireball's temperature plummeted, the proportion of light hydrogen (H) and neutrons (n) declined, while helium (He^3 and He^4) was being formed. Deuterium (D or H^2) and tritium (T or H^3) increased in amount, waned, then increased again. The scales are logarithmic. (From *The Astrophysical Journal*)

Of course, if stars condensed out of material already consisting of three parts hydrogen to one part helium by mass, then later nuclear reactions would result in a still larger fraction of helium today.

Thus, it is important to establish the hydrogen-to-helium ratio in stars, for if it should prove less than the 3:1 ratio predicted from the general relativity rate of expansion, we might well have to change our equations as suggested by Brans and Dicke. Jesse L. Greenstein of California Institute of Technology pointed out that the best evidence of helium's primeval abundance ought to come from the very oldest stars—the bright red giants and members of globular clusters—assigned to extreme Population II.

Spectroscopic studies of stellar atmospheres have yielded dependable helium measures for some twenty stars, all young ones, which are found to contain 35 per cent helium. (If the primeval value were 25 per cent, we would be forced to think that very little has been formed by nuclear reactions during the lives of these stars.) It is further true that a third of the stars in the Galaxy's extended halo show no helium, although they are classified among the oldest objects. Unfortunately, only atmospheric abundances are observed, and we do not know to what extent the core and atmosphere of a star are mixed.

Robert V. Wagoner of Cal Tech reported calculations he made with William A. Fowler and Fred Hoyle (of Cambridge, England) as to the formation of helium and heavier elements in the early stages of expansion of the universe (see Fig. 123). It was assumed that the universe had once passed through a phase in which the temperature was at least 10^{11} degrees Kelvin, sufficiently high to disrupt all elements, including helium, remaining from any previous stage (if the universe oscillated). This is an elaboration and refinement of the primeval-fireball concept. Wagoner, Fowler, and Hoyle also investigated element formation in *supermassive* stars that undergo implosion, or very rapid contraction, followed by explosion.

The behavior of such objects has been much discussed in the last few years because of their possible connection with quasars. Wagoner, Fowler, and Hoyle considered supermassive objects with masses of 10^3 to 10^{10} suns or more. Especially important are the conditions at the time the star *bounces*; that is, when it changes from implosion to explosion.

In the primeval fireball, they calculated abundances of deuterium and He^4 (Fig. 123) in agreement with Peebles for the time scale of general relativity but found that values could be much higher in supermassive stars. . . . Their conclusion suggests that elements heavier than lithium (Li^7) would not be formed in the early stages of fireball expansion, whereas under the right conditions, the bounce of supermassive objects could produce many of the heavier elements.

However, in none of the cases investigated by Wagoner, Fowler, and Hoyle were elements heavier than carbon produced in amounts similar to those found in the solar system and in Population-I stars. It would appear that these elements result primarily from nuclear reactions in the Population-I stars themselves. On the other hand, the supermassive-object abundances fit those that are observed in the oldest stars of Population II.

Finally, it was pointed out that the low helium abundance recently found in some stars, as had been reported by Greenstein, could be accounted for if the universe contains neutrinos or antineutrinos in sufficiently large numbers. In the first case, the neutrinos would react with the neutrons to produce protons and negative electrons, so the neutrons required for helium and heavier nuclei would not be available. In the second case, the reaction of the antineutrinos with protons would produce neutrons and positive electrons. Then additional reactions would form deuterium that would be later destroyed in the stars.

Many authors point out that some action, perhaps gravity, may produce a net positive *baryon number* (or, less precisely, an excess of matter over antimatter in the universe). Otherwise we would expect to find equal quantities of matter and antimatter. By the principle known as the conservation of baryon number, if matter predominates over antimatter in the universe today it must have done so in the past, and will continue to do so in the future.

Primordial Galaxies

Peebles' and Partridge's study of the formation of galaxies from the expanding fireball of the very young universe was discussed by the former at the symposium. As the primeval plasma expanded, it would cool sufficiently for protons and electrons to combine into hydrogen atoms. The temperature would then be about 4000°K, and the universe's

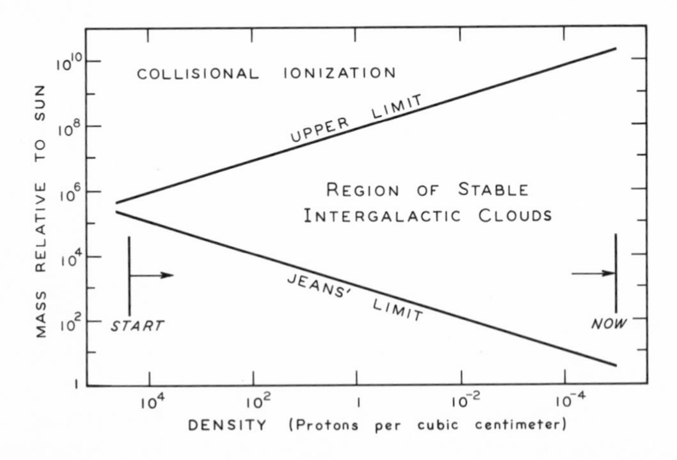

FIG. 124. As a cosmic gas cloud shares the expansion of the universe, its fate depends upon its initial mass. Above the line marked "upper limit," star formation is possible, and the cloud becomes a galaxy. Below the line marked "Jeans' Limit," clouds are unstable and break up. (Princeton University diagram by P. J. E. Peebles and R. B. Partridge)

age some 100,000 years. At the same time the mean density would have fallen to around 10^4 protons per cubic centimeter.

At this point, clouds of neutral atomic hydrogen could form from small local inhomogeneities in the expanding plasma. The line labeled "Jeans' limit" in Figure 124 shows the smallest masses that could grow from the various mean densities marked along the bottom. In later stages of the expansion, the clouds could have smaller masses, but in the beginning (labeled "Start"), 10^5 solar masses would be required.

For stars to form, these clouds must contract, but this is only possible if they can radiate away a substantial portion of their energy. An examination of numerous mechanisms of radiation suggests that none of them is at all effective in reducing an isolated cloud's total energy, even over an interval as long as the age of the universe.

But if the great primordial clouds collided, their hydrogen could become ionized, releasing electrons. Recombination of the electrons and protons would then produce light that would carry energy away and permit contraction to the stage of actual star formation. This process is most probable in the early expansion, while the density is still high, and is indicated by the area above the upward slanting line in Figure 116. This upper limit to the size of stable intergalactic clouds is related to the maximum kinetic energy which the hydrogen atoms can have and still not cause collisional ionization.

After these residual intergalactic clouds have cooled (today) they may contain enough nonluminous matter to make up the difference between the observed density of 7×10^{-31} gram per cubic centimeter for luminous matter and the critical density of 2×10^{-29} necessary for a closed universe. The existence of such intergalactic clouds might be detected by absorption in the spectra of very distant quasars at far-ultraviolet wavelengths just shorter than the Lyman-alpha hydrogen line (as red shifted from 1216 A into the visible region of the spectrum).

Peebles pointed out that near the start the masses and sizes of the original clouds were very similar to those now observed for globular clusters. Formed when the overall turbulence of the universe was low, these earliest clouds probably possessed little angular momentum, as do globular clusters today. A number of such clouds would be the units from which each galaxy was built. Some might condense to form globular clusters before passing through the protogalaxy, and their forms would be preserved thereafter, accounting for the system of globular clusters. Other gas clouds, colliding before they condense into stars, are heated by the collisions, and coalesce to form the rotating disk of the galaxy.

Peebles and Partridge have studied the question of observing very young galaxies, those so distant that we see them by light emitted billions of years ago, when they were being formed. These objects should be highly luminous in comparison with nearby galaxies similar in age to our own, and their spectra would be subject to great red shifts (z on the order of 10 to 30), hence strongest in the infrared as observed today. Taking into account numerous other sources of background radiation—terrestrial, interplanetary, galactic, and extragalactic—the Princeton scientists suggest that the integrated light of the young galaxies might be detected at wavelengths from 5 to 15 microns, from above the earth's atmosphere using infrared detectors.

Epilogue

The astronomers' universe keeps on expanding—from Milky Way to Local Group to Coma Cluster to quasars to infrared young galaxies to the radio-frequency remnants of the primordial fireball. The variety of measurements and speculations is probably characteristic of an active field in any science—and certainly confusing. The remarkable thing is that a fairly consistent model of the universe seems to be emerging from all this confusion.

After trying several alternative explanations, or "models," we find that most of them must be eliminated. Many proposals have not even been mentioned in this volume, but two of the major contenders are on the way out. Steady-State cosmology (and its continuous creation) seems to be disproved by the background 3°K radio radiation, almost certainly evidence of the "Big Bang." It would also be difficult to explain the larger numbers of quasars per unit volume of space at distances of 5 billion light years (seen as they were 5 billion years ago) if the universe remains constant in time (see pp. 249 and 283).

Although "direct proof" is yet lacking, the intermediate or short distances to quasars are nearly disproved (p. 246). While this was being written, the late George Gamow (then at Colorado) tried another (non-Doppler, noncosmological) cause of red shift—a time change in physical constants that would shift all spectrum lines toward shorter wavelengths as the universe ages. He followed an earlier speculation of Dirac that the ratio of electrical to gravitational forces between electron and proton should be proportional to the age of the universe, but instead of letting gravitation decrease (as Dicke does, p. 263), he assumed that electrical force—or charge on electron and proton—has increased. Hence the atoms in quasars, which we see as they were 5 billion years ago, had weaker electrical forces and redder emission lines than do similar atoms here and now.

Gamow retracted this proposal because of a simple, unacceptable consequence: the wavelength separation of several ionized silicon (Si-II) lines, observed in quasar spectra as absorption lines, would be changed (in addition to their general red shift). Measurements by Bahcall, Sargent, and Schmidt in a Palomar spectrum of 3C 191 show that the Si-II lines are properly spaced for the atomic fine-structure constant to be the same in 3C 191 ($z = 1.945$, distance about 7.7 billion light years) as it is here now. This failure seems to strengthen the case for cosmological distances of quasars.

Of course there are several gaps and inconsistencies remaining in the astronomers' picture: the mean density of visible matter in the universe is too low, the various shapes of galaxies are not fully understood, the structure of quasars is imperfectly determined, and their vast power supply is still a puzzle. The variable light and radio emission of quasars, and the two or more red shifts in some (p. 249) are also difficult to explain. One possibility (partly developed on p. 249) is this: there are so many galaxies between us and the quasars that we see each quasar through a closer galaxy. Absorption lines produced by interstellar gas clouds in the intervening galaxy would have a smaller red shift (as observed), and if one or the other is moving sideways, the obscuration (or gravitational deflection of light) by the nearer system could cause the light variations. However, the sidewise motions are minute compared to the velocity of light, so this explanation still requires small regions of high luminosity in a quasar. The latest observations in 1968 seem to show several "sets of red shifts" in the absorption lines of quasars, so the explanation by intervening galaxies becomes less plausible.

Looking back through earlier chapters, we see that many more notes could be added on discoveries made since this text was put together. A recent determination of the Hubble constant, using magnitudes of globular clusters around fairly distant galaxies, gives a value of 75 km/sec per megaparsec (rather than 100) and boosts the Big-Bang age of the universe to more than 13 billion years. It has been suggested that the relation between total mass and total luminosity of galaxies in groups and clusters (Fig. 101) is an indication of a "set of densities" ranging from high values typical of planets and stars to values 10^{28} times smaller that are typical of clusters of galaxies, and that this set of densities resulted from the manner in which matter condensed during the early history of the universe. Cosmic rays (p. 259) are found to contain small numbers of heavy ions, some of which may be antimatter that has traveled hundreds of millions of light years from an antimatter galaxy.

Discussion of the universe was left to this last volume of the Sky and Telescope Library of Astronomy for at least two reasons besides the obvious one: that it is a fitting climax to the series. First, recent developments have been so rapid that we waited until the last possible moment for articles to be written. Second, in order to appreciate the astronomers' universe, the reader needs all the previous volumes of the series:

Volume 1 (Wanderers in the Sky, 1965) for the mechanics of motion and relativity theory that apply to galaxies and clusters of galaxies, as well as to planets and stars.

Volume 2 (Neighbors of the Earth, 1965) for a description of the system of planets, comets, and other material near the sun—an ordinary star.

Volume 3 (The Origin of the Solar System, 1966) for an understanding of how the sun and other stars probably acquire planets, a few suitable for the evolution of living organisms.

Volume 4 (Telescopes, 1966) for a description of the instruments used by astronomers in their studies of planets, stars, galaxies, and interstellar material.

Volume 5 (Starlight, 1967) for the analysis of starlight carried out in the past century.

Volume 6 (The Evolution of Stars, 1968) for an appreciation of how the stars are ageing—slowly changing as they consume their nuclear fuel.

Volume 7 (Stars and Clouds of the Milky Way, 1968) for an "inside picture" of our own Galaxy, one of the spirals.

After Volume 8, with its sketchy coverage of the universe, there's not much left to say. Of course, continuing astronomical research in all of these areas will refine the current views, will show some details to be wrong, and may add whole new topics in a universe not yet fully explored.

THORNTON PAGE
LOU WILLIAMS PAGE

Appendix I
The Origin of *Sky and Telescope*

In March 1931, publication of a small quarterly magazine, *The Telescope,* began at Perkins Observatory of Ohio Wesleyan University in Delaware, Ohio, with the director of the observatory, Harlan T. Stetson, as editor. By July 1933 the magazine had become a larger, bimonthly periodical. After Stetson moved to the Massachusetts Institute of Technology, the Bond Astronomical Club, a society of Cambridge amateur astronomers, and Harvard College Observatory assumed sponsorship of the magazine. Loring B. Andrews became editor, and in 1937 Donald H. Menzel succeeded him. *The Telescope* carried stories of important astronomical discoveries, reviews of current astronomical work, and articles on the history of the science.

In the meantime, the first issue of the small *Monthly Bulletin of the Hayden Planetarium* (New York City) appeared, in November 1935, edited by Hans Christian Adamson. In addition to a review of the current show at the planetarium, it contained other astronomical notes and articles. The interest and encouragement of its readers led, in October 1936, to the enlargement of its size and scope. Its name was changed to *The Sky,* and while retaining its planetarium ties, it became the official organ of the Amateur Astronomers' Association in New York City, replacing the magazine *Amateur Astronomer,* which had been published from April 1929 to the spring of 1936.

The Sky grew in reputation and circulation. In February 1938 Clyde Fisher, curator-in-chief of the planetarium, became editor. On November 1, 1939, the Sky Publishing Corporation was formed, owned by Charles A. Federer, Jr., who for four years had been a planetarium lecturer. He and his wife, Helen Spence Federer, edited and published *The Sky* through its fifth volume.

Then, encouraged by Harlow Shapley, director of the Harvard College Observatory, Sky Publishing Corporation moved to Cambridge, Massachusetts, and combined *The Telescope* and *The Sky* into *Sky and Telescope,* born with the November 1941 issue. The ties with Harvard have been strong. Until the middle 1950s the magazine's offices were in the observatory—now they are located less than a mile away.

The present editorial staff includes Mr. Federer as editor-in-chief, Joseph Ashbrook as editor, and William Shawcross as managing editor. Observers' material is handled by Leif J. Robinson, books and art by Mollie Boring. Unsigned material in the magazine is prepared by this group.

During its twenty-eight years, *Sky and Telescope* has been a distinguished and increasingly well-received publication, with two overlapping purposes. It has served as a forum where amateur astronomers can exchange views and experiences, and where they are furnished with observing data. It has brought to an ever-widening circle of scientists and educated laymen detailed and reliable information on new astronomical developments, and through its pages, has introduced them to the important figures of modern astronomy.

Appendix II Astronomy Through the Ages: A Brief Chronology

ca. 3000 B.C.:	Earliest recorded Babylonian observations of eclipses, planets, and stars.
ca. 2500 B.C.:	Egyptian pyramids constructed, oriented north-south by the stars.
ca. 2000 B.C.:	Babylonian story of creation: *Enuma Elish.*
	Stonehenge built in southern England with stones lined up by the stars.
ca. 1000 B.C.:	Beginnings of Chinese and Hindu astronomical observations.
700–400 B.C.:	Greek story of creation: Hesiod's *Theogony.*
	Hebrew story of creation: *Genesis.*
	Greek philosophers Thales, Pythagoras, and Meton note regularity of celestial motions.
400–300 B.C.:	Greek philosophers Plato, Eudoxus, and Calippus develop the concept of celestial motions on spheres.
	Aristotle develops the idea of four elements and the concept that heavy things fall, light ones rise.

300–100 B.C.: Aristarchus proposes that the earth moves.
Eratosthenes measures the size of the earth.
Hipparchus makes accurate observations of star positions.

ca. A.D. 150: Ptolemy's *Almagest* summarizes the geocentric theory; the planets' motions are explained by epicycles and other motions in circles.

ca. 1400: Ulugh-Beg, in Samarkand, reobserves star positions.

1530: Copernicus, in Poland, postulates that the earth and planets move around the sun because this involves fewer circular motions. This revolutionary idea later rouses strong opposition.

ca. 1600: Tycho Brahe measures the motions of the planets accurately; Kepler uses these measurements to show that the orbits of planets are ellipses rather than combinations of circles.
Galileo uses the first telescope to observe the moons of Jupiter and the crescent shape of Venus, supplying strong support for the Copernican idea. Galileo also establishes that falling weights would all be accelerated in the same way if there were no air to hold the lighter ones back.

1680: Newton combines Kepler's and Galileo's findings, together with observations of moon and comets, into the fundamental laws of mechanics and gravitation. He also studies light, its colors, and spectrum. By this time, pendulum clocks are in use.

1690: Halley notes the periodic reports of a large comet every seventy years and concludes they refer to one object moving in a long, thin ellipse around the sun.

1755: Kant postulates that the sun and planets were formed by the coagulation of a cloud of gas like the spiral nebulae.

1780: William Herschel builds large telescopes, discovers the planet Uranus, and explains the Milky Way as a flat disk of stars around the sun.

1700–1800: Mathematical astronomy flourishes, involving many Europeans—Cassini, Bradley, d'Alembert, Laplace, Lagrange, and others—who apply Newton's mechanics to celestial motions with remarkable precision.

1840: The first astronomical photograph (of the moon) obtained by J. W. Draper. By 1905 photography is well established for accurate observations with telescopes ranging up to 40 inches in aperture, photographing stars 100,000 times fainter than those visible to the naked eye.

1800–1900: Navigation has become a precise and important practical application of astronomy. The accurate observations of star positions show that annual parallax is due to the earth's orbital motion around the sun, confirming the Copernican idea and providing a method of measuring distances to the stars. Other small motions show that the stars are moving.

1850–1900: The laboratory study of light together with physical theory shows that spectrum analysis can be used to determine temperature and chemical composition of a light source.

1843: Doppler explains the effect of motion on the spectrum of a light source.

1877: Schiaparelli observes "canals" on Mars.

1900: Chamberlin and Moulton speculate that the planets were formed after another star passed close to the sun.

1904–20: Einstein establishes the theory of relativity.

Large reflecting telescopes are built at the Mount Wilson Observatory in California.

1910–30: Russell and Eddington establish the theory of stellar structure.

1917–30: Shapley and Oort establish the size, shape, and rotation of the Milky Way Galaxy.

1930: Discovery of Pluto.

1910–40: Slipher and Hubble find that other galaxies are moving away from ours. De Sitter, Eddington, Lemaître, and others explain this recession by application of relativity theory.

1930–60: Bethe, Gamow, and others in the U.S. apply the results of nuclear physics to explaining the source of stellar energy. This is followed by the work of many astrophysicists on evolution of the stars from large interstellar gas clouds.

Von Weizsäcker, Kuiper, Urey, and others develop a theory of the origin of the solar system from a large gas cloud.

1947–60: Instruments are shot above the atmosphere in the U.S. for astronomical observations.

	Large radio telescopes are built in the U.S., Australia, and England.
1957:	Sputnik I, the first artificial satellite of the earth, is launched by Soviet scientists.
1959:	First space probe to hit the moon is launched by Soviet scientists.
1961:	First manned space flight around the earth by Soviet astronaut Yuri Gagarin.
1962:	X-ray sources detected in the Milky Way and beyond.
1963–64:	Radio telescopes locate quasi-stellar sources ("quasars"), found to have large optical red shifts.
1964–65:	First close-up photographs of the lunar surface obtained by U.S. space probes Ranger 7 and Ranger 8.
1965:	Photographs of Mars, taken at about 11,000 miles distance by Mariner 4, show a cratered surface.
1966:	First soft landing on the moon by the Russian Luna 9.
1967:	Physical analysis of moon's surface carried out by Surveyor 3 and Surveyor 5.
1968:	Radio telescopes locate regularly pulsating sources ("pulsars") with periods of about 1 sec, estimated to be at distances of a few hundred light years. U.S. space probe Apollo 8 carries three men out to the moon, around it, and back to earth.

Appendix III Notes on the Contributors

ABELL, GEORGE O. (1927–), astronomy professor at U.C.L.A. since 1956 and lecturer at Griffith Planetarium in Los Angeles; worked on the Palomar Sky Survey, 1953–1956; well known for his *Catalogue of Clusters of Galaxies*; author of *Exploration of the Universe*. ("The Local Group of Galaxies")

ASHBROOK, JOSEPH (1918–), astronomer specializing in variable stars; on the Yale faculty from 1946 to 1953, when he joined the editorial staff of *Sky and Telescope*. ("The Nucleus of the Andromeda Nebula," "Spiral Structure in Galaxies")

BOWYER, C. STUART (1934–), on the faculty of the Department of Astronomy, University of California (Berkeley); his chief interest in astronomy is high-energy astrophysics; formerly in the Department of Space Science, Catholic University of America, at the U.S. Naval Research Laboratory, and at the National Bureau of Standards. ("Galactic X-Ray Astronomy")

COLES, ROBERT R. (1907–), lecturer and curator at the Hayden Planetarium in New York City from 1939 to 1951 and its chairman from 1951 to 1953. ("Extragalactic Cepheids Studied")

FAULKNER, D. J. (1937–), Queen Elizabeth II Fellow at Mount Stromlo Observatory, Canberra, Australia; in 1963 and 1964 in the physics department of University College, London; his chief interests in astronomy are the gaseous nebulae and stellar interiors. ("Current Studies of the Magellanic Clouds")

GREEN, LOUIS C. (1911–), professor of astronomy at Haverford College, where he has been since 1941; in 1932 and 1933 he taught astronomy at Rutgers University and from 1937 to 1941 at Allegheny College; a Guggenheim Fellow in Munich, Germany, 1955 and 1956; a researcher in theoretical astrophysics. ("Dallas Conference on Super Radio Sources," "Relativistic Astrophysics," "Observational Aspects of Cosmology," "The Fourth 'Texas' Symposium")

HESS, VIKTOR F. (1883–1964), Austrian physicist; in 1936 awarded the Nobel Prize for his discovery of cosmic rays in 1914; assistant at the Institute of Radium Research of the Viennese Academy of Science, 1920–1931; professor of physics at University of Innsbruck, Austria, 1931–1939, and at Fordham University, New York City, 1939–1964. ("Mysterious Messengers from beyond the Solar System")

HODGE, PAUL W. (1934–), astronomer, now at University of Washington in Seattle; formerly at University of California (Berkeley) and at the Smithsonian Astrophysical Observatory in Cambridge, Massachusetts; known for his work on the distances of galaxies and the evolution of stars; author of *Galaxies and Cosmology*. ("The Sculptor and Fornax Dwarf Galaxies")

HOFFLEIT, DORRIT (1907–), conducted the "News Notes" page of *Sky and Telescope* from 1941 through 1956, while associated with Harvard College Observatory; since June 1957 director of the Maria Mitchell Observatory, Nantucket, Massachusetts, and astronomer at Yale University Observatory. ("Composite Photographs of the Whirlpool Nebula")

KREIMER, EVERED (1923–), repairer of electronic equipment; specialist in the techniques of astronomical photography; member of the

Phoenix (Arizona) Observatory Association. ("A Messier Album," with John H. Mallas)

Lynds, Beverly T. (1929–), astronomer, on the faculty of the University of Arizona since 1962; a research associate at the National Radio Astronomy Observatory, 1959–1961; her chief astronomical interest is the interstellar medium; coauthor of *Elementary Astronomy*. ("Spiral Patterns in Galaxies")

Mallas, John H. (1927–), a computer expert and visual observer of galaxies, nebulae, clusters, and double stars; former member of the Kansas City (Kansas) and Phoenix astronomy clubs. ("A Messier Album," with Evered Kreimer)

Mayall, N. U. (1906–), director of the Kitt Peak National Observatory, Tucson, Arizona; at Lick Observatory, 1933–1960; his chief fields of study are galactic nebulae, globular clusters, and red shifts and internal motions of galaxies. ("Inclinations of Spectrum Lines in Spirals")

Mumford, George S. (1928–), astronomer on the faculty of Tufts University, Medford, Massachusetts; since January 1965 he has conducted the "News Notes" department of *Sky and Telescope*. (Author of many short items in this volume)

Page, Thornton (1913–), Fisk Professor of Astronomy at Wesleyan University, Middletown, Connecticut; at Yerkes Observatory and University of Chicago, 1938–1940, 1946–1951; NAS Research Associate at the Smithsonian Astrophysical and Harvard Observatories, 1965–1967; editor of *Stars and Galaxies* and of the *Sky and Telescope* Library of Astronomy. ("The Evolution of Galaxies," "Galaxies and Quasars: Research Reported at the IAU Congress in Prague, August 1967")

Struve, Otto (1897–1963), director of the Yerkes Observatory of the University of Chicago, 1932–1949; founder and director (1938–1949) of McDonald Observatory; the last of a family that produced four generations of renowned astronomers in Russia; he was one of the leading figures in American astronomy in this century, widely known for his research on stellar spectra, and the author of many research papers and books; his last book was *Astronomy of the 20th Century*. ("Some Thoughts on Olbers' Paradox," "The Clouds of Magellan," "The Distance Scale of the Universe," "Ages of the Stars")

Thackeray, A. D. (1910–), director of Radcliffe Observatory, Pretoria, South Africa, where he has been since 1948; 1937–48, chief assistant, Solar Physics Observatory, Cambridge University; Commonwealth Fund Fellow at Mount Wilson Observatory, 1934–36; his chief interests in astronomy are stellar spectroscopy and radial velocities of stars; author of *Astronomical Spectroscopy.* ("Supernova in Messier 83")

WITHERELL, PERCY W. (1877–), graduate of M.I.T.; past president of the Bond Club at Harvard College Observatory; for many years treasurer of the American Association of Variable Star Observers, and now its auditor. ("Man and His Expanding Universe")

Glossary

Powers of ten are used to save space in writing very large or very small numbers: 10^9 means 1,000,000,000, or one billion; 10^{21} is 1 followed by 21 zeros; 10^{-6} is $1/10^6$, or 0.000001, or one millionth; and so on.

Throughout this book the following abbreviations are used:

A = angstroms
a.u. = astronomical unit
b = brightness, or galactic latitude
B = blue magnitude
c = velocity of light, 3×10^{10} cm/sec or 186,000 mi/sec
cm = centimeter
D = distance from us
δ = declination
$\Delta\lambda$ = change in wavelength
E = energy, or elliptical galaxy
e.V. = electron-volt
f = frequency, or focal length
f.u. = radio flux unit
G = gravitational constant
H = hydrogen atom, or the constant in Hubble's law, or H-line of calcium in spectra
Hz = Hertz = 1 cycle per second
$H\alpha$, $H\beta$, $H\gamma$, . . . = Balmer lines of hydrogen
Irr = irregular galaxy
k = Boltzmann constant
°K = Kelvin degrees of temperature
km = kilometer (0.61 mi.)
kpc = kiloparsec
l = galactic longitude
L = intrinsic luminosity
λ = wavelength
Λ = cosmological constant
l.y. = light year
$Ly\alpha$, $Ly\beta$, $Ly\gamma$ . . . = Lyman lines of hydrogen
m = meter, or mass, or apparent magnitude
M = mass, or absolute magnitude
MHz = megaHertz = megacycles/sec
Mpc = megaparsec
mi = mile (1.7 km)
p = proton, or hydrogen ion
psc = parsec (3.26 light years)
r = radius, or distance
S = spiral galaxy
SB = barred spiral galaxy
sec = second of time
t = time
T = temperature
θ = angle
U = ultraviolet magnitude
v = velocity, or speed
V = visual, or yellow, magnitude
x,y = rectangular coordinates
z = red shift = $\Delta\lambda/\lambda$
° = degree of arc (angle) = 60′
′ = minute of arc = 60″
″ = second of arc

acceleration Speeding up or slowing down; more precisely, a change in amount or direction of velocity.

angle Angular distance measured in degrees, minutes, and seconds of arc: all the way around the sky is 360°; 1° = 60′ (minutes of arc), 1′ = 60″ (seconds of arc).

angstrom Unit of length used to measure wavelengths of light, abbreviated A; 1 centimeter (about 0.4 inch) is 100 million A.

astronomical unit The distance from sun to earth (about 93 million miles), abbreviated *a.u.*, and used as a unit of distance in the solar system.

astrophysics Study of physical conditions in planets, stars, nebulae, galaxies, and regions between them.

atmosphere The outer, gaseous layers of a star (or planet), through which light can pass, and where the absorption or emission lines in the spectrum are formed.

atomic energy Energy released in nuclear reactions which change the total mass of material by packing nuclear particles in a different arrangement.

BD (followed by a number) The *Bonner Durchmusterung* catalogue number, which serves to identify a star.

background radiation Radio waves, equal in intensity from all directions, believed to be the remnants of the "Big Bang" 10 billion years ago, since cooled by expansion of the universe to a temperature of 3°K. See p. 283.

Balmer lines A series of spectral lines produced by hydrogen atoms. The first (red) line at wavelength 6563 A is called $H\alpha$ (hydrogen-alpha), the second (green, 4861 A) $H\beta$, the third (4341 A) $H\gamma$, and so on, ending at the Balmer limit at 3650 A.

binary A double star or pair of stars close together in the sky. *Visual* binaries can be seen as two separate stellar images in a telescope. An *eclipsing* binary looks like one star but varies periodically in brightness. A *spectroscopic* binary varies periodically in radial velocity (measured by Doppler shift in its spectrum).

Boltzmann equation When atoms are in equilibrium at temperature, T, the relative numbers in energy levels (of energy E) are proportional to $e^{-E/kT}$, where k is Boltzmann's constant and e the base of natural logarithms.

3C (followed by a number) The *Third Cambridge Catalogue* number, which serves to identify small radio sources, some of them quasars.

CTA (followed by a number) The Cal Tech number which serves to identify small radio sources.

carbon cycle A series of nuclear reactions involving carbon, nitrogen, and hydrogen at very high temperature. It results in four hydrogen atoms forming one helium atom and releasing energy.

catalogue A list of stars or other celestial objects, giving their positions on the sky and some other characteristic, such as brightness.

celestial sphere The "sphere of sky" with moon, stars, and planets "attached," half of which appears to arch over us on a clear, dark night. Maps of the sky are maps of this sphere. The position of a star on the celestial sphere indicates a direction in space.

cluster A group of stars (or galaxies) relatively close together in space.

color index Difference between photographic and visual magnitude, ranging from —0.5 for blue stars to +1.5 and more for red stars.

constellation A group of bright stars in one region of the sky, or the area on the sky occupied by such a group, named, in most cases, for ancient Greek mythological figures.

coordinate A distance or angle characterizing the position or location of an object. Coordinates x and y are generally used on a plane surface, longitude and latitude on the earth's surface, right ascension and declination, or galactic longitude and galactic latitude, on the sky. Three coordinates are needed in space.

coudé focus The place where a large telescope with extra mirrors forms an image that remains fixed when the telescope is moved.

declination Angle in the sky north or south of the celestial equator; a coordinate analogous to latitude on the earth's surface.

diffraction The bending of light around an obstruction of any size, such as the edge of the moon or the edges of small particles.

distance modulus Difference between apparent magnitude and absolute magnitude; a measure of the ratio of luminosity to brightness, or of the distance of an object (on the assumption that space is "clean," so that brightness is proportional to $1/\text{distance}^2$.) See p. 60.

diurnal motion Daily apparent motion of celestial objects westward due to the eastward rotation of the earth at 15° per hour.

Doppler shift A slight shift in wavelength in the spectrum of a light source moving toward or away from the observer. The shift is toward shorter wavelength (more violet color) for an approaching source and toward longer wavelength (redder color) for a receding source (p. 185).

dyne A very small unit of force, about 1/1000 the weight of a gram, the force necessary to speed up 1 gram by 1 cm/sec in 1 sec.

Einstein shift An increase in the wavelength (reddening) of light emitted by atoms in a strong gravitational field, as predicted by Einstein's general relativity theory. For a spherical mass, M, of radius R, the increase in wavelength (λ) from the surface is $\Delta\lambda = 2.13 \times 10^{-6}\lambda\, M/R$.

electrons The small, negatively charged particles in the outer region of any atom, in motion about its nucleus. Sometimes electrons are removed from atoms and move independently, as in a stream forming an electric current.

element A chemically pure substance consisting of only one type of atom.

ellipse an oval-shaped closed curve, precisely defined by the equation in rectangular coordinates $x^2/a^2 + y^2/b^2 = 1$.

elliptical galaxy A smooth-looking galaxy with an elliptical image, designated "E-type," or "E," galaxy. See Fig. 26.

emission line A sharp excess of one color (one wavelength of light or radio waves) in a spectrum; generally characteristic of low-density gas.

emulsion The light-sensitive coating on film or glass plate which is "developed" to show a negative image of the light that fell on the film or plate.

energy Capacity for doing work—that is, exerting a force through some distance. In many processes energy is changed from one form to another (radiation, heat, chemical energy, potential energy) without being created or destroyed. In nuclear reactions mass can be converted into energy.

energy level The energy corresponding to the different shells or orbits an electron can occupy in the Bohr model of the atom. The jump of an electron to a shell having an orbit of smaller energy (closer to the nucleus) is accompanied by radiation emission. In order to move an electron to an orbit farther from the nucleus, energy must be supplied.

erg A unit of energy; the work done in pushing 1 gram so that it speeds up from rest to 1.414 cm/sec.

evolution of stars The formation of stars from clouds of gas and dust by initial collapse, increase in temperature, and subsequent nuclear reactions that convert a small fraction of the original mass to energy and produce new elements in the material. See p. 112.

flux The amount of energy (usually in the form of radiation: light or radio waves) passing through 1 square meter per second. Flux on a telescope mirror from one celestial object is a measure of the brightness of that object.

focal ratio The "speed" of a lens or telescope mirror, expressed (inversely) as the ratio of focal length, f, to aperture (lens size). A telescope of focal ratio $f/4$ has an aperture equal to one quarter of its focal length and can photograph a nebula or other extended object in one ninth of the exposure time required with an $f/12$ telescope.

forbidden line A spectral line that is normally emitted only by rarefied gases in interstellar space.

frequency Number of periodic changes (cycles) per second. The *period* is the reciprocal of the frequency. Radio waves have frequencies of thousands of cycles per second (kilocycles per second, or kilo-Hertz) to many millions (megacycles per second, or *Mc/s*, or *MHz*), and wavelengths of c/f, where c is the velocity of light and radio waves, 3×10^{10} cm/sec, and f is the frequency in cycles per second, or Hertz.

galactic latitude Angular distance of an object north or south of the galactic equator measured along a great circle passing through that object and the celestial poles.

galactic longitude Angular distance of an object from the galactic center, measured eastward along the galactic equator to the great circle passing through the galactic poles and through the object.

galaxy A vast disk-shaped assemblage of stars, gas, and dust. The sun is located in the Milky Way Galaxy, a spiral of type Sb. Other types (Sc, Sa, SBc, SBb, SBa, SO, E7, E3, EO, Irr) are shown in Figures 24-27.

gauss A unit of magnetic-field strength; the force on a unit magnetic pole is 1 dyne where the field is 1 gauss.

grating A set of parallel narrow slits which deflect light passing through them by the process of diffraction. A coarse grating used in front of a telescope lens gives several images of each bright star. A fine diffraction grating produces the spectrum of colors in light passing through it or reflected from a reflection grating.

great circle A line drawn completely around a sphere, dividing it into halves (as the equator does). Segments of great circles on a sphere correspond to straight lines on a plane; the shortest distance between two points on a sphere is along a great circle drawn through them.

HD (followed by a number) The *Henry Draper Catalogue* number, referring to stars with spectral types determined at the Harvard College Observatory.

H-R diagram A plot of brightness versus color (or temperature, or spectral type), both on a logarithmic or magnitude scale, the brightness adjusted for distance. Each star appears as a point on the H-R diagram, and many of these points (for a group of stars) fall near a diagonal line called the *main sequence*.

halo A spherical cloud of old Population-II stars that surrounds the Milky Way Galaxy or other spiral galaxies.

helium A gas formed of single atoms each about four times the mass of the hydrogen atom.

Hertz A unit of frequency, abbreviated Hz; one cycle per second (cps).

Hubble constant The increase of radial velocity, v, with distance, D, in the Hubble law of red shifts, or the constant H in the equation $v = HD$. It was first estimated by Hubble as 529 km/sec for each million parsecs, then reduced to 75 and now estimated as 100 km/sec per megaparsec or 18 mi/sec per million light years. See p. 134.

hydrogen alpha (Hα) The most prominent spectral line of the Balmer series, at wavelength 6563 A; an H-II region is detected by its H-α emission.

IC (Index Catalogue) The supplement to Dreyer's *New General Catalogue* (NGC) of star clusters and nebulae.

image tube An electronic device that increases the brightness of an image focused on its "photocathode," from which electrons are ejected, multiplied, and focused electrically on a phosphor screen.

inertia The tendency of massive bodies to remain at rest or continue moving in a straight line.

infrared The "color" of invisible light with longer wavelength than red light and shorter wavelength than radio waves.

intensity Energy received per second, usually in the form of light or radio waves, but also as sound waves or such particles as cosmic rays. The intensity of an absorption line measures the gap in energy near one wavelength, or frequency, in a spectrum.

interferometer An instrument that measures the small difference in direction of light arriving from two sources by the interference of the light waves or radio waves. See p. 203. It can also be used to measure the wavelength of light with high precision.

intrinsic luminosity The brightness a star would have if it were at a standard distance; abbreviated L and often expressed in terms of the sun's luminosity (which is 4×10^{33} ergs/sec). Thus, a galaxy of intrinsic luminosity 10^{10} suns radiates 4×10^{43} ergs/sec. *Absolute magnitude* also measures luminosity (on a logarithmic scale); if L is measured in suns, absolute magnitude (M) is about $5 - 2.5 \log. L$.

ion An atom or group of atoms with one or more electrons removed (positive ion) or added (negative ion).

ionization The process of forming ions, or the number of ions in a given region (usually the number per cubic centimeter).

isotopes Two atoms of the same type (element) with masses differing by one or more neutron masses. Unstable isotopes undergo *radioactive decay*. Atoms of stable isotopes normally remain unchanged.

isotropic Equal in all directions. Our view of the universe is generally considered to be isotropic.

kinetic energy Energy of motion, $\frac{1}{2}mv^2$, equal to the work done in pushing a mass m until it moves at speed v.

light year A large unit of distance, about 10^{13} kilometers or 6×10^{12} miles. It is the distance that light travels in one year (3.17×10^7 sec) at a speed of 3×10^5 km/sec. See p. 10.

lines Gaps in the spectrum of sunlight (colors missing) are called *Fraunhofer lines*; gaps in the spectra of light from other stars are called *absorption lines*. The stronger ones are known by letters, such as H and K lines; all of them can be designated by wavelength. Nebulae and some stars have *emission lines*, brighter patches in their spectra. See **spectrum.**

luminosity The inherent brightness of a star in terms of the sun's brightness, if the two were at the same distance from us. See p. 60.

Lyman alpha A line in the far-ultraviolet part of the spectrum (1215 angstroms), strongly absorbed and easily emitted by hydrogen atoms. Lyman beta, Lyman gamma, and others, as well as Lyman continuum, are similar but at shorter wavelengths.

M (followed by a number) The *Messier Catalogue* number which serves to identify bright galaxies, nebulae, and star clusters (see p. 29).

magnetic field A region near a magnet where another magnet would be acted upon by a force; also a region near moving electric charges where a magnet is affected similarly.

magnifying power Magnification of an image by a telescope-eyepiece combination. "Power 50X" means that an object like the moon appears 50 times larger in the eyepiece than to the unaided eye.

magnitude An indication of the brightness of a celestial object. The brightest stars are "of the first magnitude" and the faintest stars visible to the naked eye are of the sixth magnitude (mag. 6). With telescopes the scale has been extended to over mag. 20. Every 5 magnitudes corresponds to 100 times fainter.

main sequence See **H-R diagram.**

megacycle One million cycles, a unit of radio frequency, abbreviated Mc or MHz.

minute of arc An angle equal to 1/60 of a degree; therefore an angular distance in the sky which is 1/(60)(360) of a circle; written 1′.

monochromatic Light of one selected wavelength.

NGC (followed by a number) The *New General Catalogue* number, which serves to identify clusters, nebulae, and galaxies.

nebula A vast cloud of gas between the stars.

neutrino A small particle with no electric charge and almost zero mass, produced in some of the nuclear reactions inside stars.

neutron A neutral particle with approximately the same mass as a proton; a component of atomic nuclei.

neutron star A super-dense star of mass about equal to that of the sun, diameter 5 or 10 miles, composed mostly of neutrons with a very thin hydrogen atmosphere. The high density (10^{15} gm/cm^3) is theoretically possible because neutrons can be packed very closely. No such stars have yet been observed.

nova A star that suddenly increases its light output 10,000 times or more and then fades slowly.

nuclear processes Interactions between the nuclei of atoms that generally release vast quantities of energy as one kind of atom is changed into another.

parallax The very small change in direction of a star caused by the motion of the earth around the sun. It can be measured only for the nearer stars, on photographs taken six months apart, and is used to obtain the distance (1/parallax = distance in parsecs; 3.26/parallax = distance in light years). See Fig. 1.

parsec A large unit of distance equal to 3.26 light years; the distance at which a star has a parallax of 1 second of arc (1″).

period The time interval of one complete circuit of an orbit or one complete rotation of a rotating body or the complete cycle of any periodic change.

photodiode and **photomultiplier** Photoelectric cells.

photometer An instrument designed to measure the intensity of light falling on it. It is described as visual, photographic, or photoelectric, depending on the detector used.

planet A nonluminous body moving around the sun or a star in a nearly circular orbit, and shining by reflected light.

planetary nebula A shell of gas ejected from and expanding around an extremely hot star; many appear as a ring because we are viewing it through the thin dimension on the front and rear parts of the shell.

plasma A cloud of ions.

polarized light Light that consists of waves that vibrate across the beam of light in one direction. Ordinary light is generally unpolarized, and the vibrations are in all directions. The polarization may be partial or complete (100 per cent).

pole A point on the earth where its axis of rotation cuts the surface. The *celestial poles* in the sky are the directions of the two ends of the earth's axis and are located at the centers of the stars' circular motions during any night.

pole of the galaxy A point 90° from the center line of the Milky Way. There are two such poles at opposite points in the sky, analogous to the two poles of the earth, each 90° from the earth's equator.

Population I Consists of young stars with normal metal content, mostly in the outer parts of the disk of a spiral galaxy and associated with clouds of gas and dust. **Population II** consists of older stars with subnormal metal content near the nucleus of a galaxy or in globular clusters.

position angle The direction of a line in the sky, measured from 0° (direction north) through 90° (direction east).

proper motion The very small change in direction of a star due to its velocity across the line of sight as seen from near the sun. It is expressed as a fraction of a second of arc per year, and can be measured on photographs of the star taken many years apart.

proton A hydrogen atom with its one electron removed.

quantum A discrete amount of energy; the smallest amount that can be radiated or absorbed by matter. The size depends upon the wavelength and is larger for the shorter wavelengths.

quasar A quasi-stellar radio source, identified with an optical starlike object that has a large red shift. The radio source is small and generally double. See p. 138.

radial velocity The motion of a star or other light source along the line of sight, detected by *Doppler shift* of lines in the spectrum; red shift indicates recession (positive radial velocity). See p. 185.

radio astronomy The study of astronomical bodies by use of radio telescopes.

radiometer A radio receiver designed to measure the radio energy received per second by the antenna.

red shift The fractional increase in wavelength, $\Delta\lambda/\lambda$, in spectra of distant galaxies, usually denoted by z and considered to be a Doppler shift resulting from velocity of recession. See p. 185.

refractor A telescope in which a large lens (rather than a concave mirror) forms images of stars and other celestial objects.

resolving power The ability of a telescope to distinguish two stars very close together in the sky; the ability of a spectrograph to distinguish two lines very close together in a spectrum.

right ascension One coordinate of a star in the sky, measured eastward like longitude on earth, starting from a point called the vernal equinox.

Schmidt camera A telescope or camera consisting of a spherical mirror with a "correcting plate" (thin lens) in front of it, generally with large focal ratio and able to take a wide-angle photograph. See Fig. 11.

second A unit of time (1/3600 of an hour), abbreviated "sec" or "s." A second of arc (1″) is a (different) unit of angle (1/3600 of a degree). In 1 sec the earth rotates 15″ eastward and a star seems to move 15″ toward the west.

seeing Changing fuzziness and a slight jumping of a star image in a telescope is caused by air currents in the earth's atmosphere and is called "poor seeing."

sidereal time Time that gains on normal time by 4 minutes each day, or 24 hours in 1 year. "Sidereal" refers to stars, which appear to move 360° around the sky in 24 hours sidereal time (23 hours and 56 minutes of normal time), the true rotation period of the earth.

spectral type One of the designations *O, B, A, F, G, K, M,* based on the pattern of lines in a star's spectrum. Surface temperature is high in "early-type" *O* and *B* stars and lower in "late-type" *K* and *M* stars.

spectrogram Photograph of a spectrum.

spectrograph An instrument attached to a telescope to obtain photographs of stellar spectra.

spectrum The various colors of light from a source spread out in the sequence from red to violet (long wavelength to short wavelength), as in a rainbow. Invisible wavelengths extend from the red to infrared and radio waves, and from blue-violet to ultraviolet and X rays.

spiral galaxy A galaxy with spiral arms, generally mottled, or showing dark dust lanes. Two types, designated "S" (normal spiral) or "SB" (barred spiral) are shown in Figures 24 and 25.

Stark effect Broadening of spectral lines caused by strong electric fields on the absorbing atoms. In stellar atmospheres electric fields are usually produced by ions and electrons. Stark broadening is proportional to the ion density or ion pressure, and is a good indicator of total atmospheric pressure and of the star's luminosity.

synchrotron radiation Usually of radio wavelength, synchrotron radiation is produced by electrons (or other charged particles) moving in a magnetic fields. The radio spectrum has a slope of about —0.7 on a logarithmic plot of flux density (intensity) against frequency. See p. 211 and p. 215.

transmutation A change of one type of atom (element) into another by nuclear reactions or radioactive decay.

triangulation Determining the distance to an inaccessible object by measuring its direction from the two ends of a base line of known length. See Fig. 1.

turbulence Whirls or vortices in a fluid (liquid or gas).

ultraviolet The "color" of invisible light with wavelengths shorter than those of visible light (less than about 4000 angstroms).

variable star A star that changes in brightness due to pulsations of its surface, or flares, or eclipse by a companion star.

vernal equinox The point in the sky where the sun crosses the celestial equator on about March 21 each year; used as the origin of the coordinates, right ascension and declination.

wavelength The distance between the crests (or troughs) of regular waves. Visible light of various colors has wavelengths ranging from about 1/10,000 inch to about twice that length. See **spectrum.**

white dwarf A small hot star of high density and low luminosity; the remains of a large bright star that blew up.

width of a spectral line Neither absorption lines nor emission lines are at exactly one wavelength; many have a spread of several angstroms.

X rays Light of too short wavelength to be visible; it can penetrate some distance into most materials. X rays from the sun or other, more distant, sources are entirely absorbed in the earth's atmosphere.

Zeeman effect The splitting of spectral lines caused by a strong magnetic field; where the magnetic field is weak, the spectral lines appear slightly widened.

Suggestions for Further Reading

ABELL, G. O. *Exploration of the Universe*. New York: Holt, Rinehart and Winston, 1964.
BONDI, H. *The Universe at Large*. Garden City, New York: Doubleday (Anchor Books), 1960.
HODGE, P. W. *Galaxies and Cosmology*. New York: McGraw-Hill, 1966.
HUBBLE, E. P. *The Realm of the Nebulae*. New Haven, Connecticut: Yale University Press, 1936. (Dover paperback edition, 1958.)
KAHN, F. D., AND PALMER, H. P. *Quasars*. Cambridge, Massachusetts: Harvard University Press, 1967.
ÖPIK, E. J. *The Oscillating Universe*. New York: New American Library (Mentor Books), 1960.
PAGE, T., ed. *Stars and Galaxies*. Englewood Cliffs, New Jersey: Prentice-Hall, 1962.
PAYNE-GAPOSCHKIN, CECILIA. *Introduction to Astronomy*. New York: Prentice-Hall, 1954.
SCHATZMAN, E. *The Origin and Evolution of the Universe*. New York: Basic Books, 1965.
SHAPLEY, H. *Galaxies* (2nd ed.). Cambridge, Massachusetts: Harvard University Press, 1960.
SINGH, J. *Great Ideas and Theories of Modern Cosmology*. New York: Dover, 1961.
STRUVE, O., AND ZEBERGS, V. *Astronomy of the 20th Century*. New York: Macmillan, 1962.
WHITROW, G. J. *The Structure and Evolution of the Universe*. New York: Harper (Torchbooks), 1959.

ADVANCED

BONDI, H. *Cosmology*. New York: Cambridge University Press, 1960.
McVITTIE, G. C. *General Relativity and Cosmology*. Urbana, Illinois: University of Illinois Press, 1965.
The Structure and Evolution of Galaxies (Solvay Institute Proceedings). New York: Interscience, 1965.

Index